Recent Progress in Bioimaging and Targeted Therapy

Recent Progress in Bioimaging and Targeted Therapy

Guest Editors

Huangxian Ju
Ying Liu
Huanghao Yang
Zong Dai

MDPI

Basel • Beijing • Wuhan • Barcelona • Belgrade • Novi Sad • Cluj • Manchester

Guest Editors
Huangxian Ju
State Key Laboratory of Analytical Chemistry for Life Science
Nanjing University
Nanjing
China

Ying Liu
State Key Laboratory of Analytical Chemistry for Life Science
Nanjing University
Nanjing
China

Huanghao Yang
College of Chemistry
Fuzhou University
Fuzhou
China

Zong Dai
School of Biomedical Engineering
Sun Yat-sen University
Shenzhen
China

Editorial Office
MDPI AG
Grosspeteranlage 5
4052 Basel, Switzerland

This is a reprint of the Special Issue, published open access by the journal *Targets* (ISSN), freely accessible at: https://www.mdpi.com/journal/targets/special_issues/VYT439U1V6.

For citation purposes, cite each article independently as indicated on the article page online and as indicated below:

Lastname, A.A.; Lastname, B.B. Article Title. *Journal Name* **Year**, *Volume Number*, Page Range.

ISBN 978-3-7258-5049-5 (Hbk)
ISBN 978-3-7258-5050-1 (PDF)
https://doi.org/10.3390/books978-3-7258-5050-1

Contents

About the Editors

Huangxian Ju

Huangxian Ju is the director of State Key Laboratory of Analytical Chemistry for Life Science, Chemistry Department, Nanjing University. His research interests focus on analytical biochemistry, biosensing and molecular diagnosis. He has authored 108 patents (65 approved), 7 English and 6 Chinese books and 20 chapters, and has published 994 papers with a h-index of 112 (Google Scholar h-index 124 with more than 58,800 citations).

Ying Liu

Ying Liu is a PI in State Key Laboratory of Analytical Chemistry for Life Science, Chemistry Department, Nanjing University. Her research interests include DNA assembly for bioimaging and therapy, as well as NIR-II in vivo imaging. She has 40+ publications in *J. Am. Chem. Soc.*, *Angew. Chem. Int. Ed.*, and *Nat. Commun.*, etc., and 4 patents authorized in recent years.

Huanghao Yang

Huanghao Yang is the vice president of Fuzhou University and a professor in MOE Key Laboratory for Analytical Science of Food Safety and Biology, College of Chemistry. His research interests focus on design of functional luminescent probes and applications in biosensing and optical bioimaing. He has published 200+ SCI papers (e.g., *Nature*, *Nat. Methods*, *J. Am. Chem. Soc.*, *Angew. Chem. Int. Ed.*) with a H-index of 104, and authored 30 patents.

Zong Dai

Zong Dai is a PI in the School of Biomedical Engineering, Sun Yat-sen University. His research interests focus on biomedical analysis and biosensing, particularly in bioimaging and molecular diagnosis and nucleic acid analysis. He has published 90+ SCI papers (e.g., *J. Am. Chem. Soc.*, *Angew. Chem. Int. Ed.*, *Adv. Mater.*), and authored 15 patents. He is also a member of multiple academic committees and journal editorial boards.

Preface

As Guest Editors of this Reprint, we are pleased to present a curated collection highlighting transformative advances at the dynamic intersection of bioimaging and targeted therapy. This powerful synergy represents a paradigm shift in precision medicine, enabling unprecedented real-time, non-invasive visualization of biological processes across molecular, cellular, and tissue scales, while simultaneously facilitating spatiotemporally controlled therapeutic interventions. Such integration promises to redefine diagnostics and treatment, particularly for complex diseases like cancer, where early detection and the minimization of off-target toxicity are paramount. Our impetus for this reprint stems from the critical need to accelerate the translation of groundbreaking technological innovations from the laboratory bench into tangible clinical impact, effectively bridging fundamental discovery with improved patient outcomes.

This reprint comprehensively showcases cutting-edge developments spanning the field. Core themes include the development of novel in vivo imaging strategies and sophisticated biosensing platforms for the sensitive detection of diverse disease biomarkers—ranging from small molecules and nucleic acids to circulating tumor cells and tumor microenvironments. Central to these advances is the innovative design and fabrication of targeted imaging probes, leveraging the unique properties of organic molecules, inorganic nanomaterials, DNA nanostructures, and stimuli-responsive polymers. Concerted efforts to enhance imaging performance—improving signal-to-noise ratios, resolution, specificity, and targeting efficiency—are crucial for success. These enhanced imaging capabilities directly empower advanced therapeutic modalities, including chemotherapy, photodynamic therapy (PDT), photothermal therapy (PTT), gene therapy, and immunotherapy, increasingly integrated within multifunctional theranostic platforms that combine diagnosis and treatment. Equally vital is the exploration of novel therapeutic targets and signaling pathways, essential for maximizing therapeutic efficacy while rigorously diminishing systemic side effects.

This reprint of eight seminal contributions (two research articles, six comprehensive reviews) is meticulously curated for researchers and clinicians driving innovation in nanomedicine, chemical biology, biomedical engineering, and translational medicine. It serves as an essential resource for scientists pioneering next-generation probe design, imaging methodologies, and therapeutic delivery systems, as well as for clinicians seeking deeper insights into applying these technologies for precise diagnosis, image-guided surgical intervention, optimized treatment planning, and real-time efficacy monitoring. The featured papers vividly exemplify the field's interdisciplinary vigor, covering diverse and impactful topics such as photo-cleavable polycation/upconversion nanoparticle systems for efficient siRNA delivery and cancer therapy, activatable nanoscintillator probes for in vivo telomerase activity detection, advances in small-molecule fluorescent and organic afterglow probes for diagnostics, machine learning applications in cell detection, rational design of cancer-targeting peptides, engineered bacteriophage for theranostics, and novel influenza vaccine strategies based on conserved epitopes. Collectively, these works demonstrate remarkable progress in engineering exquisite specificity and control into both diagnostic imaging and therapeutic intervention.

We extend our sincere gratitude to all contributing authors for their exceptional work, which forms the cornerstone of this reprint. We also thank the Editors of Targets for their support and the expert reviewers for their invaluable critiques that significantly strengthened the manuscripts. It is our hope that this reprint will inspire further innovation and collaboration, ultimately harnessing the full potential of bioimaging-guided targeted therapy to improve patient care globally.

Huangxian Ju, Ying Liu, Huanghao Yang, and Zong Dai
Guest Editors

targets

MDPI

Article

Photo-Cleavable Polycations-Wrapped Upconversion Nanoparticles for Efficient siRNA Delivery and Cancer Therapy

Yuling He [1,2,*], Shuwen Guo [3], Huangxian Ju [2] and Ying Liu [2,*]

1 Shaanxi Key Laboratory of Brain Disorders, Institute of Basic and Translational Medicine, Xi'an Medical University, No. 1 Xinwang Road, Xi'an 710021, China
2 State Key Laboratory of Analytical Chemistry for Life Science, School of Chemistry and Chemical Engineering, Nanjing University, Nanjing 210023, China
3 Key Laboratory of Analytical Chemistry for Life Science of Shaanxi Province, School of Chemistry and Chemical Engineering, Shaanxi Normal University, Xi'an 710100, China
* Correspondence: heyuling0925@163.com (Y.H.); yingliu@nju.edu.cn (Y.L.)

Abstract: RNA interference (RNAi) therapy is a promising approach for cancer therapy. However, due to the weak binding affinity between a carrier and small interference RNA (siRNA) and complicated tumor environment, efficient loading and release of siRNA still remain challenging. Here, we design photo-cleavable polycations-wrapped upconversion nanoparticles (PC-UCNPs) for spatially and temporally controllable siRNA delivery. The PC-UCNPs are synthesized by in situ reversible addition−fragmentation chain transfer (RAFT) polymerization of photo-cleaved 5-(2-(dimethylamino)ethoxy)-2-nitrobenzyl acrylat (MENA) monomer and poly(oligo(ethylene oxide) methyl ether acrylate (OEMA) mononer through a chain transfer agent that anchored on the surface of silica-coated upconversion nanoparticles (UCNPs@SiO_2). After reacting with CH_3I, siRNA and hyaluronic acid (HA) are adsorbed on the particle surface to prepare PC-UCNPs/siRNA/HA. The reaction with cell-secreted hyaluronidase (HAase) achieves the intracellular delivery of PC-UCNPs/siRNA/HA, and 980 nm laser irradiation causes siRNA release, which effectively improves the gene silencing efficiency in vitro and suppresses tumor growth in vivo; therefore, these processes have a promising potential application in precision medicine.

Keywords: polycations; Upconversion Nanoparticles (UCNPs); photo-cleavage; siRNA delivery

1. Introduction

Gene interference therapy using exogenous small interfering RNA (siRNA) to selectively silence gene expression and inhibit protein transcription has become a promising cancer therapeutic approach [1–3]. However, the key challenge for the further application of RNA interference (RNAi) therapy still remains in the efficient delivery and release of these small, fragile biomolecules. Currently, many types of responsive siRNA delivery systems have been developed with pH [4,5], small molecule [6], light irradiation [7,8] and temperature [9] as stimuli, which enhances the controllability for gene release and the efficiency of gene transfection.

Light irradiation is a promising strategy for the on-demand release of payload [10–12]. Recently, near-infrared (NIR) light has attracted much attention due to its minimal photodamage on living tissues, low auto-fluorescence background, non-photobleaching and deep penetration [13–15]. The lanthanide-doped upconversion nanoparticles (UCNPs) can absorb long-wavelength NIR light and generate short-wavelength UV and visible light emissions [16,17]. The integration of UCNPs with photoresponsive molecules, including o-nitrobenzy [18], spiropyran [19] and azobenzene [20], have been studied for drug delivery, gene delivery and in vitro/in vivo imaging [18,21]. Although NIR-regulated UCNPs delivery systems have been reported for the successful release of siRNA [22,23], surface-loaded siRNA usually lacks protection, impairing loading efficiency and resulting in siRNA degradation or leakage during the delivery process [24,25]. Therefore, a NIR-responsive siRNA delivery system with a satisfactory loading capacity, stability and release efficiency is needed with urgency.

With the positive charge and convenience for functionalization, cationic polymers could enhance siRNA loading efficiency [26–29]. Here, we designed NIR-cleavable polycations-encapsulated UCNPs for efficient siRNA delivery and tumor therapy. UCNPs are coated with SiO_2 and functionalized with chain transfer reaction (CTA) initiator 2-(((propanethio) carbonothioyl)thio)acetic acid. Reversible addition−fragmentation chain transfer (RAFT) polymerization is then carried out on the UCNPs surface with photo-responsive 5-(2-(dimethylamino)ethoxy)-2-nitrobenzyl acrylate (MENA) and poly(oligo(ethylene oxide) methyl ether acrylate (OEMA) as monomers. After reacting with CH_3I, the as-obtained PC-UCNPs subsequently absorb siRNA and hyaluronic acid (HA) via electrostatic interactions to obtain PC-UCNPs/siRNA/HA (Scheme 1a). The outer layer of negatively charged HA not only extends blood circulation but also achieves siRNA delivery specificity by recognizing the CD44 receptor on the tumor cell membrane. The degradation of HA by hyaluronidase (HAase) exposes inner cationic polymers and promotes intracellular delivery and lysosome escape. Upon subsequent 980 nm laser irradiation, the UV emission of UCNPs cleaves the photosensitive o-nitrobenzyl segment and detaches the polymer coating from the UCNPs surface for efficient siRNA release (Scheme 1b). This strategy demonstrates the effective silencing of target gene expression and suppression of tumor growth; therefore, it should become a universal strategy for efficient NIR-assisted gene therapy.

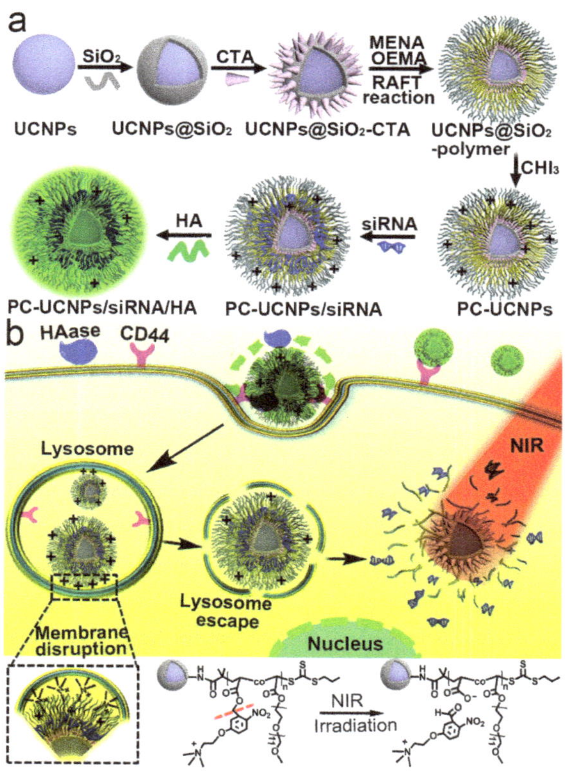

Scheme 1. Schematic illustration of (**a**) synthesis and (**b**) intracellular delivery of PC-UCNPs/siRNA/HA with NIR responsive siRNA release.

2. Results and Discussion

2.1. Synthesis and Characterization of UCNPs@SiO_2-Polymer

UCNPs core $NaYF_4$:Yb,Tm was synthesized using the previously reported solvent thermal method [30] and coated with an $NaYF_4$ shell to prevent the influence of surface defects and enhance particle luminescence [31]. The as-obtained core–shell-structured UCNPs $NaYF_4$:Yb,Tm@$NaYF_4$ was further coated with a thin silica layer using a reverse-phase microemulsion method to obtain UCNPs@SiO_2 [32]. The synthesized UCNPs with an average size of 29 $\pm$ 2.82 nm is shown in Figure 1a.

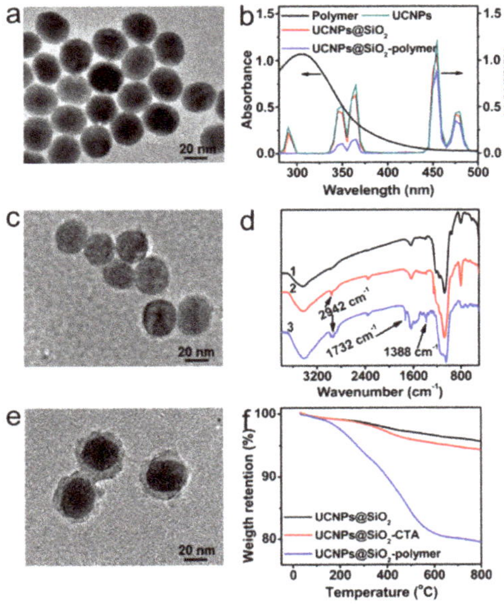

Figure 1. TEM image of (**a**) UCNPs, (**c**) UCNPs@SiO_2, and (**e**) UCNPs@SiO_2-polymer, (**b**) Upconversion luminescence spectra of UCNPs (green line), UCNPs@SiO_2 (red line), PC-UCNPs (blue line) and UV–Vis absorption spectrum of polymer (black line), (**d**) FT-IR spectra of (1) UCNPs@SiO_2, (2) UCNPs@SiO_2-CTA and (3) UCNPs@SiO_2-polymer, (**f**) Thermogravimetric analysis of UCNPs@SiO_2, UCNPs@SiO_2-CTA and UCNPs@SiO_2-polymer.

Emission peaks at 290 nm, 345 nm and 363 nm correspond to $^1I_6 \rightarrow {}^3H_6$, $^1I_6 \rightarrow {}^3F_4$ and $^1D_2 \rightarrow {}^3H_6$ transitions of Tm^{3+}, respectively, and 455 nm and 475 nm correspond to $^1D_2 \rightarrow {}^3F_4$ and $^1G_4 \rightarrow {}^3H_6$ transitions of Tm^{3+}, respectively (Figure 1b, green line). SiO_2 coating increased the size of UCNPs@SiO_2 to 31 $\pm$ 2.1 nm (Figure 1c), which had little effect on UCNPs emissions (Figure 1b, red line). The chain transfer agent, 2-(((propanethio) carbonothioyl)thio)acetic acid (CTA) (Figure S1, Supplementary Materials), was conjugated to UCNP@SiO_2 via the amide bond. The as-obtained UCNPs@SiO_2-CTA demonstrated a CTA characteristic absorption peak at 305 nm in the UV-Vis spectrum (Figure S2) and –CH_2– characteristic absorption peak at 2942 cm^{-1} corresponding to C–H stretching vibration in the FT-IR spectrum (Figure 1d, UCNPs@SiO_2-CTA). Reversible addition−fragmentation chain transfer (RAFT) polymerization of 5-(2-(dimethylamino)ethoxy)-2-nitrobenzyl acrylate (MENA) (Figures S3–S5) and Poly(oligo(ethylene glycol)) methyl ether acrylate (OEMA) was conducted on the surface of UCNP@SiO_2-CTA as the RAFT agent (Figure S6), which obtained P(OEMA-co-MENA)-conjugated UCNPs (UCNPs@SiO_2-polymer) with a size of 44 $\pm$ 1.7 nm and a thickness corresponding to a polymer shell, 6 $\pm$ 0.9 nm (Figure 1e). The molecular mass of as-obtained P(OEMA-co-MENA) on UCNPs surface was measured via Gel permeation chromatographic (GPC) analysis as 11.6 kDa (Figure S7). The characteristic absorbance of UCNPs@SiO$_2$-polymer in the range of 280–440 nm overlapped with the emission of UCNPs@SiO$_2$ at 290 nm, 345 nm and 375 nm (Figure 1b, black line, red line), which led to a substantial decrease in UCNPs emission peak at 290 nm, 345 nm and 375 nm (Figure 1b, blue line). Moreover, the FT-IR spectrum of UCNPs@SiO$_2$-polymer showed the new absorbance peak at 1732 cm^{-1} corresponding to stretching vibrations of the ester group and at 1388 cm^{-1} corresponding to stretching vibrations of -NO$_2$ compared with UCNPs@SiO$_2$-CTA (Figure 1d, UCNPs@SiO$_2$-polymer). In addition, thermogravimetric analysis of UCNPs@SiO$_2$-polymer showed about 19% of weight loss (Figure 1f, UCNPs@SiO$_2$-polymer). These results demonstrated that RAFT polymerization was successfully initiated on the surface of UCNPs@SiO$_2$-CTA.

2.2. Preparation of PC-UCNPs/siRNA/HA and In Vitro Verification of HA Degradation and siRNA Release

To load negatively charged siRNA via electrostatic interaction, the tertiary amine groups from UCNPs@SiO$_2$-polymer were reacted with CH$_3$I to form quaternary ammonium salts. Compared with UCNPs@SiO$_2$-polymer, the as-obtained PC-UCNPs had an increased zeta potential of 58 ± 2.2 mV, while its hydrodynamic diameter remained unchanged (Figure S8). A series amount of siRNA was loaded on PC-UCNPs, and the weight ratio of PC-UCNPs/siRNA was optimized as 1/1, as indicated by the significantly retarded siRNA band in agarose gel retardation assay (Figure 2a). HA was further coated with PC-UCNPs/siRNA to obtain PC-UCNPs/siRNA/HA. After siRNA loading, the as-obtained PC-UCNPs/siRNA showed a zeta potential of 26 ± 2.3 mV and a hydrodynamic diameter of 61 ± 1.9 nm (Figure 2b, PC-UCNPs/siRNA). Continuous surface modification of HA decreased zeta potential to −23 ± 2.6 mV due to carboxylate groups of HA and increased the hydrodynamic diameter of particles to 70 ± 2.5 nm (Figure 2b, PC-UCNPs/siRNA/HA).

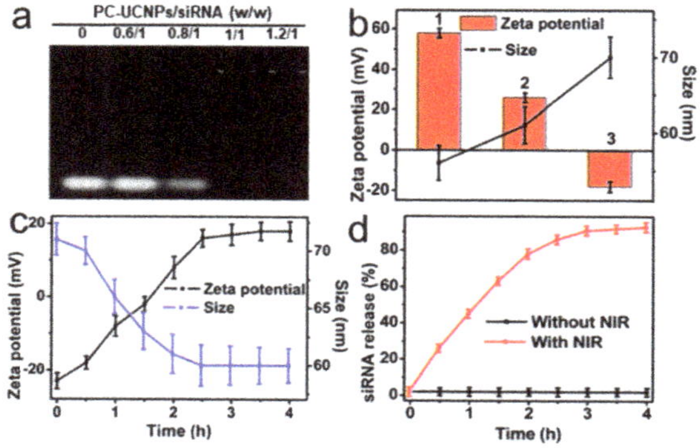

Figure 2. (**a**) Agarose gel retardation assay for PC-UCNPs/siRNA at various weight ratios (w/w). (**b**) Zeta potential and DLS analysis of (1) PC-UCNPs, (2) PC-UCNPs/siRNA, (3) PC-UCNPs/siRNA/HA. (**c**) Zeta potential and DLS analysis of PC-UCNPs/siRNA/HA in response to HAase at different time points. (**d**) The release curve of siRNA from PC-UCNPs/siRNA/HA as a function of time in response to NIR irradiation. The error bars indicate means ± SD (n = 4).

The serum stability of the delivery carrier is important to prolong circulation time and elevate tumor targeting [33]. The hydrodynamic diameter and polydispersity index (PDI) of PC-UCNPs/siRNA/HA remained unchanged when incubated with DMEM containing 10% fetal bovine serum (Figure S9), indicating the satisfactory stability of PC-UCNPs/siRNA/HA.

The negatively charged surface of PC-UCNPs/siRNA/HA can effectively extend systemic circulation time [34] and respond to internal tumor microenvironments for surface charge conversion to facilitate intracellular delivery. Due to the surface-adsorbed HA, PC-UCNPs/siRNA/HA was recognized by the CD44 membrane receptor. The HA coating layer was then degraded by the overexpressed HAase enzyme in the tumor cell [35], which exposed the positively charged surface of PC-UCNPs/siRNA to achieve the intracellular delivery and endosomal escape of PC-UCNPs/siRNA/HA. According to the HA degradation period, the zeta potential of PC-UCNPs/siRNA/HA continued its increase towards 25 ± 2.4 mV, indicating the gradual decomposition of negatively charged HAs and exposure of the positively charged PC-UCNPs/siRNA surface (Figure 2c, zeta potential). Along with the degradation of HA, the hydrodynamic size of PC-UCNPs/siRNA/HA continued its decrease to 62 ± 2.7 nm (Figure 2c, size), which was similar to that of PC-UCNPs/siRNA (Figure 2b, PC-UCNPs/siRNA, size), indicating the complete degradation of HA from the UCNPs surface. SiRNA release was achieved via external NIR irradiation, and siRNA-loaded PC-UCNPs/siRNA/HA with NIR irradiation showed rapid release, and the release percentage saturated at around 92% at 3 h (Figure 2d). On the contrary, PC-UCNPs/siRNA/HA showed a negligible release of siRNA (2%) in the absence of NIR irradiation (Figure 2d), indicating a satisfactory precise control of siRNA release via NIR light irradiation.

2.3. Internalization of PC-UCNPs/siRNA/HA and Intracellular siRNA Release

After the degradation of the HA coating layer, the highly cationic surface of PC-UCNPs/siRNA could facilitate the lysosomal escape of PC-UCNPs/siRNA. Acridine orange (AO) assay was conducted to monitor the lysosomal membrane permeabilization of HepG2 cells. HepG2 cells were incubated with PBS, UCNPs@SiO_2-polymer, PC-UCNPs/HA, and PC-UCNPs/siRNA/HA, respectively, and stained with AO. Strong green fluorescence was observed from PBS-treated HepG2 cells and UCNPs@SiO_2-polymer-treated HepG2 cells (Figure 3a, PBS, UCNPs@SiO_2-polymer), while declined green fluorescence and increased red fluorescence was observed from PC-UCNPs/HA and PC-UCNPs/siRNA/HA-treated HepG2 cells (Figure 3a, PC-UCNPs/HA, PC-UCNPs/siRNA/HA), proving the contribution of cationic surface to lysosomal membrane permeabilization. A lysosomal colocalization experiment was further performed to trace the internalization and lysosomal escape of PC-UCNPs/siRNA/HA by CLSM images. FAM labelled siRNA (siRNA-FAM) was used to prepare PC-UCNPs/siRNA-FAM/HA, which demonstrated good overlap of green fluorescence of FAM and red fluorescence of Lysotracker Red after 2 h incubation, confirming efficient internalization of PC-UCNPs/siRNA-FAM/HA (Figure 3b, 2 h). After 6 h of incubation, the FAM fluorescence had spread to the entire cytoplasm, while the lysosome probe showed weaker fluorescence due to the acidity dependence of Lysotracker red (Figure 3b, 6 h) [36], indicating the efficient lysosomal escape of PC-UCNPs/siRNA-FAM/HA.

To verify the successful release of siRNA from PC-UCNPs to cytoplasm, PC-UCNPs/siRNA-FAM/HA was incubated with HepG2 cells for 4 h, and UCNPs emission at 454 nm and 475 nm and FAM emission at 529 nm was monitored via CLSM. Separation of FAM fluorescence (green channel) and UCNPs fluorescence (red channel) was observed for PC-UCNPs/siRNA-FAM/HA-treated HepG2 cells after 40 min of NIR light irradiation (Figure 4a, NIR(+)), indicating the efficient release of siRNA from PC-UCNPs. On the contrary, no fluorescence separation was observed for PC-UCNPs/siRNA-FAM/HA-treated HepG2 cells in the absence of NIR irradiation (Figure 4a, NIR(−)).

HA-mediated endocytosis specificity was further confirmed by CLSM images. Strong intracellular FAM fluorescence was observed for PC-UCNPs/siRNA-FAM/HA incubated HepG2 cells (Figure 4b, HepG2), while excess HA pretreated HepG2 cells showed little fluorescence after incubation with PC-UCNPs/siRNA/HA (Figure 4b, HA pretreated HepG2). Moreover, the semiquantitative analysis of intracellular FAM-siRNA fluorescence also showed higher intensity for HepG2 cells than HepG2 cells pretreated with HA (Figure S10). These results indicated good targeting specificity for HA-mediated cancer cell internalization.

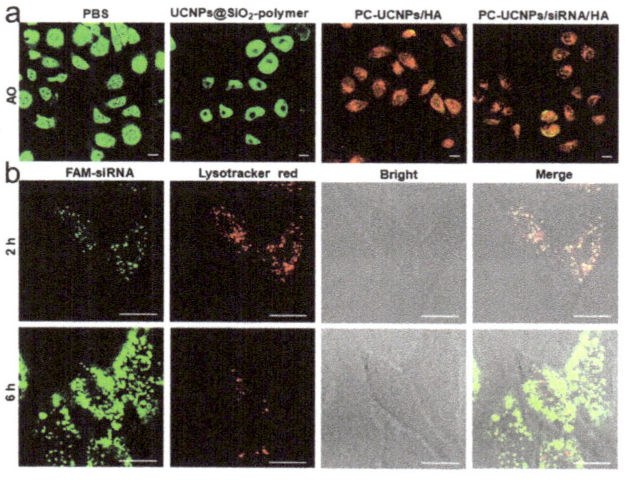

Figure 3. Confocal microscopic images of (**a**) AO-stained HepG2 cells treated with PBS, UCNPs@SiO_2-polymer, PC-UCNPs/HA and PC-UCNPs/siRNA/HA (with 10 nM siRNA loading, Scale bar: 20 μm), (**b**) colocalization of lysosome and PC-UCNPs/siRNA/HA when incubated with HepG2 cells for different durations (With 10 nM siRNA loading, Scale bar: 8 μm).

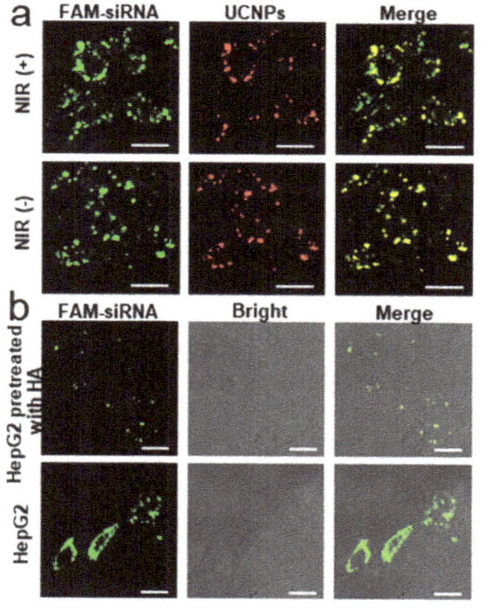

Figure 4. Confocal microscopic images of (**a**) co-localization of UCNPs at 454 nm and 475 emission and FAM-siRNA at 529 nm for PC-UCNPs/siRNA/HA-incubated HepG2 cells in the presence and absence of NIR irradiation (Scale bar: 25 μm), (**b**) PC-UCNPs/siRNA/HA-incubated HepG2 cells and PC-UCNPs/siRNA/HA-incubated HepG2 cells that pretreated free HA (Scale bar: 10 μm).

2.4. *In Vitro Therapeutic Effect of PC-UCNPs/siRNA/HA*

The polo-like family of serine/threonine protein kinases (PLKs) is the key to cell cycle regulation and proliferation, and PLK1 can promote oncogenic transformation; however, inhibiting PLK1 expression can cause mitotic catastrophe, cell cycle arrest and apoptosis [37, 38]. Therefore, PLK1 siRNA was chosen as the sample siRNA and loaded to PC-UCNPs to verify the therapeutic effect of PC-UCNPs/siRNA/HA. HepG2 cells were incubated with PC-UCNPs/siPLK1/HA for 6 h, irradiated with 980 nm light for 2 h (6 min intervals for every 10 min of light exposure to avoid heating), and continuously cultured for 24 h, and PLK1 mRNA expression and PLK1 protein expression were measured via qRT-PCR and enzyme-linked immunosorbent assay (ELISA), respectively. The PLK1 mRNA expression was down-regulated to 31% for the PC-UCNPs/siPLK1/HA-incubated HepG2 cells during NIR irradiation (Figure 5a, column 4), while PC-UCNPs/siPLK1/HA-incubated HepG2 cells showed 92% of mRNA expression in the absence of NIR irradiation (Figure 5a, column 2), and PC-UCNPs/siRNA'(nontherapeutic siRNA)/HA-incubated HepG2 cells showed 89% of mRNA expression during NIR irradiation (Figure 5a, column 3). The PLK1 protein expression in HepG2 cells with different treatments showed similar tendency with 36% of PLK1 protein expression for PC-UCNPs/siPLK1/HA-treated HepG2 cells during NIR irradiation (Figure 5b, column 4) and very limited suppressions of PLK1 protein expression for both PC-UCNPs/siPLK1/HA-treated HepG2 cells without NIR light irradiation (Figure 5b, column 2) and PC-UCNPs/siRNA' (nontherapeutic siRNA)/HA-treated HepG2 cells with NIR light irradiation (Figure 5b, column 3). In addition, PC-UCNPs/siPLK1/HA-treated Hela cells during NIR irradiation showed similar tendencies in mRNA expressions and PLK1 protein expressions (Figure S11a,b). These results demonstrated the satisfactory gene silencing efficiency and protein expression suppression of PC-UCNPs/siRNA/HA.

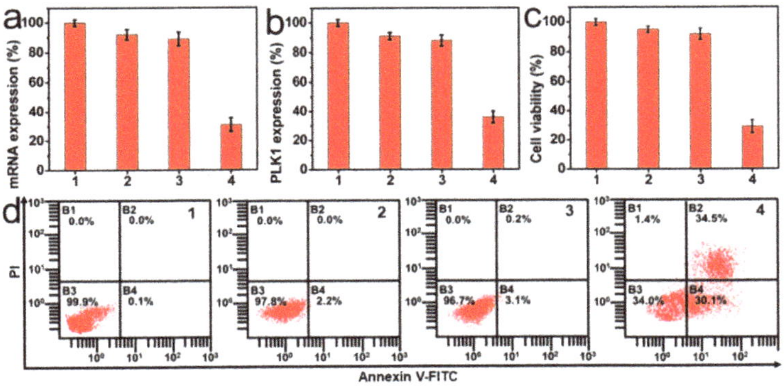

Figure 5. Gene silencing and cell viability. Expression levels of (**a**) PLK1 mRNA expression via qRT-PCR and (**b**) PLK1 protein expression via ELISA. Relative HepG2 cell viability determined via (**c**) MTT and (**d**) Annexin V-fluorescein isothiocyanate (FITC)/propidium iodide (PI) apoptotic kit. (1) Untreated HepG2 cells, (2) PC-UCNPs/siPLK1/HA-treated HepG2 cells without NIR irradiation, (3) PC-UCNPs/siRNA'/HA treated HepG2 cells with NIR irradiation and (4) PC-UCNPs/siPLK1/HA treated HepG2 cells with NIR irradiation. The error bars indicate means ± SD ($n = 5$).

The therapeutic effect of PC-UCNPs/siRNA/HA was evaluated using a standard MTT assay. HepG2 cells treated with PC-UCNPs/siPLK1/HA showed 29% of cell viability under NIR irradiation (Figure 5c, column 4), while cell viability was 96% for PC-UCNPs/siPLK1/HA-treated HepG2 cells in the absence of NIR irradiation (Figure 5c, column 2) and 94% for PC-UCNPs/siRNA'/HA-treated HepG2 cells under NIR irradiation (Figure 5c, column 3). Meanwhile, PC-UCNPs/siPLK1/HA-treated Hela cells showed 49% cell viability during NIR irradiation (Figure S11c). These results indicated the high therapeutic efficiency of PC-UCNPs/siRNA/HA. Flow cytometric assay results also showed the highest apoptosis rate of 64.6% for PC-UCNPs/siRNA/HA-treated HepG2 cells under NIR irradiation (Figure 5d). In addition, PC-UCNPs/siRNA/HA demonstrated satisfactory biocompatibility, with an 87% cell viability even at high concentrations, e.g., 200 µg/mL (Figure S11). Moreover, UCNPs/siRNA/HA showed a hemolysis percentage of 3.6% at a concentration of 1000 µg/mL (Figure S12), further indicating good biocompatibility.

2.5. In Vivo Therapeutic Efficiency of PC-UCNPs/siRNA/HA

Considering the overexpression of the CD44 receptor on the surface of the membrane, HepG2 cell tumor-bearing nude mice could be optimal tumor models [34]. The in vivo antitumor efficiency of PC-UCNPs/siPLK1/HA was evaluated using HepG2 cell tumor-bearing nude mice. PC-UCNPs/siPLK1/HA-injected mice after 2 h of NIR irradiation (6 min intervals for every 10 min of light exposure to avoid heating) showed the most effective inhibition of tumor growth compared with the PBS-treated mice group, and the PC-UCNPs/siPLK1/HA-treated mice group without NIR irradiation (Figure 6a), indicating the prominent in vivo antitumor capability of PC-UCNPs/siPLK1/HA. There was no noticeable body weight change in the mice during the treatment period (Figure 6b), indicating satisfactory biocompatibility and therapeutic specificity of PC-UCNPs/siPLK1/HA. All mice were euthanized on day 14, and the tumors were collected, weighed, sliced and stained using H&E and TUNEL to observe apoptosis via a CLSM image. PC-UCNPs/siPLK1/HA-treated mice group with NIR light irradiation also had the smallest tumor weight and size (Figure 6c,d) with the maximum disappearance of tumor nucleus in H&E images and the highest level of cell apoptosis in TUNEL images (Figure 6e). The in vivo applicability of PC-UCNPs/siRNA/HA was only evaluated for HepG2 cell tumor bearing mice and demonstrated satisfactory therapeutic efficiency. By conjugating with tumor targeting molecules, such as folic acid and aptamer AS1411, the as-presented PC-UCNPs/siRNA/HA nanoparticles could be widely suitable for a variety of tumor models.

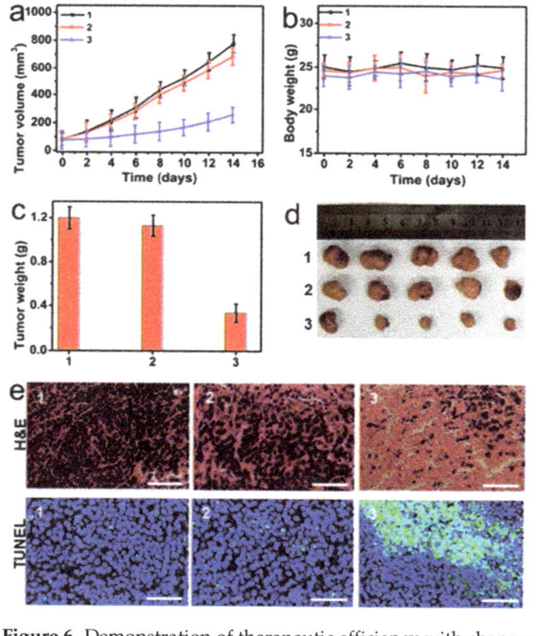

Figure 6. Demonstration of therapeutic efficiency with change of (**a**) tumor volume, (**b**) body weight of mice as a function of time, (**c**) tumor weight, (**d**) representative photos, (**e**) H&E staining and TUNEL staining of collected tumors for mice groups treated with PBS (1), PC-UCNPs/siPLK1/HA without NIR irradiation (2), PC-UCNPs/siPLK1/HA with NIR irradiation (3). The error bars indicate means ± SD (*n* = 5) (Scale bar, 100 μm).

3. Experimental

3.1. Materials and Reagents

Acryloyl chloride, 5-Hydroxy-2-nitrobenzaldehyde (98%), 2-(Dimethylamino) ethyl chloride hydrochloride (87%), Carbon disulfide (CS_2), 1-Propanethiol, 2-Bromo-2-methylpropionic acid, 1-

Ethyl-(3-dimethyllaminopropyl) carbodiie hydrochlide (EDC·HCl), N-Hydroxysuccinimide (NHS), (3-aminopropyl) triethoxysilane (APTES), Poly(oligo(ethylene glycol)) methyl ether acrylate (OEMA, Mn = 550 g/mol), 2,2′-Azoisobutyronitrile (AIBN) were purchased from J&K Chemical Co., Ltd. (Beijing, China). Anhydrous yttrium chloride (YCl_3) (99.9%), anhydrous ytterbium chloride ($YbCl_3$) (99.9%) and anhydrous thulium chloride ($TmCl_3$) (99.9%) were purchased from Sigma-Aldrich (Burlington, MA, USA). Tetraethylorthosilicate (TEOS), Sodium hydroxide (NaOH), Potassium phosphate (K_3PO_4), Ammonium fluoride (NH_4F), Cyclohexene, Oleic acid (OA), 1-Octadecene (ODE), Triethylamine (TEA), CO-520 surfactant, Tetraethyl orthosilicate (TEOS), (3-aminopropyl)triethoxysilane, Sodium borohydride ($NaBH_4$) were purchased from Alfa Aesar (Shanghai, China). Dulbecco's modified eagle's medium (DMEM), 4,6-Diamidino-2-phenylindole (DAPI), Trypsin, Penicillin streptomycin, 3-[4,5-dimethylthiazol-2-yl]-2,5-diphenyltetra-zoliumbromide (MTT), Fetal bovine serum (FBS), Phosphate buffered saline (PBS, pH 7.4), 6 x RNA loading buffer and Annexin V-FITC apoptosis detection kit were purchased from KeyGEN BioTECH (Nanjing, China). HepG2 human liver cancer cell lines were purchased from cell bank of Chinese Academy of Sciences. RNA extraction kit, PrimeScript RT reagent kit and SYBR premix EX Taq kit were purchased from Takara (Dalian, China). PLK1 ELISA kit was purchased from Jin Yibai Biological Technology (Shanghai, China). siRNA targeting PLK1 mRNA sequence (sense strand: 5′-AUAUUCGACUUUGGUUGCCTT-3′, antisense strand: 5′-GGCAACCAAAGUCGAAUAUTT-3′) and FAM fluorescein-labelled negative control siRNA sequence (FAM-siRNA) (sense strand: 5′-FAM-UUCUCCGAACGUGUCACGUTT-3′, antisense strand: 5′-ACGUGACACGUUCGGAGAATT-3′) were purchased from GenePharma Co., Ltd. (Shanghai, China). All other reagents and solvents were of analytical grade and used directly.

3.2. Apparatus

UV−Vis absorption spectra were recorded on a Nanodrop-2000C UV-Vis spectrophotometer (Nanodrop, Madison, WI, USA). Fourier-transform infrared (FT-IR) spectra were recorded on a Nicolet 6700 spectrophotometer (Nicolet Plastique, Nicolet, QC, Canada). Thermogravimetric analysis (TGA) was performed on a Setaram TGA 92 instrument (Setaram, Caluire-et-Cuire, France). Upconversion emission spectra were measured with a ZolixScan ZLX-UPL spectrometer (Beijing, China) with an external continuous-wave laser (980 nm) as the excitation source. Transmission electron microscopy (TEM) micrographs were acquired with JEM-2100 transmission electron microscope (JEOL Ltd., Tokyo, Japan), and confocal fluorescence images were acquired on TCS SP5 confocal laser scanning microscope (Leica, Frankfurt, Germany). Flow cytometric analysis was performed on a Coulter FC-500 flow cytometer (Beckman-Coulter, Pasadena, CA, USA). MTT and ELISA assays were performed on a Hitachi/Roche System Cobas 6000 (Bio-Rad, Hercules, CA, USA). Real-time PCR was carried out using a CFX96 touch real-time PCR detection system (Bio-Rad, Hercules, CA, USA). Zeta potential and dynamic light scattering (DLS) were measured via a Zetasizer Nano-ZS (Malvern Instruments, Malvern, UK) and ZetaPlus 90 Plus/BI-MAS (Brook Hhaven, New York, NY, USA), respectively.

3.3. Synthesis of 2-(((propanethio) carbonothioyl)thio)acetic Acid (CTA)

The 1-Propanethiol (548 mg, 7.19 mmol) was added to 10 mL acetone containing K_3PO_4 (1.53 g, 7.19 mmol) with magnetic stirring. After stirring for 10 min, CS_2 (1.64 g, 21.56 mmol) was added, and the reaction continued for 10 min. Additionally, 2-bromo-2-methylpropionic acid (1 g, 5.99 mmol) was then added and stirred overnight. The solvent was removed, and the residue was extracted using CH_2Cl_2 and 1 M HCl aqueous solution. The collected organic phase was then washed with water and dried using anhydrous Na_2SO_4, and the CH_2Cl_2 was removed via rotary evaporation. The crude product was further purified via recrystallization with hexane to obtain a yellow product (3.07 g, 65% yield). Moreover, ^{1}H-NMR ($CDCl_3$, 400 MHz, 298 K) δ (ppm): 3.27 (t, 2H), 1.78–1.67 (m, 8H), 1.00 (t, 3H) (Figure S1).

3.4. Synthesis of 5-(2-(dimethylamino)ethoxy)-2-nitrobenzyl Acrylate (MENA) Monomer

Synthesis of 5-hydroxy-2-nitrobenzyl alcohol (Compound 1): 5-hydroxy-2-nitrobenzaldehyde (500 mg, 3 mmol) in 10 mL methanol was syringed into the Schleck flask in an ice water bath under N_2 protection, $NaBH_4$ (226 mg, 5 mmol) was then slowly added, and the solution was stirred at 0 °C for 24 h. The reaction was cautiously quenched by the addition of a 10% HCl solution, adjusted to a pH of 2, and extracted using ethyl acetate. Organic layer was collected, washed with saturated NaCl solution, dried with anhydrous $MgSO_4$, filtered, and concentrated by a rotary evaporator. The crude product was further purified via silica gel column chromatography (hexane/ethyl acetate, $v/v = 1/1$). In addition, 5-hydroxy-2-nitrobenzyl alcohol (Compound 1) was obtained as a yellowish solid (705 mg, 97% yield). ^{1}H NMR (DMSO-d_6, 400 MHz, 298 K) δ (ppm): 10.88 (s, 1H, Ar–OH), 8.04, 7.24 and 6.78 (m, 3H, Ar), 5.5 (s, 1H, –CH_2OH), 4.81 (s, 2H, –CH_2OH) (Figure S3).

Synthesis of 5-(2-(dimethylamino)ethoxy)-2-nitrobenzyl alcohol (Compound 2): the mixture of compound 1 (500 mg, 2.96 mmol), 2-(dimethylamino)ethyl chloride hydrochloride (381 mg, 3.05 mmol) and NaOH (260 mg, 6.5 mmol) were added into a Schleck flask with both 15 mL toluene and 3 mL ethanol under N_2 protection. The reaction mixture was refluxed at 120 °C for 48 h and poured into 30 mL water after cooling. The organic layer was isolated, washed with saturated NaCl solution, dried over anhydrous Na_2SO_4, and concentrated by rotary evaporator. The as-obtained crude product was further purified via silica gel column chromatography using methanol as the eluent, and 5-(2-(dimethylamino)ethoxy)-2-nitrobenzyl alcohol (Compound 2) was obtained as a yellowish solid (572 mg, 65% yield). ^{1}H NMR ($CDCl_3$, 400 MHz, 298 K) δ (ppm): 8.16, 7.26, and 6.93 (m, 3H, Ar), 5.00 (s, 2H, –CH_2OH), 4.24 (t, 2H, –$OCH_2CH_2N(CH_3)_2$), 2.86 (t, 2H, –$OCH_2CH_2N(CH_3)_2$), 2.43 (s, 6H, –$OCH_2CH_2N(CH_3)_2$) (Figure S4).

Synthesis of 5-(2-(dimethylamino)ethoxy)-2-nitrobenzyl acrylate (MENA): Compound 2 (500 mg, 2.08 mmol) was added to the mixture solution of 7 mL anhydrous THF and 333 μL TEA at 0 °C under N_2 protection. Acryloyl chloride (198 μL, 2.185 mmol) in 500 μL anhydrous THF was then added dropwise. After reaction for 5 h at room temperature, the insoluble salts were removed via filtration. The filtrate was concentrated and further purified using silica gel column chromatography (hexane/ethyl acetate, $v/v = 3/1$) as the eluent. Additionally, 5-(2-(dimethylamino)ethoxy)-2-nitrobenzyl acrylate (MENA) was obtained as a yellowish solid (380 mg, 73% yield). ^{1}H NMR ($CDCl_3$, 400 MHz, 298 K) δ (ppm): 8.22, 7.09, and 6.94 (m, 3H, Ar), 6.54 (d, 1H, $CHCOOCH_2$–), 6.26 (m, 1H, $CHCOOCH_2$–), 5.95 (d, 1H, $CHCOOCH_2$–), 5.64 (2, 2H, Ar-CH_2O-), 4.21 (t, 2H, –$OCH_2CH_2N(CH_3)_2$), 2.86 (t, 2H, –$OCH_2CH_2N(CH_3)_2$), 2.42 (s, 6H, –$OCH_2CH_2N(CH_3)_2$) (Figure S5).

3.5. Synthesis of Core-Shell UCNPs: $NaYF_4$:Yb,Tm@$NaYF_4$

The UCNPs core $NaYF_4$:Yb,Tm was synthesized according to the previous report [39]. Briefly, YCl_3 (0.78 mmol), $YbCl_3$ (0.2 mmol), and $TmCl_3$ (0.02 mmol) were mixed with 6 mL OA and 15 mL ODE, heated to 150 °C and stirred for 2 h to remove oxygen and water. After cooling down to 30 °C, NH_4F (4 mmol) and NaOH (2.5 mmol) in 10 mL methanol were added dropwise into the mixture solution within 15 min. The reaction solution was then heated to 45 °C for 30 min, 90 °C for 30 min and 310 °C for 1 h under N_2 protection. The reaction solution was then cooled to room temperature and precipitated with the mixture of ethanol and acetone to obtain UCNPs core and re-dispersed in cyclohexane for further use.

For the synthesis of core–shell structured UCNPs $NaYF_4$:Yb,Tm@$NaYF_4$, YCl_3 (0.08 mmol) were mixed in ODE (15 mL) and OA (6 mL), heated to 150 °C, stirred for 2 h and subsequently cooled down to 75 °C. The above-obtained UCNPs core $NaYF_4$:Yb,Tm was added into the reaction mixture, and the reaction mixture was heated to 90 °C to remove the cyclohexane. After cooling to room temperature, NH_4F (4 mmol) and NaOH (2.5 mmol) in 10 mL methanol were added and continuously stirred at 45 °C for 30 min, at 90 °C for 30 min and at 290 °C for 45 min under N_2 protection to obtain core–shell structured UCNPs

$NaYF_4$:Yb,Tm@$NaYF_4$. The as-obtained UCNPs $NaYF_4$:Yb,Tm@$NaYF_4$ were precipitated with ethanol and re-dispersed in 10 mL of cyclohexane for future use. The morphology of as-prepared UCNPs was characterized by TEM, and its properties were characterized via fluorescence spectrometer.

3.6. Preparation of UCNPs@SiO_2-CTA

The UCNPs@SiO_2-NH_2 was synthesized via the reversed-phase microemulsion method according to the previous report with a slight modification [40]. Briefly, 500 μL CO-520 surfactant was added into 1 mg/mL of above-obtained UCNPs in 20 mL cyclohexane and sonicated for 10 min. A total of 100 μL $NH_3 \cdot H_2O$ (wt 30%) was added into the above mixture solution and continuously sonicated for 10 min until it formed a transparent solution. Then, 36 μL TEOS was added into the mixture solution, reacted for 3 h and centrifuged. The as-obtained UCNPs@SiO_2 was redispersed in ethanol, 400 μL of (3-aminopropyl)triethoxysilane and 70 μL of $NH_3.H_2O$ were added and stirred at 80 °C for 8 h. The as-obtained UCNPs@SiO_2-NH_2 was centrifuged and redispersed in DMSO for further use.

The mixture of CTA (20 mg, 0.08 mmol), NHS (9.7 mg, 0.08 mmol) and EDC·HCl (16.9 mg, 0.088 mmol) was added into 10 mL DMSO. After reaction for 4 h, the above-obtained UCNPs@SiO_2-NH_2 was added into the reaction solution, continuously reacted for 24 h, and centrifuged to obtain UCNPs@SiO_2-CTA. The property of UCNPs@SiO_2-CTA was characterized using UV-Vis spectrum.

3.7. Preparation of Photo-Cleavable Polycations-Wrapped UCNPs (PC-UCNPs)

The above obtained UCNPs@SiO_2-CTA (1.2 g), poly(oligo(ethylene glycol)) methyl ether acrylate (OEMA, 4.56 g, 9.6 mmol), MENA (744 mg, 2.4 mmol) and AIBN (12 mg) were added to a glass ampoule with anhydrous 1,4-dioxane (19.2 mL). The ampoule was degassed through three freeze–thaw cycles, sealed under vacuum, maintained in an oil bath preheated to 70 °C and kept for polymerization reaction for 24 h. The polymerization reaction was quenched with liquid nitrogen, and the reaction mixture was centrifuged at 8000 rpm for 5 min and washed with methanol. To endow positive charge to the as-obtained UCNPs@SiO_2-polymer, it (0.30 g) was dispersed in CH_2Cl_2 (30 mL), slowly added to CH_3I (0.1 g, 2 mmol) via a dropping funnel, and kept in the dark for 24 h at room temperature. The as-obtained PC-UCNPs was collected after centrifugation, dispersed in 10 mL PBS and stored at 4 °C for further use. The obtained PC-UCNPs were characterized via TEM, DLS and TG.

The gel permeation chromatographic (GPC) analysis of polycations coating of P(OEMA-co-MENA) was performed by dispersing above-obtained PC-UCNPs (200 mg) in hydrofluoric acid (HF, 5 mL), stirring it for 6 h at 50 °C and centrifuging it at 8000 rpm for 5 min. The supernatant was then collected, dialyzed (molecular weight cut off, 3.5 kDa) and freeze-dried. The obtained powder (1 mg) was mixed with THF (1 mL, mass spectrum grade) and analyzed by GPC using THF as eluent and polystyrene as reference polymer.

3.8. Preparation of siRNA-Loaded and HA-Wrapped PC-UCNPs (PC-UCNPs/siRNA/HA)

siRNA was loaded into the above-obtained PC-UCNPs via electrostatic interaction. After mixing 2 μg of siRNA in RNase-free DEPC with 2 μL of 6 X siRNA loading buffer, PC-UCNPs were added to PBS with a series of mass ratio (PC-UCNPs/siRNA (*w*/*w*) of 0, 0.6/1, 0.8/1, 1/1, 1.2/1), incubated at room temperature for 30 min and applied for the agarose gel retardation assay. Hyaluronic acid (HA) was then wrapped to the as-obtained PC-UCNPs/siRNA via electrostatic interaction to obtain PC-UCNPs/siRNA/HA. HA (50 mM) in PBS buffer was added into the prepared PC-UCNPs/siRNA with a mass ratio of 1/1, incubated for 10 min at room temperature and centrifuged at 8000 rpm for 5 min to obtain PC-UCNPs/siRNA/HA. The surface modification process was then characterized using zeta potential.

3.9. In Vitro Verification of HA Degradation and siRNA Release

PC-UCNPs/siRNA/HA was incubated with 0.5 mg/mL of HAase solution (HEPES buffer, pH 6.5) at 37 °C, and the degradation of HA was characterized by DLS and zeta potential over prearranged time intervals. To obtain the release profile of siRNA, the as-treated solutions were exposed under 980 nm of irradiation (2 W/cm^{-2}) for different durations, and the as-obtained solutions were centrifuged at predetermined time intervals. The supernatant containing released siRNA was measured via UV-Vis spectroscopy.

3.10. Cell Culture

The human hepatocellular carcinoma cells (HepG2, cancer cell) were incubated with DMEM medium containing 10% FBS and 1% penicillin-streptomycin (10,000 U/mL) at 37 °C in a humidified atmosphere containing 5% CO_2.

3.11. Intracellular Delivery of PC-UCNPs/siRNA/HA

To confirm the intracellular delivery specificity, HepG2 cells (CD44 receptor overexpression) were firstly pretreated with 2 mg/mL free HA at 37 °C for 1 h and then incubated with FAM-siRNA-loaded PC-UCNPs (PC-UCNPs/siRNA-FAM/HA, 5 μg/mL with siRNA 10 nM) for 6 h to observe FAM fluorescence using CLSM. Moreover, the fluorescence of directly PC-UCNPs/siRNA-FAM/HA-incubated HepG2 was observed.

3.12. Lysosomal Membrane Permeabilization of HepG2 Cells

To prove the lysosomal membrane permeabilization of HepG2 cells, HepG2 cells were incubated with PBS, UCNPs@SiO-polymer (in the absence of cationic quaternary ammonium groups), PC-UCNPs/HA, PC-UCNPs/siRNA/HA (5 μg/mL with siRNA 10 nM), respectively, for 6 h. The as-treated cells were stained with acridine orange (AO) (100 μL, 5 μg/mL) for 30 min, washed with PBS twice, and AO fluorescence was taken via CLSM.

3.13. Lysosomal Escape of siRNA

To verify the lysosomal escape of siRNA, HepG2 cells were incubated with 5 μg/mL of PC-UCNPs/siRNA-FAM/HA (siRNA 10 nM) for 2 h and 6 h, respectively. The as-treated cells were washed with PBS, incubated with 75 nM LysoTracker Green for 30 min, and observed via CLSM.

3.14. MTT Assay

To perform MTT assay, HepG2 cells (4000 cells/well) were seeded in a 96-well plate and incubated overnight in 200 μL of DMEM containing 10% FBS and 1% penicillin-streptomycin, the incubation medium was then replaced with 200 μL of fresh DMEM containing serial concentrations (40, 80, 120, 160 and 200 μg/mL) of PC-UCNPs/siRNA/HA and continuously incubated for 24 h. A total of 20 μL MTT (5 mg/mL in PBS) was subsequently added per well and continuously incubated for 4 h. After removing cell culture medium, 80 μL of DMSO was added to dissolve the crystals precipitate. The optical density was measured at 490 nm with a Bio-Rad microplate reader. Moreover, PBS-treated HepG2 cells were set as control group. The relative cell viability (%) was calculated using $(A_{test}/A_{control}) \times 100\%$.

SiRNA' with scrambled sequence and no biotoxicity was loaded to PC-UCNPs, and the as-obtained PC-UCNPs/siRNA'/HA was used to investigate the biotoxicity of PC-UCNPs/siRNA'/HA (siRNA 5 nM). Therapeutic siPLK1 was loaded to PC-UCNPs, and the as-obtained PC-UCNPs/siPLK1/HA (siPLK1 5 nM) was used to evaluate the therapeutic effect of PC-UCNPs/siPLK1/HA. PC-UCNPs/siPLK1/HA-incubated HepG2 cells were irradiated with the 980 nm laser (2 W/cm^2) for 2 h (6 min intervals for every 10 min of light exposure to avoid heating) and continuously incubated for 24 h for MTT assay. At the same time, Hela cells were also used to evaluate the therapeutic effect of PC-UCNPs/siPLK1/HA.

3.15. Hemolysis Assay of PC-UCNPs/siRNA/HA

Blood samples from healthy mice were collected in a container filled with heparin and centrifuged at 3000 rpm for 6 min to obtain red blood cells (RBCs). After washing and diluting with saline, RBCs (200 μL) were mixed with PC-UCNPs/siRNA/HA (800 μL) in saline at different concentrations (200, 400, 600, 800 and 1000 μg/mL) and incubated at 37 °C. After incubation for 4 h, the mixture solution was centrifuged, and the supernatant absorbance at 570 nm (A_{sample}) was measured via UV−vis. Moreover, RBCs were mixed with saline and water as a negative control (A_{saline}) and a positive control (A_{water}), respectively. The hemolysis percentage was calculated via $(A_{sample} - A_{saline})/(A_{water} - A_{saline}) \times 100\%$.

3.16. Gene Silencing Assay

HepG2 cells were seeded into a 6-well plate at a density of 5×10^5 cells/well and cultured at 37 °C overnight. The old medium was displaced with fresh medium containing PC-UCNPs/siPLK1/HA at siPLK1 concentrations of 250 nM, incubated for 4 h and exposed under 980 nm laser (2 W/cm^2) for 2 h (6 min intervals for every 10 min of light exposure to avoid heating). The as-treated HepG2 cells were continuously cultured for 24 h, and the expression levels of PLK1 mRNA and protein were evaluated using RT-PCR and ELISA kit following the manufacturer's instructions. The control group was set as siRNA′ (scrambled sequence)-loaded PC-UCNPs/siRNA′/HA-treated HepG2 cells with NIR irradiation and PC-UCNPs/siPLK1/HA-treated HepG2 cells without NIR irradiation for the comparison of gene-silencing effect. Moreover, Hela cells were also adopted to evaluate the gene-silencing effect of PC-UCNPs/siPLK1/HA.

3.17. Cell Apoptosis Assay

HepG2 cells were seeded into a 6-well plate at a density of 2×10^5 cells/well and incubated with PC-UCNPs/siPLK1/HA at siPLK1 concentrations of 250 nM at 37 °C for 4 h, irradiated upon with a 980 nm laser for 2 h and continuously incubated for 24 h. Then, the as-treated cells were washed with PBS, stained with a mixture of 5.0 μL Annexin V-FITC and 5.0 μL propidium iodide for 10 min and measured with flow cytometry over FL1 (Annexin V-FITC) and FL3 (PI) channels. Control groups were set as siRNA′ (scrambled sequence)-loaded PC-UCNPs/siRNA′/HA-treated HepG2 cells with NIR irradiation and PC-UCNPs/siPLK1/HA-treated HepG2 cells without NIR irradiation.

3.18. In Vivo Antitumor Efficiency

Pathogen-free female BALB/c nude mice were purchased from KeyGEN BioTECH (Nanjing, China). All experiment procedures were approved by the Model Animal Research Center of KeyGEN BioTECH. To build the tumor model in mice, HepG2 cells were only selected as models since their overexpressed CD44 receptor on the surface of membrane.1.0×10^7 HepG2 cells were subcutaneously injected to establish HepG2 tumor xenograft mouse model. After the tumor volumes grew to 50 mm^3, the tumor-bearing mice were randomly divided into three groups, with five mice in each group, and intratumorally injected, respectively, with 50 μL of PBS (1 group), PC-UCNPs/siPLK1/HA (2 groups) at a dose of 1 mmol siPLK1 per mouse. At 2 h post-injection, one of the PC-UCNPs/siPLK1/HA-injected mouse group was exposed to a 980 nm laser irradiation (2 W/cm^2) for 2 h (6 min intervals for every 10 min of light exposure to avoid heating). The injection and irradiation processes were repeated every other two days, a total of six times. The tumor sizes were measured every 2 days with caliper. The tumor volumes were calculated as $V = (L \times W^2)/2$, where L and W are the length and width of the tumor, respectively. To evaluate the safety of PC-UCNPs/siPLK1/HA in mice, the mice weights were also recorded during the therapeutic process. After treatment for 14 days, all mice were sacrificed, and the obtained tumors were then collected, weighed, photographed and stained using H&E and TUNEL.

4. Conclusions

Photo-cleavable polycations wrapping upconversion nanoparticles for the spatial and temporal control of siRNA delivery and release were presented. Photo-cleavable 5-(2-(dimethylamino)ethoxy)-2-nitrobenzyl acrylate and Poly(oligo(ethylene glycol)) methyl ether acrylate were polymerized in situ on the chain transfer agent CTA functionalized UCNPs@SiO_2 surface via RAFT polymerization, which was loaded with therapeutic PLK1 siRNA and HA for NIR-controlled in vivo siRNA release and tumor therapy. The as-presented PC-UCNPs/siPLK1/HA showed good inhibition of cell viability and tumor growth and should become a promising platform for precise drug delivery and tumor therapy.

Supplementary Materials: The following supporting information can be downloaded at: https://www.mdpi.com/article/10.3390/targets1010006/s1, Figure S1: Synthetic route and ^{1}H NMR spectrum (in $CDCl_3$) of chain transfer agent CTA; Figure S2: UV-Vis spectra of UCNPs@SiO_2, UCNPs@SiO_2-CTA and CTA; Figure S3: Synthetic route and ^{1}H NMR spectrum (in DMSO-d_6) of 5-hydroxy-2-nitrobenzyl alcohol (Compound 1).; Figure S4: Synthetic route and ^{1}H NMR spectrum (in $CDCl_3$) of 5-(2-(dimethylamino)ethoxy)-2-nitrobenzyl alcohol (Compound 2); Figure S5: Synthetic route and ^{1}H NMR spectrum (in $CDCl_3$) of 5-(2-(dimethylamino)ethoxy)-2-nitrobenzyl acrylate (DMNA); Figure S6: Synthetic route of UCNPs@SiO_2-polymer via RAFT polymerization; Figure S7: Gel permeation chromatographic (GPC) analysis of P(OEOA-co-DMNA) from the surface of UCNP@SiO_2-polymer using THF as eluent; Figure S8: Zeta potential and DLS analysis of UCNPs@SiO_2-polymer and PC-UCNPs; Figure S9: DLS analysis and polydispersity index (PDI) of PC-UCNPs/siRNA/HA incubated with DMEM containing 10% fetal bovine serum for different times; Figure S10: Semiquantitative analysis of intracellular FAM fluorescence intensity for PC-UCNPs/siRNA/HA incubated HepG2 cells and HA pre-treated HepG2 cells; Figure S11: Gene silencing and cell viability. Expression levels of (a) PLK1 mRNA expression via qRT-PCR, (b) PLK1 protein expression via ELISA and (c) relative Hela cell viability via MTT. (1) Untreated Hela cells, (2) PC-UCNPs/siPLK1/HA treated Hela cells without NIR irradiation, (3) PC-UCNPs/siRNA'/HA treated Hela cells with NIR irradiation, (4) PC-UCNPs/siPLK1/HA treated Hela cells with NIR irradiation. The error bars indicate means $\pm$ SD (n = 5).; Figure S12: Relative cell viabilities of HepG2 cells treated with PC-UCNPs/siRNA/HA at the series concentrations; Figure S13: Hemolysis percentages of red blood cells (RBCs) incubated with 0, 200, 400, 600, 800, 1000 μg/mL PC-UCNPs/siRNA/HA. The data error bars indicate means $\pm$ SD (n = 4).

Author Contributions: Y.H., S.G. and Y.L. are responsible for executing the methodology. Y.H. and S.G. handle the software and validation. The formal analysis, investigation, data curation, and writing of the original draft are carried out by Y.H., Y.L. and H.J. are responsible for the writing, review, editing, and acquisition of funding. All authors have read and agreed to the published version of the manuscript.

Funding: We gratefully acknowledge the National Natural Science Foundation of China (21974064, 22022405, 22374073, 22101310, 82302278) and the State Key Laboratory of Analytical Chemistry for Life Science (5431ZZXM2204, 5431ZZXM2307).

Institutional Review Board Statement: The study was approved by the Model Animal Research Center of KeyGEN BioTECH (project identification code IACUC2021040501) on 5 April 2021.

Informed Consent Statement: Not applicable.

Data Availability Statement: The raw/processed data reported in this work are available upon request.

Conflicts of Interest: The authors declare no conflict of interest.

References

1. Conde, J.; Ambrosone, A.; Hernandez, Y.; Tian, F.; McCully, M.; Berry, C.C.; Baptista, P.V.; Tortiglione, C.; de la Fuente, J.M. 15 years on siRNA delivery: Beyond the state-of-the-art on inorganic nanoparticles for RNAi therapeutics. *Nano Today* **2015**, *10*, 421–450. [CrossRef]

2. Ozcan, G.; Ozpolat, B.; Coleman, R.L.; Sood, A.K.; Lopez-Berestein, G. Preclinical and clinical development of siRNA-based therapeutics. *Adv. Drug Deliv. Rev.* **2015**, *87*, 108–119. [CrossRef] [PubMed]
3. Kanasty, R.; Dorkin, J.R.; Vegas, A.; Anderson, D. Delivery materials for siRNA therapeutics. *Nat. Mater.* **2013**, *12*, 967–977. [CrossRef]
4. Liu, X.; Appelhans, D.; Voit, B.J. Hollow capsules with multiresponsive valves for controlled enzymatic reactions. *Am. Chem. Soc.* **2018**, *140*, 16106–16114. [CrossRef]
5. Wang, C.; Wang, X.; Du, L.; Dong, Y.; Hu, B.; Zhou, J.; Shi, Y.; Bai, S.; Huang, Y.; Cao, H.; et al. Harnessing pH-sensitive polycation vehicles for the efficient siRNA delivery. *ACS Appl. Mater. Interfaces* **2021**, *13*, 2218–2229. [CrossRef] [PubMed]
6. Li, R.; Huang, X.; Lu, G.; Feng, C. Characterization of a library of vitamin A-functionalized polymethacrylate-based nanoparticles for siRNA delivery. *Polym. Chem.* **2021**, *12*, 911–925.
7. Rapp, T.L.; DeForest, C.A. Targeting drug delivery with light: A highly focused approach. *Adv. Drug Deliv. Rev.* **2021**, *171*, 94–107. [CrossRef]
8. Lin, J.; Hu, J.; Wang, W.; Liu, K.; Zhou, C.; Liu, Z.; Kong, S.; Lin, S.; Deng, Y.; Guo, Z. Thermo and light-responsive strategies of smart titanium-containing composite material surface for enhancing bacterially anti-adhesive property. *Chem. Eng. J.* **2021**, *407*, 125783. [CrossRef]
9. Yang, X.; Chen, X.; Wang, Y.; Xu, G.; Yu, L.; Ding, J. Sustained release of lipophilic gemcitabine from an injectable polymeric hydrogel for synergistically enhancing tumor chemoradiotherapy. *Chem. Eng. J.* **2020**, *396*, 125320. [CrossRef]
10. Moroz-Omori, E.V.; Satyapertiwi, D.; Ramel, M.-C.; Høgset, H.; Sunyovszki, I.K.; Liu, Z.; Wojciechowski, J.P.; Zhang, Y.; Grigsby, C.L.; Brito, L.; et al. Photoswitchable gRNAs for spatiotemporally controlled CRISPR-Cas-based genomic regulation. *ACS Cent. Sci.* **2020**, *6*, 695–703. [CrossRef]
11. Li, S.; Xia, B.; Javed, B.; Hasley, W.D.; Melendez-Davila, A.; Liu, M.; Kerzner, M.; Agarwal, S.; Xiao, Q.; Torre, P.; et al. Direct visualization of vesicle disassembly and reassembly using photocleavable dendrimers elucidates cargo release mechanisms. *ACS Nano* **2020**, *14*, 7398–7411. [CrossRef]
12. Yang, S.; Pieters, P.A.; Joesaar, A.; Bögels, B.W.A.; Brouwers, R.; Myrgorodska, I.; Mann, S.; de Greef, T.F.A. Light-activated signaling in DNA-encoded sender–receiver architectures. *ACS Nano* **2020**, *14*, 15992–16002. [CrossRef] [PubMed]
13. Wang, S.-B.; Zhang, C.; Ye, J.-J.; Zou, M.-Z.; Liu, C.-J.; Zhang, X.-Z. Near-infrared light responsive nanoreactor for simultaneous tumor photothermal therapy and carbon monoxide-mediated anti-inflammation. *ACS Cent. Sci.* **2020**, *6*, 555–565. [CrossRef] [PubMed]
14. Li, Y.; Feng, P.; Wang, C.; Miao, W.; Huang, H. Black phosphorus nanophototherapeutics with enhanced stability and safety for breast cancer treatment. *Chem. Eng. J.* **2020**, *400*, 125851. [CrossRef]
15. Li, X.; Hu, H.; Shi, Y.; Liu, Y.; Zhou, M.; Huang, Z.; Li, J.; Ke, G.; Chen, M.; Zhang, X.-B. PtSnBi Nanoplates Enable Photoacoustic Imaging-Guided Highly Efficient Photothermal Tumor Ablation. *Chem. A Eng. J.* **2023**, *29*, e202203227. [CrossRef]
16. Wang, R.; Zhou, L.; Wang, W.; Li, X.; Zhang, F. In vivo gastrointestinal drug-release monitoring through second near-infrared window fluorescent bioimaging with orally delivered microcarriers. *Nat. Commun.* **2017**, *8*, 14702. [CrossRef] [PubMed]
17. Yang, Y.; Chen, Y.; Pei, P.; Fan, Y.; Wang, S.; Zhang, H.; Zhao, D.; Qian, B.-Z.; Zhang, B.-Z. Fluorescence-amplified nanocrystals in the second near-infrared window for in vivo real-time dynamic multiplexed imaging. *Nat. Nanotechnol.* **2023**, *6*, 1–10. [CrossRef]
18. Yan, B.; Boyer, J.C.; Branda, N.R.; Zhao, Y. Near-infrared light-triggered dissociation of block copolymer micelles using upconverting nanoparticles. *J. Am. Chem. Soc.* **2011**, *133*, 19714–19717. [CrossRef]
19. Wang, X.; Hu, J.; Liu, G.; Tian, J.; Wang, H.; Gong, M.; Liu, S. Reversibly switching bilayer permeability and release modules of photochromic polymersomes stabilized by cooperative noncovalent interactions. *J. Am. Chem. Soc.* **2015**, *137*, 15262–15275. [CrossRef]
20. Chen, G.; Ma, B.; Xie, R.; Wang, Y.; Dou, K.; Gong, S. NIR-induced spatiotemporally controlled gene silencing by upconversion nanoparticle-based siRNA nanocarrier. *J. Control. Release* **2018**, *282*, 148–155. [CrossRef]
21. Xie, S.; Du, Y.; Zhang, Y.; Wang, Z.; Zhang, D.; He, L.; Qiu, L.; Jiang, J.; Tan, W. Aptamer-based optical manipulation of protein subcellular localization in cells. *Nat. Commun.* **2020**, *11*, 1347. [CrossRef]
22. Wang, M.; Han, Y.; Yu, X.; Liang, L.; Chang, H.; Yeo, D.C.; Wiraja, C.; Wee, M.L.; Liu, L.; Liu, X.; et al. Upconversion Nanoparticle Powered Microneedle Patches for Transdermal Delivery of siRNA. *Adv. Healthcare Mater.* **2020**, *9*, 1900635. [CrossRef]
23. Zhang, Y.; Ren, K.; Zhang, X.; Chao, Z.; Yang, Y.; Ye, D.; Dai, Z.; Liu, Y.; Ju, H. Photo-tearable tape close-wrapped upconversion nanocapsules for near-infrared modulated efficient siRNA delivery and therapy. *Biomaterials* **2018**, *163*, 55–66. [CrossRef] [PubMed]
24. Wu, M.; Meng, Q.; Chen, Y.; Du, Y.; Zhang, L.; Li, Y.; Zhang, L.; Shi, J. Large-pore ultrasmall mesoporous organosilica nanoparticles: Micelle/precursor co-templating assembly and nuclear-targeted gene delivery. *Adv. Mater.* **2015**, *27*, 215–222. [CrossRef]
25. Lai, W.F.; Rogach, A.L.; Wong, W.T. Molecular design of upconversion nanoparticles for gene delivery. *Chem. Sci.* **2017**, *8*, 7339–7358. [CrossRef] [PubMed]
26. Kim, H.S.; Son, Y.J.; Mao, W.; Leong, K.W.; Yoo, H.S. Atom transfer radical polymerization of multishelled cationic corona for the systemic delivery of siRNA. *Nano Lett.* **2018**, *18*, 314–325. [CrossRef] [PubMed]
27. Martins, J.N.; Raimundo, B.; Rioboo, A.; Folgar-Cameán, Y.; Montenegro, J.; Basílio, N. Photoswitchable Calixarene Activators for Controlled Peptide Transport across Lipid Membranes. *J. Am. Chem. Soc.* **2023**, *145*, 13126–13133. [CrossRef] [PubMed]

28. Yonezawa, S.; Koide, H.; Asai, T. Recent advances in siRNA delivery mediated by lipid-based nanoparticles. *Adv. Drug Deliv. Rev.* **2020**, *154–155*, 64–78. [CrossRef]
29. Li, F.; Wang, M.; Guan, S.; Huang, Z.; Liu, S.; Li, X.; Jiang, X.; Luo, Q.; Xu, J.; Liu, J. Cucurbit[8]uril-based supramolecular polymer nanocapsules as an effective siRNA delivery platform for gene therapy. *Polym. Chem.* **2019**, *10*, 5659–5664. [CrossRef]
30. Haase, M.; Schafer, H. Upconverting nanoparticles. *Angew. Chem. Int. Ed.* **2011**, *50*, 5808–5829. [CrossRef]
31. Wang, Y.; Tu, L.; Zhao, J.; Sun, Y.; Kong, X.; Zhang, H. Upconversion Luminescence of β-NaYF4: Yb3+, Er3+@β-NaYF4 Core/Shell Nanoparticles: Excitation Power Density and Surface Dependence. *J. Phys. Chem. C* **2009**, *113*, 7164–7169. [CrossRef]
32. Xia, A.; Chen, M.; Gao, Y.; Wu, D.; Feng, W.; Li, F. Gd3+ complex-modified NaLuF4-based upconversion nanophosphors for trimodality imaging of NIR-to-NIR upconversion luminescence, X-Ray computed tomography and magnetic resonance. *Biomaterials* **2012**, *33*, 5394–5405. [CrossRef] [PubMed]
33. Malcolm, D.W.; Varghese, J.J.; Sorrells, J.E.; Ovitt, C.E.; Benoit, D.S.W. The Effects of Biological Fluids on Colloidal Stability and siRNA Delivery of a pH-Responsive Micellar Nanoparticle Delivery System. *ACS Nano* **2018**, *12*, 187–197. [CrossRef] [PubMed]
34. He, Y.; Guo, S.; Wu, L.; Chen, P.; Wang, L.; Liu, Y.; Ju, H. Near-infrared boosted ROS responsive siRNA delivery and cancer therapy with sequentially peeled upconversion nano-onions. *Biomaterials* **2019**, *225*, 119501. [CrossRef]
35. Gu, T.-T.; Li, C.; Xu, Y.; Zhang, L.; Shan, X.; Huang, X.; Guo, L.; Chen, K.; Wang, X.; Ge, H.; et al. Stimuli-responsive combination therapy of cisplatin and Nrf2 siRNA for improving antitumor treatment of osteosarcoma. *Nano Res.* **2020**, *13*, 630–637. [CrossRef]
36. Park, S.-J.; Park, W.; Na, K. Tumor Intracellular-Environment Responsive Materials Shielded Nano-Complexes for Highly Efficient Light-Triggered Gene Delivery without Cargo Gene Damage. *Adv. Funct. Mater.* **2015**, *25*, 3472–3482. [CrossRef]
37. Liu, F.; Park, J.-E.; Qian, W.-J.; Lim, D.; Gräber, M.; Berg, T.; Yaffe, M.B.; Lee, K.S.; Burke, T.R. Serendipitous alkylation of a Plk1 ligand uncovers a new binding channel. *Nat. Chem. Biol.* **2011**, *7*, 595–601. [CrossRef]
38. Barr, F.A.; Sillje, H.H.; Nigg, E.A. Polo-like kinases and the orchestration of cell division. *Nat. Rev. Mol. Cell Biol.* **2004**, *5*, 429–440. [CrossRef]
39. Liu, B.; Li, C.; Yang, P.; Hou, Z.; Lin, J. 808-nm-Light-excited lanthanide-doped nanoparticles: Rational design, luminescence control and theranostic applications. *Adv. Mater.* **2017**, *29*, 1605434. [CrossRef]
40. Sivakumar, S.; Diamente, P.R.; van Veggel, F.C. Silica-Coated Ln3+-Doped LaF3 Nanoparticles as Robust Down- and Upconverting Biolabels. *Chem. Eur. J.* **2006**, *12*, 5878–5884. [CrossRef]

targets

Article

An Activatable Nanoscintillator Probe for Detecting Telomerase Activity and Screening Inhibitors In Vivo

Baoliu Chen [1], Junduan Dai [1], Sijie Song [1], Xianzhe Tang [1], Yuheng Guo [1], Ting Wu [1], Mengnan Wu [1], Chaojie Hao [1], Xiaofeng Cheng [1], Xucong Lin [2], Yijie Bian [3], Zhaowei Chen [1,*] and Huanghao Yang [1,*]

1 MOE Key Laboratory for Analytical Science of Food Safety and Biology, New Cornerstone Science Laboratory, Fujian Provincial Key Laboratory of Analysis and Detection Technology for Food Safety, and State Key Laboratory of Photocatalysis on Energy and Environment, College of Chemistry, Fuzhou University, Fuzhou 350108, China
2 Engineering Technology Research Center on Reagent and Instrument for Rapid Detection of Product Quality and Food Safety in Fujian Province, College of Chemistry, Fuzhou University, Fuzhou 350108, China
3 College of Biological Science and Engineering, Fuzhou University, Fuzhou 350108, China; bianyijie@fzu.edu.cn
* Correspondence: chenzw@fzu.edu.cn (Z.C.); hhyang@fzu.edu.cn (H.Y.)

Abstract: Telomerase represents an essential molecular machinery for tumor occurrence and progression and a potential therapeutic target for cancer treatment. Sensitive and reliable analysis of telomerase activity is of significant importance for the diagnosis and treatment of cancer. In this study, we developed a telomerase-activated nanoscintillator probe for deep-tissue and background-free imaging of telomerase activity and screening telomerase inhibitors in tumor-bearing living mice models. The probe was constructed by modifying lanthanide-doped nanoscintillators with aptamer-containing DNA anchor strands which hybridized with quencher labelled–oligonucleotide strands and telomerase primers. The X-ray-induced fluorescence of the probe was quenched originally but turned on upon telomerase-catalyzed extension of the primer. Benefiting from exceptional tissue penetrating properties and negligible autofluorescence of X-ray excitation, this probe enabled direct detection of telomerase activity in vivo via fluorescence imaging. Furthermore, with the direct, readable fluorescent signals, the probe enabled the screening of telomerase inhibitors in living cells and whole-animal models in the native states of telomerase. This strategy would inspire the development of low autofluorescence and deep tissue bioimaging probes for disease diagnosis and drug development in high-level living settings.

Keywords: bioimaging; nanoscintillator; telomerase; drug screening; cancer

1. Introduction

Telomerase is a ribonucleoprotein reverse transcriptase that synthesizes telomeric DNA repeats at chromosome termini with an internal RNA template to compensate for the 'end-replication problem' [1,2]. Usually, telomerase activity is strictly repressed in most normal cells but greatly activated in a majority of tumor cells (85–90%) because the maintenance of telomere length by overexpressed telomerase is essential for tumors to escape the replicative senescence and crisis [3]. In this sense, telomerase has received enormous attention as an appealing molecular marker for cancer diagnosis and anticancer drug discovery [4–6]. This has spurred the development of polymerase chain reaction-based telomerase repeat amplification protocol (PCR-TRAP), the most classic method for analyzing telomerase activity in cell and tissue lysates [7,8]. To minimize the artifacts caused by PCR, modified TRAP methods have been developed by combining with electrochemical [9,10] and optical technologies [11]. However, these methods still face interferences from cell lysates, not to mention the requirement of time-consuming procedures for sample preparation. Alternatively, given the high sensitivity, straightforwardness, and non-invasiveness of fluorescence imaging, various fluorescent probes have been attempted to directly visualize telomerase activity in living cells [12–14] and even in vivo [15,16]. Nevertheless, these probes are not without shortcomings, including the photobleaching of fluorophores, autofluorescence from biological specimens, and limited tissue penetration depth of UV/visible (UV/vis) excitation light sources [17,18]. Along this line, the development of a more reliable fluorescent probe overcoming these issues would allow better interrogation of the information on telomerase in its native physiological state.

Towards this goal, here, we developed an activatable nanoscintillator probe for tracking telomerase activity and screening telomerase inhibitors in vivo. Nanoscintillators are a class of luminescent nanomaterials that can absorb and convert ionizing radiation (X-rays, γ-rays, etc.) into UV/vis/near-infrared photons [19]. Of note, the exceptional penetration depth of the excitation sources (i.e., X-rays), avoidance of autofluorescence, and optical stability have enabled broad applications of nanoscintillators in the biomedical field [20–22]. In this study, we leveraged core–shell $NaLuF_4$:Tb/Gd@$NaYF_4$ nanoscintillators (NSs) and telomerase-responsive DNA molecular devices (DMDs) to construct a telomerase probe. As shown in Scheme 1, in the telomerase-responsive DMDs (Table S1), a long single-strand DNA (A-strand, 66 base, red) is covalently immobilized on the nanoscintillator surface and has three functional domains: an AS1411 aptamer sequence at the 3′-terminal for targeting cancer cells, the middle sequence for carrying telomerase primer (TP, blue), and the telomeric repeat-complementary sequence at the 5′-end for binding BHQ1 (black hole quencher 1)-labeled strand (BHQ1-strand, green). Initially, the fluorescence of nanoscintillators is quenched due to the proximity of BHQ1 to their surface. Once encountering telomerase, the extension of the TP strand displaces the BHQ1-labeled strand, turning on the fluorescence of nanoscintillators for signal readout. With systemic studies, we demonstrated the application of this probe for detecting telomerase activity in solution, living cells, and tumors, and screening telomerase inhibitors in animal models.

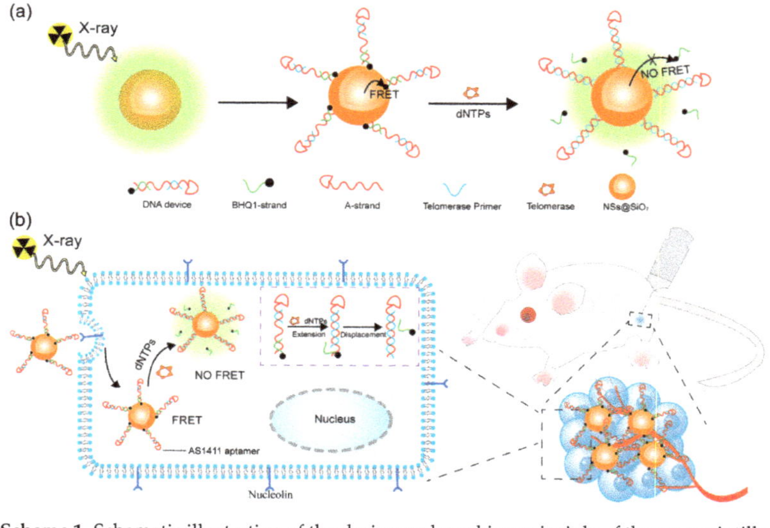

Scheme 1. Schematic illustration of the design and working principle of the nanoscintillator probe for detecting telomerase in vivo. (**a**) The X-ray-induced fluorescence of the nanoscintillators (NSs) is quenched by BHQ1 in the beginning due to the fluorescence resonance energy transfer (FRET) effect; upon telomerase-catalyzed extension of the primer, strand-displacement leads to the detachment of the BHQ1-strand from the surface of NSs and the FRET effect is inhibited, whereby the fluorescence of NSs is recovered for signal readout. (**b**) Specific recognition and internalization of probe into tumor cells for in vivo monitoring telomerase activity.

2. Materials and Methods

Additional information on reagents and materials, characterizations, experimental details, DNA sequences, and cell culture was given in the Supporting Information.

Synthesis of $NaLuF_4$:Tb(15%)/Gd(25%) nanoparticles. In a typical experiment, a mixture of $Ln(CH_3COO)_3{\cdot}4H_2O$ (0.5 mmol; Ln = Lu, Gd, Tb at the desired ratio), 4 mL oleic acid (OA), and 16 mL 1-octadecene (ODE) were added to a 50 mL two-neck round-bottom flask. Then, the resultant mixture was heated to 160 °C under vacuum for 45 min. After cooling the solution to room temperature, 10 mL methanol containing 1.25 mmol NaOH and 2 mmol NH_4F were added to the reaction system and stirred for 30 min at 50 °C, followed by heating to 100 °C for another 10 min under vacuum. Thereafter, the solution was heated to 300 °C at a rate of 20 °C/min and kept at 300 °C for 60 min under a nitrogen atmosphere while stirring. After cooling to room temperature, the resultant nanoparticles were collected by precipitation with ethanol and centrifugation at 8000 rpm and washed with ethanol and cyclohexane (1:1, *v*/*v*) three times. Finally, the obtained nanoparticles were redispersed in 5 mL cyclohexane.

Synthesis of $NaLuF_4$:Tb(15%)/Gd(25%)@$NaYF_4$ (NSs). In a typical procedure, 0.5 mmol $Y(CH_3COO)_3{\cdot}4H_2O$, 5 mL OA and 7.5 mL ODE were added to a two-neck round-bottom flask, and then the resultant mixture was heated to 160 °C under vacuum for 50 min. After cooling the solution to room temperature, the core $NaLuF_4$:Tb(15%)/Gd(25%) nanoparticles dispersed in 4 mL cyclohexane were added to the reaction system, and the suspension was heated at 80 °C for 20 min to evaporate the cyclohexane. After cooling to room temperature, 10 mL methanol containing 1.25 mmol NaOH and 2 mmol NH_4F were added to the reaction system and stirred at 50 °C for 30 min, and then the resultant mixture was heated to 80 °C for another 20 min under vacuum to evaporate the methanol. The suspension was then heated to 290 °C at a rate of 20 °C/min and kept at 290 °C for 90 min under a nitrogen atmosphere while stirring. After cooling to room temperature, the nanoparticles were collected by precipitation with ethanol and centrifugation at 8000 rpm and washed with ethanol and cyclohexane (1:1, *v*/*v*) three times. Finally, the nanoparticles were redispersed in 5 mL cyclohexane and stored at 4 °C for further use.

Synthesis of silica-coated NSs (NSs@SiO_2) particles. 1-hexanol (2 mL), Triton X-100 (2 mL), and cyclohexane (10 mL) were mixed and stirred until the mixture became clear. Then, 300 µL deionized water and NSs solution dispersed in cyclohexane (4 mL 0.01 M) were sequentially added to the solution. The resultant mixture was stirred for 10 min at room temperature before the addition of 80 µL tetraethyl orthosilicate. After further vigorously stirring the suspension for 20 min, ammonium hydroxide (160 µL) was added, and the mixture was stirred at room temperature for 20 h. In the end, NSs@SiO_2 particles were collected by centrifugation at 8000 rpm and washed with ethanol three times.

Preparation of carboxyl group-functionalized NSs@SiO_2 (NSs@SiO_2-COOH) particles. NSs@SiO_2 particles (180 mg) were first dispersed in isopropyl alcohol (120 mL), 3-aminopropyltriethoxysilane (200 µL) was then added to the suspension, and the resultant suspension was refluxed in an oil bath at 85 °C for 10 h. The resultant nanoparticles (NSs@SiO_2-NH_2) were collected by centrifugation at 10,000 rpm for 10 min and washed with deionized water and ethanol. Sequentially, to prepare NSs@SiO_2-COOH particles, NSs@SiO_2-NH_2 particles (120 mg) were dispersed in *N*,*N*-dimethylformamide, and succinic anhydride (190 mg) was added to the suspension, which was further stirred at room temperature for 20 h. Thereafter, the resulting nanoparticles were collected by centrifugation at 10,000 rpm and washed with deionized water three times.

Preparation of the probe. NSs@SiO_2-COOH (2 mg) particles were dispersed in 2-(*N*-morpholino) ethanesulfonic acid (MES) buffer (pH 6.0, 10 mM), then 60 µL *N*-(3-(dimethylamino) propyl)-*N*-ethylcarbodiimide (30 mg/mL) and 120 µL *N*-hydroxysuccinimide (50 mg/mL) were added to the suspension. After stirring for 30 min, the pH of the suspension was adjusted with PBS to neutral, and then amino functionalized A-strand (40 µL, 0.1 M) was added, followed by stirring at room temperature for 18 h. In the end, the resultant nanoparticles were collected by centrifugation at 10,000 rpm, washed with water (pH 7.4) for three times, and re-dispersed in

Tris-HCl buffer (10 mM Tris, pH 7.4, 250 mM NaCl, 100 mM KCl) for further use. Then, 1 mL of A-strand-grafted NSs@SiO_2 (1 mg/mL) dispersed in Tris-HCl buffer was incubated with 8 nmol of BHQ1-strand and 8 nmol of telomerase primer at room temperature for 12 h under shaking. The probe was collected by centrifugation at 10,000 rpm and washed with Tris-HCl buffer three times.

Telomerase Extraction. Telomerases were extracted from Hela cells with CHAPS buffer. Firstly, Hela cells (1.0×10^6) were collected and transferred in a centrifuge tube when Hela cells were in the exponential growth stage. Secondly, Hela cells were washed with iced PBS two times. Thirdly, the cells were transferred in the cold CHAPS lysis buffer (0.5% (w/v) CHAPS, 10 mM Tris, pH 7.5, 1 mM ethylene glycol-bis(β-aminoethylether)-*N*,*N*,*N*′,*N*′-tetraacetic acid, 0.1 mM phenylmethanesulfonyl fluoride, 10% glycerol, and 1 mM $MgCl_2$) and incubated at 4 °C for 30 min. Then, the lysate was centrifugated (12,000 rpm) at 4 °C for 20 min. The supernatant was collected and stored at −80 °C for further use.

Detection of telomerase activity in vitro. To determine the telomerase-responsive character of the probe, the time-dependent fluorescence variation kinetics of the probe was performed. The probe (0.5 mg/mL) and deoxynucleotide solution mixture (dNTPs; 200 μM) were added into TRAP buffer containing telomerase extracts from 100,000 Hela cells, and then the mixture was incubated at 37 °C for different times (10, 20, 30, 40, 60, 100, and 120 min) and time-dependent variation in fluorescence intensity was recorded. For comparison, we used PBS to replace the telomerase extracts to record the changes in potential background fluorescence.

To investigate the relationship between X-ray-induced fluorescence signals and telomerase activity, dNTPs (200 μM) and different amounts of telomerase extracts equivalent to different numbers (0, 100, 500, 1000, 2000, 5000, 10,000, 20,000, 50,000, 100,000) of Hela, HepG2, or MCF-7 cells were added to TRAP buffer containing the probe (0.5 mg/mL) and then the resultant suspensions were incubated at 37 °C for 40 min before recording the X-ray-induced fluorescence of each group. The change in the fluorescence intensity at λ = 546 nm of the probe was plotted against the number of cells. The limit of detection (LOD) was calculated using the equation LOD = $3\sigma/S$, where σ is the standard deviation of ten consecutive measurements of the blank, and S is the slope of the linear part of the plot.

To verify the specificity of the probe towards telomerase, a series of biomolecules (BSA, thrombin, GSH, trypsin, lysozyme, and glucose) were respectively added into TRAP buffers containing 0.5 mg/mL of the probe. Then, the resultant mixtures were incubated at 37 °C for 40 min, and the fluorescence was recorded at the end. To study the stability of the probe, the probe was incubated with PBS buffer, Dulbecco's modified Eagle's medium (DMEM), DMEM containing 10% fetal bovine serum (FBS), and saline (0.9% NaCl solution) at 37 °C, respectively, for a series of time (0, 1, 4, 8, 12, 24, 36, and 48 h). At these indicated time points, the X-ray-induced fluorescence and hydrodynamic size of the probe were recorded.

Detection of telomerase activity in living cells. To assess telomerase activity in living cells with the probe, 5×10^5 Hela cells were first seeded in a 6-well plate. After 24 h incubation at 37 °C, the medium was removed, and the cells were washed with PBS, followed by the addition of fresh medium containing the probe (0.5 mg/mL), and the cells were incubated for different times (0, 1, 2, 3, 4, and 5 h). Thereafter, the cells were washed with PBS and digested by trypsin, and the cell suspensions of each group were collected for recording the X-ray-induced fluorescence intensity.

To detect the telomerase activity in different cancer cells, MCF-7, Caco-2, HepG2, and A549 cells were seeded in 6-well plates and cultured for 24 h; then, the culture medium was replaced with fresh medium containing the probe (0.5 mg/mL). After incubating for 6 h, cells were washed with PBS and digested by trypsin; the cell suspensions were collected for recording the X-ray-induced fluorescence intensity and imaging, respectively. Imaging was performed using the IVIS Lumina III Series system (PerkinElmer).

For intracellular telomerase inhibitor screening, Hela cells (5×10^5) were seeded in a 6-well plate at 37 °C for 24 h, followed by replacing the culture medium with fresh culture medium containing different concentrations of epigallocatechin gallate (EGCG; 0, 5, 10, 15, 20, and 25 μM), 3′-Azido-3′-deoxythymidine (AZT; 25 μM), curcumin (25 μM), Doxorubicin (Dox; 25 μM), and 2-[(E)-3-naphthalen-2-ylbut-2-enoylamino]benzoic acid (BIBR1532; 25 μM) for 6 h. Then, the cells were washed with PBS and fed with fresh culture medium containing the probe (0.5 mg/mL) for 4 h. Thereafter, the cells were digested by trypsin for recording the X-ray-induced fluorescence intensity and imaging, respectively.

Imaging of telomerase activity and screening telomerase inhibitors in vivo. All animal experiments were approved by the Institutional Animal Care and Use Committee of Fuzhou University (Protocol Number 013-20210218-006 and the date of approval is 18 February 2021). Nude mice (6–8 weeks, ~18–20 g) were purchased from Shanghai SLAC Laboratory Animal Co., Ltd. (Shanghai, China). To establish human cervix tumors in mice, 1×10^6 Hela cells were dispersed in 50 μL PBS buffer and then subcutaneously injected into the right hind leg of athymic nude mice. When the tumor volumes reached a size of 150–350 mm^3, the probe or the mis-probe lacking telomerase primer dispersed in saline was injected intratumorally. Then, fluorescence imaging of tumors in the nude mice was performed at 0, 1, 2, 3, 4, and 5 h after administration of the probe (15 mg/kg, 50 μL, in 0.9% saline) or control probe (15 mg/kg, 50 μL, in 0.9% saline) using the IVIS Lumina III Series system (PerkinElmer, Waltham, MA, USA). Mice receiving intratumor injections of saline were set as the control. For screening telomerase inhibitors in vivo, 6 h after Hela tumor-bearing mice treated with AZT (50 mg/kg), EGCG (50 mg/kg), curcumin (50 mg/kg), BIBIR1532 (50 mg/kg), or Dox (50 mg/kg) by intratumor injection, the mice were treated with the probe for 4 h, and then X-ray-induced fluorescence imaging was recorded.

Statistical analysis. The obtained data were expressed as the mean value ± standard deviation (SD), and the statistical significance between groups was analyzed with one-way ANOVA and Tukey's post hoc tests using the software GraphPad Prism. Reported p values were classified as follows: **** $p < 0.0001$, *** $p < 0.001$, ** $p < 0.01$, and * $p < 0.05$. A value of $p < 0.05$ was considered statistically significant.

3. Results and Discussion

3.1. Synthesis and Characterization of the Probe

The detailed synthetic process of the probe is depicted in Figure S1. To construct the probe, lanthanide-doped core–shell nanoscintillators ($NaLuF_4$:Tb(15%)/Gd(25%)@$NaYF_4$) with sizes around 25 nm were first prepared according to our previous work [23] and then deposited with a c.a. 2 nm thick silica shell (NSs@SiO_2, Figure S2 and Figure 1a–c). The X-ray excited fluorescence spectra of NSs and NSs@SiO_2 particles showed that four characteristic emission bands at 489, 546, 584, and 612 nm, assigned to the $^5D_4 \rightarrow {}^7F_J$ (J = 6, 5, 4, 3) transitions of the Tb^{3+} luminescence center (Figure 1d). The deposition of silica shell showed no obvious impact on the emission characteristics of the NSs. Thereafter, the obtained NSs@SiO_2 particles were sequentially functionalized with amino and carboxyl groups (Figure S3), followed by grafting the A-strand. Finally, the TP and BHQ1-strand were hybridized into their complementary segments in the A-strand to form the telomerase-responsive DMDs. The conjugation of the DMDs was confirmed by the characteristic absorption bands of DNA at 260 nm and BHQ1 centered at 546 nm in the UV–vis absorption spectrum (Figure 1e) and a negatively charged zeta potential (Figure S4). The X-ray-induced fluorescent spectra of the probe showed that the emission of NSs was strongly quenched (Figure 1f) because the overlap of emission bands with the absorption spectrum of BHQ1 led to fluorescence resonance energy transfer (FRET). These results indicated the successful modification of DMDs to the surface of NSs@SiO_2. Finally, the content of the DNA device in the probe was measured to be about 0.499 nmol per mg probe by measuring the conjugated number of BHQ1-strands (Figure S5).

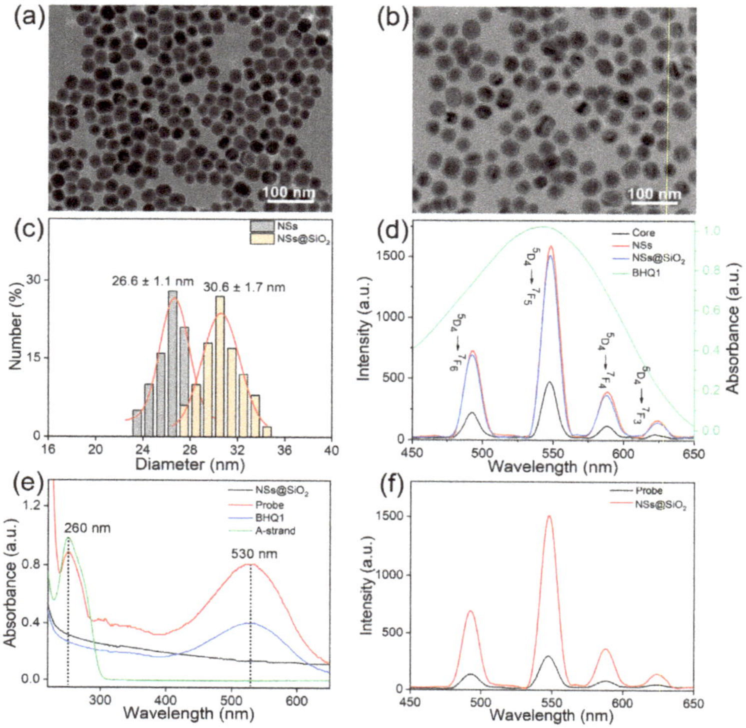

Figure 1. Synthesis and characterization of the probe. Representative TEM images of the as-prepared (**a**) NSs and (**b**) NSs@SiO_2 particles. (**c**) Size distributions of NSs and NSs@SiO_2 particles. (**d**) The normalized UV-vis absorption spectrum of BHQ1-strand (green line) and the X-ray-induced fluorescence spectra of the core nanoparticles ($NaLuF_4$:Tb(15%)/Gd(25%), black line), the core–shell NSs (red line), and NSs@SiO_2 particles (blue line). (**e**) UV-vis absorbance spectrum of NSs@SiO_2 particles, the probe, BHQ1, and A-strand DNA. (**f**) X-ray-induced fluorescence spectra of NSs@SiO_2 particles and the probe.

3.2. Telomerase Detection In Vitro

The strand-displacement behavior of DMDs by telomerase-triggered TP elongation was first validated by polyacrylamide gel electrophoresis (Figure S6). Then, the X-ray-induced fluorescence of the probe under different conditions was measured to confirm its feasibility for telomerase activity sensing. After incubating the probe with a mixture of deoxynucleotides (dNTPs) and telomerase extracts from 100,000 Hela cells at 37 °C for 2 h, the X-ray-induced fluorescence intensity of the probe at 546 nm was significantly increased compared with that of the nontreated probe. In contrast, neither the extracts from the normal cell line (L02, normal human liver with limited expression of telomerase) nor the heat-inactivated Hela cell extracts (95 °C for 20 min) induced obvious fluorescence enhancement. Moreover, minimal fluorescence enhancement was observed for a nonspecific probe lacking the TP strand (denoted as mis-probe) (Figure 2a). These results proved that telomerase could catalyze the extension of the TP strand and, thereafter, trigger the release of the BHQ1-strand via strand displacement. Concomitantly, the detachment of the BHQ1-strand from the surface of the NSs@SiO_2 reduced the FRET efficiency, recovering the fluorescence of NSs for signal readout.

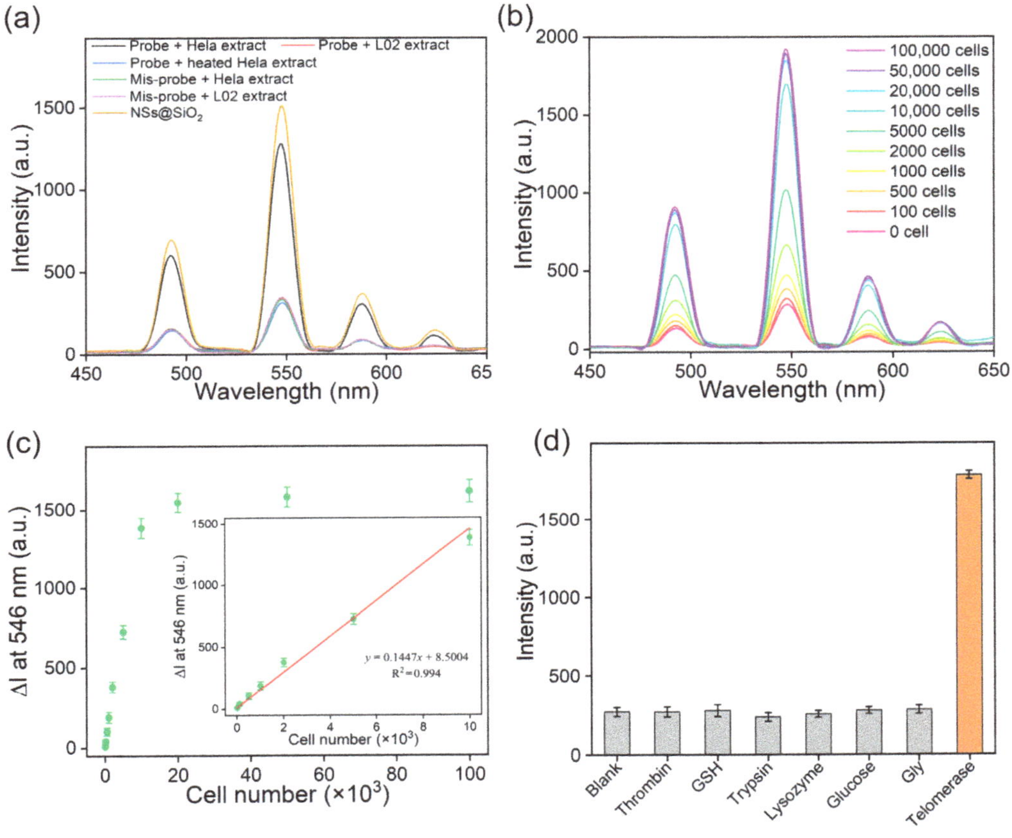

Figure 2. Sensitivity and selectivity of the probe. (**a**) X-ray-induced fluorescence spectra of suspensions containing the probe, dTNPs, and different analysts as indicated. Experiments conducted with the mis-probe were used as the control. The spectrum of an equal amount of $NSs@SiO_2$ particles was shown for better comparison. (**b**) X-ray-induced fluorescence spectra of the probe upon incubation with telomerase extracts from different numbers of Hela cells. (**c**) The relationship between fluorescence intensity enhancement (ΔI) at 546 nm and Hela cell numbers. Inset: linear plot of the ΔI versus Hela cell numbers. (**d**) Fluorescence intensity at 546 nm of the probe after incubating with telomerase and different control analytes as indicated. Error bars represented the standard deviations of three independent experiments.

Next, we measured the time-dependent fluorescence variation kinetics of the probe by incubating it with dNTPs and telomerase extracted from 100,000 Hela cells at 37 °C. As shown in Figure S7, the X-ray-induced fluorescence of the probe enhanced over time and reached a plateau within 40 min. Based on these results, 40 min of reaction time was chosen in the following experiments. Then, we tested the sensitivity of the probe by incubating it with telomerase extracts from different numbers of cells. As shown in Figure 2b, the intensity of X-ray-induced fluorescence increased gradually with increasing cell numbers, and there was a good linear relationship between the fluorescence enhancement (ΔI) at 546 nm and the telomerase activity equivalent to 0–10,000 Hela cells (Figure 2c). Remarkably, the limit of detection was calculated to be 37 Hela cells, which was comparable to the detection limit of most of the previously reported methods [24,25]. Furthermore, similar trends in the fluorescence changes were also observed for the probe after being treated with telomerase extracts from different cancer cells, such as HepG2 (Figure S8) and MCF-7 cells (Figure S9), and the detection limits were measured to be 51 and 48 cells, respectively. These results suggested that this scintillating probe provided an alternative tool for the sensitive detection of telomerase activity.

We next evaluated the selectivity and stability of the probe. The selectivity of the probe was verified by incubating it with various biomolecules, including thrombin, glutathione (GSH), trypsin, lysozyme, glucose, and glycine. As shown in Figure 2d, the X-ray-induced fluorescence had no significant change upon incubating with the control analytes. In sharp contrast, a distinct fluorescence enhancement was observed in the presence of telomerase. These results demonstrated good sensing specificity of the developed probe for detecting telomerase even in complex biological media. Furthermore, we measured the stability of the probe by incubating the probe in PBS buffer, DMEM, DMEM plus 10% FBS, and saline for 48 h, which showed that the fluorescence intensity and hydrodynamic size of the probe changed negligibly (Figure S10). These results substantialized the stability of the probe in biorelevant aqueous environments.

3.3. Analysis of Intracellular Telomerase Activity

The probe was then employed for the monitoring of intracellular telomerase activity. To show whether the AS1411 aptamer in the probe could favor its uptake by cancer cells, we employed Hela cells as the experimental cell line and L02 cells as a control. To directly visualize the cellular uptake of the probe under confocal laser scanning microscopy (CLSM), we replaced the BHQ1 with Cy3. As shown in Figure S11, after incubating with Hela cells, green fluorescence could be observed in the cytoplasm after 1 h of incubation, and the fluorescence intensity enhanced gradually with increasing incubation time and plateaued around 3 h. In contrast, although the fluorescence signal of Cy3 could be detected in the normal cells (L02), the intensity was much lower than that in Hela cells at each time point. These results indicated that the probe could be preferably internalized by Hela cells because AS1411 aptamer in stand A could specifically recognize and bind to nucleolin that is overexpressed on the surface of cancer cells [26,27]. We next evaluated the cytotoxicity of the probe by incubating it with Hela and L02 cells at various concentrations; the cellular viabilities of both cell lines were almost not affected up to 200 μg/mL (Figure S12). These results indicated that the probe had a satisfactorily low cytotoxicity.

Before utilizing the probe for directly detecting intracellular telomerase activity, we assessed the autofluorescence of cell suspensions under X-ray excitation. As shown in Figure 3a, compared to the strong background fluorescence generated under UV light (365 nm) excitation, the autofluorescence was reduced by more than 99% upon excitation by X-ray. These results confirmed that the scintillator-based probe was reliable for analyzing target biomolecules such as telomerase in complex biological environments. We next leveraged the probe for tracking telomerase activity in living cells. After treating Hela cells with the probe at different times, cells were digested, and the fluorescence intensity of cell suspensions was measured. As shown in Figure 3b, the fluorescence intensity at 546 nm gradually increased over incubation time, and the fluorescence enhancement reached a plateau around 4 h. By contrast, no significant enhancement in X-ray-induced fluorescence was observed in L02 cells over the same time. In additional experiments using the mis-probe, no obvious changes in X-ray-induced fluorescence were recorded in either Hela cells or L02 cells, indicating that the enhancement in fluorescence in Hela cells was due to the specific response of the probe to telomerase. Moreover, similar changes in X-ray-induced fluorescence were also observed in other four kinds of cancer cells (including MCF-7, Caco-2, HepG2, and A549 cells) after being treated with the probe (Figure 3c), showcasing the generality of the probe for sensing telomerase activity in different cancer cell lines. Compared with assays sensing telomerase activity from cellular extracts, directly detecting telomerase activity in living cells with our probe would more reliably reflect the activity of telomerase and related biological processes in the native state [28].

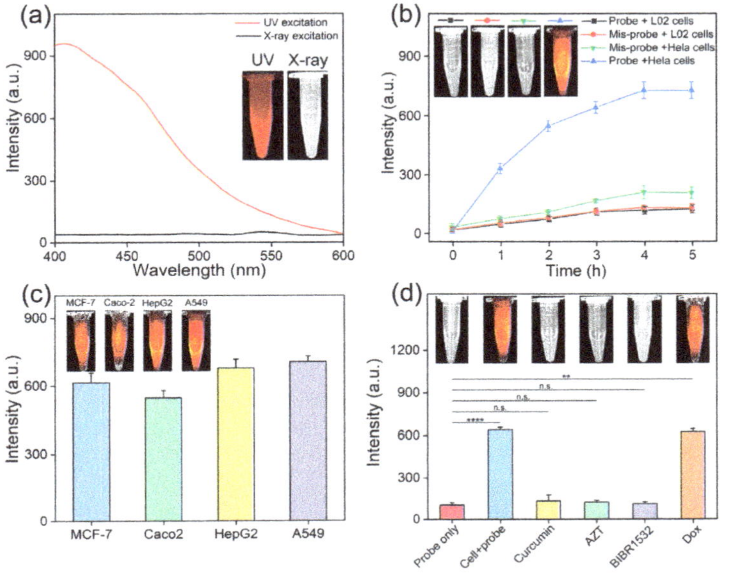

Figure 3. Sensing telomerase activity in living cells. (**a**) Fluorescence spectra of Hela cell suspensions under UV and X-ray excitations; inset showed the corresponding photographs of the suspensions under different excitations. (**b**) Changes in the intensity of X-ray-induced fluorescence of the suspensions of Hela cells and L02 cells after incubating with the probe and mis-probe for different times. Inset photographs showing the fluorescence images of the cell suspensions. (**c**) The fluorescence intensity and fluorescence images of suspensions of MCF-7, Caco-2, HepG2, and A549 cells after incubating with the probe for 4 h. (**d**) The fluorescence intensity and fluorescence images of Hela cells after being treated with different drugs and the probe. The probe alone and nontreated Hela cells plus probe (cell + probe) were set as controls. Error bars represented the standard deviations of three independent experiments. Statistical significance was calculated via ANOVA analysis with Tukey's post hoc test. ** $p < 0.01$, **** $p < 0.0001$. n.s., not significant.

3.4. Screening Telomerase Inhibitors in Living Cells

Telomerase inhibitors could be promising therapeutic drugs for cancer treatment, as most tumors rely on the reactivation of telomerase to maintain telomeres to preserve the proliferation potential while the telomerase activity of most normal cells is repressed [29,30]. Hence, it is of great significance to develop sensitive and selective tools for screening telomerase inhibitors. To confirm whether our probe could also be employed for evaluating the activity of inhibited telomerase and thus for screening telomerase inhibitors, EGCG was first studied as a representative inhibitor. As shown in Figure S13, 500,000 Hela cells were treated with different concentrations of EGCG (0, 5, 10, 15, 20, and 25 μM) for 6 h before incubating with the probe. The fluorescence signals gradually declined upon increasing the doses of EGCG and reached a plateau above 20 μM, indicating that the probe could read out the dose-dependent inhibition of EGCG toward telomerase activity. Next, we employed the probe to screen telomerase inhibitors in living cells. Curcumin, AZT, and BIBR1532 were used as model drugs. These drugs inhibit telomerase activity by down-regulating human telomerase reverse transcriptase expression [31–34]. Dox was chosen as a control because its action is not directly associated with the telomerase activity. Afterward, these drugs were incubated with 500,000 Hela cells for 6 h, and the cells were then treated with the probe. As shown in Figure 3d, negligible fluorescence enhancement was observed for groups treated with curcumin, AZT, and BIBR1532, which suggested that the telomerase activity was suppressed by these drugs. In contrast, the fluorescence signal in the Dox group was similar to the control group treated with the probe only and much higher than that of the telomerase inhibitor-treated groups. These results demonstrated that the probe could be used for the monitoring of telomerase activity in drug-treated cancer cells and would be applicable for facilitating telomerase inhibitor screening in living cells.

3.5. In Vivo Imaging of Telomerase Activity

The use of visible or UV light as the excitation source for most in vivo imaging applications suffers from high tissue autofluorescence and low tissue penetration depth [17], which could affect the accuracy and reliability of signal outputs. The exceptional tissue penetrating and background-free characteristics of X-rays might solve these problems [19]. To this end, we first compared the autofluorescence background in Hela tumor-bearing mice under UV light and X-ray irradiations. As shown in Figure S14, there was almost no autofluorescence background in X-ray-excited imaging graphs, while an obvious fluorescence background signal was detected upon UV (365 nm) illumination. Next, the tissue penetration depth of X-ray-induced fluorescence imaging was studied by a tissue phantom study. Typically, capillary tubes were first filled with a telomerase-pretreated probe, and then pork of different thicknesses was placed between the capillary tubes and the X-ray source. As shown in Figure S15, X-ray-induced fluorescence imaging of the activated probe was clearly presented by X-ray excitation, even with a tissue thickness of 2.5 cm. These results indicated that the use of an X-ray-induced fluorescence imaging probe for detecting telomerase in vivo would essentially eliminate the autofluorescence background and achieve deep tissue imaging.

Encouraged by these results, we then employed the probe to directly visualize telomerase activity in Hela tumor-bearing athymic nude mice. The probe, mis-probe, or saline was injected into tumors, where mice treated with saline and the mis-probe were set as the control groups. The probe and mis-probe were dispersed in 0.9% saline solution (15 mg/kg, 50 μL). As illustrated in Figure 4a, the fluorescence signal in the tumor of the probe group gradually increased over time and reached a maximum within 4 h post-injection under X-ray excitation. In contrast, the tumor fluorescence of the saline group and the mis-probe groups remained undetected and showed negligible variation over the same time. Consistently, quantitative analysis showed that the intratumoral fluorescence signal of the probe group revealed significantly much higher enhancement during 4 h post-injection, but no obvious intensity changed in the control groups (Figure 4b). These results indicated the feasibility of the probe for monitoring telomerase activity in vivo.

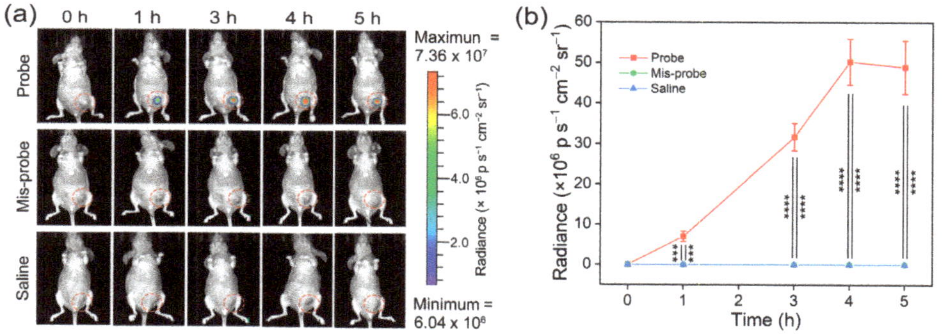

Figure 4. Imaging telomerase activity in vivo. (**a**) X-ray-induced fluorescence imaging of telomerase activity in Hela tumor-bearing athymic nude mice at different time points after intratumoral injection of the probe. Mice receiving intratumor injection of saline and the mis-probe were utilized as the control groups. (**b**) Quantitative analysis of changes in the average fluorescence intensity in the tumor regions after receiving different treatments as shown in (**a**); p, photons; sr, steradian. Error bars represented the standard deviation of five mice. Statistical significance was calculated via ANOVA analysis with Tukey's post hoc test. *** $p < 0.001$, **** $p < 0.0001$.

To evaluate the biosafety of the probe in vivo, the body weight of mice was measured after administration of the probe. As shown in Figure S16, there was no significant change in the body weight of mice over 20 days post-injection. Additionally, standard hematoxylin–eosin (H&E) staining of the tissue sections of major organs (liver, spleen, heart, lungs, and kidneys) indicated that there were no obvious signs of pathological damage in comparison with the saline group (Figure S17). Moreover, the blood biochemistry tests showed that all blood paraments of hematological biomarkers (Figure S18) and organ function biomarkers (Figure S19) exhibited no statistical difference from the group treated with saline. These results validated the biosafety of the probe for bioimaging applications.

3.6. Telomerase Inhibitor Screening In Vivo

Although we have demonstrated that the probe could detect the activity of telomerase in inhibitor-treated living cells, two-dimensional cell culture condition cannot replicate the complexity of the three-dimensional tumor tissue architectures, and it is hard to reflect the interaction between the tumor microenvironment and cancer cells [35,36]. On the other hand, tumor-beating animal models could closely recapitulate human pathology, providing a favorable preclinical tool for screening drugs before clinical testing [37]. Of note, the quality of information obtained has a strong correlation with the output measure [38].

Regarding that our probe could directly provide fluorescence readout of telomerase activity in tumor-bearing mice, we next studied the capacity of the probe for screening telomerase inhibitors in vivo. We chose curcumin (50 mg/kg body weight), AZT (50 mg/kg body weight), BIBR1532 (50 mg/kg body weight), EGCG (50 mg/kg body weight), and Dox (50 mg/kg body weight) as the model drugs. Hela-tumor-bearing mice of different groups were intratumorally injected with these drugs for 6 h before administration of the probe. Mice treated with PBS before administration of the probe were set as the blank control group. As shown in Figure 5a,b, the X-ray-induced fluorescence in the tumors of the groups treated with telomerase inhibitors (curcumin, EGCG, AZT, BIBR1532) was obviously lower than that of the blank control group. In comparison with the fluorescence variation in the control group, the X-ray-induced fluorescence changed negligibly in the group treated with Dox. These results indicated that in comparison with the conventional output method of screening drugs that evaluated the animal death or tumor volumes, our probe could provide an intuitional readout and allow more reliable and rapid measurement of the effects of telomerase-related drugs. This activatable nanoscintillator probe would provide a powerful tool for identifying telomerase-related anticancer drugs and could serve as a highly efficient platform for drug discovery. Moreover, it holds great potential to expand the concept of this design to develop novel tools for screening other target-related drugs.

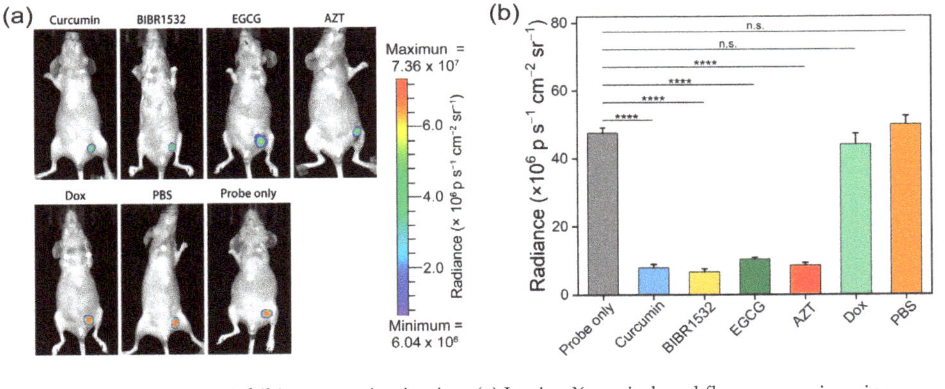

Figure 5. Telomerase inhibitor screening in vivo. (**a**) In vivo X-ray-induced fluorescence imaging of the Hela tumor-bearing athymic nude mice after intratumoral injection of different drugs in PBS

(50 μL, 50 mg/kg) and the probe (15 mg/kg, 50 μL, in 0.9% saline solution). Mice receiving PBS (50 μL) and the probe were used to exclude the possible effect of the drug vehicle. (**b**) Quantitative analysis of changes in the average fluorescence intensity in the tumor regions after receiving different treatments as shown in (**a**); p, photons; sr, steradian. Error bars represented the standard deviation of five mice. Statistical significance was calculated via ANOVA analysis with Tukey's post hoc test. **** $p < 0.0001$. n.s., not significant.

4. Conclusions

In summary, we have developed a novel probe by combining X-ray-excited nanoscintillators and telomerase-responsive DNA molecular devices for tracking telomerase activity and screening inhibitors in vivo. The proposed strategy outperformed the gold standard method of detecting telomerase activity measurement using PCR in terms of flexibility as it enabled telomerase activity sensing beyond cell lysates. The negligible autofluorescence background and high tissue penetration under X-ray excitation allowed the probe to directly visualize telomerase activity in living cells and tumor-bearing mice by producing turn-on fluorescence. More importantly, it is applicable to screen telomerase inhibitors in whole-animal models with easy-to-read fluorescent output signals. Overall, this study would inspire the development of advanced probes targeting different biomarkers for cancer diagnosis and anticancer drug discovery.

Supplementary Materials: The following supporting information can be downloaded at https://www.mdpi.com/article/10.3390/targets1010004/s1. Figure S1: Schematic illustration of the synthetic process of the probe. Figure S2: Synthesis and characterization of $NaLuF_4$:Tb/Gd nanoparticles. Figure S3: FTIR spectra of the oleate capped $NaLuF_4$:Tb/Gd@$NaYF_4$ (NSs), NSs@SiO_2, NSs@SiO_2-NH_2, and NSs@SiO_2-COOH particles. Figure S4: Zeta potential of NSs@SiO_2 particles (A), NSs@SiO_2-NH_2 particles (B), NSs@SiO_2-COOH particles (C), NSs@SiO_2-A-strand particles (D), and probe (E). Figure S5: Plot of the absorbance of BHQ1-strand at 546 nm versus the concentration of BHQ1-strand. Figure S6: Gel electrophoresis of the telomerase-responsive DNA devices (DMDs) and the telomerase-triggered telomerase primer extension and strand displacement. Figure S7: Fluorescence changes at 546 nm of the probe after incubation with telomerase extracts and PBS for different reaction times. Figure S8: (a) X-ray-induced fluorescence spectra of the probe after incubation with telomerase extracts from different numbers of HepG2 cells; (b) the relationship between fluorescence intensity enhancement (ΔI) at 546 nm and HepG2 cell numbers; inset: linear plot of the fluorescence enhancement versus HepG2 cell numbers. Figure S9: (a) X-ray-induced fluorescence spectra of the probe after incubation with telomerase extract from different numbers of MCF-7 cells; (b) the relationship between fluorescence intensity enhancement (ΔI) at 546 nm and MCF-7 numbers; inset: linear plot of the fluorescence enhancement versus MCF-7 cell numbers; Figure S10: The stability of the probe in different solutions after incubation at 37 °C for different times, and NSs@SiO_2 in these solutions was set as the control. Figure S11: Confocal laser scanning microscopy (CLSM) fluorescence images of (a) Hela cells and (b) L02 cells after treated with the probe for 0, 1, 2, 3, and 4 h. Figure S12: Cell viabilities of Hela and L02 cells after treated with different concentrations of the probe for 24 h. Figure S13: Inhibition of telomerase activity by EGCG. Figure S14: Comparison of the autofluorescence background under excitations of UV (365 nm) and X-ray source. Figure S15: Evaluation of the tissue penetration depth of X-ray. Figure S16: Body weight changes in healthy mice after intravenous injection of the probe and saline at different times. Figure S17: H&E-stained tissue sections of the major organs (heart, liver, spleen, lung, and kidney) after intravenous injection of the probe and 0.9% saline for 20 days. Figure S18: Hematology assays of healthy mice at days 10 and 20 after intravenous injection of the probe. Figure S19: Biochemical blood analysis of healthy mice at days 10 and 20 after intravenous injection of the probe; Table S1: DNA sequences in this work.

Author Contributions: Conceptualization, B.C., Y.B., Z.C. and H.Y.; methodology, B.C., J.D., S.S., X.T., Y.G., T.W., M.W., C.H., X.C., X.L., Y.B. and Z.C.; investigation, B.C., J.D., S.S., X.T., Y.G., T.W., M.W., C.H., X.C., X.L., Y.B. and Z.C.; resources, Y.B., Z.C. and H.Y.; writing—original draft preparation, B.C., Y.B., Z.C. and H.Y.; writing—review and editing, Y.B., Z.C. and H.Y.; supervision, Z.C. and H.Y.; project administration, Z.C. and H.Y.; funding acquisition, Z.C. and H.Y. All authors have read and agreed to the published version of the manuscript.

Funding: This work was funded by the National Key Research and Development Program of China (2020YFA0210800, H.Y.), the National Natural Science Foundation of China (22027805, H.Y.; 22107019, Z.C.; 22277011, Z.C.), and the Major Project of Science and Technology of Fujian Province (2020HZ06006, H.Y.). No APC was charged.

Institutional Review Board Statement: The animal study protocol was approved by the Institutional Animal Care and Use Committee of Fuzhou University (Protocol Number 013-20210218-006 and the date of approval is 18 February 2021).

Informed Consent Statement: Not applicable.

Data Availability Statement: All data generated or analyzed during this study are included in this published article (and its Supplementary Information files) or are available from the authors upon request.

Conflicts of Interest: The authors declare no conflict of interest.

References

1. Ghanim, G.E.; Fountain, A.J.; van Roon, A.-M.M.; Rangan, R.; Das, R.; Collins, K.; Nguyen, T.H.D. Structure of human telomerase holoenzyme with bound telomeric DNA. *Nature* **2021**, *593*, 449–453. [CrossRef] [PubMed]
2. Roake, C.M.; Artandi, S.E. Regulation of human telomerase in homeostasis and disease. *Nat. Rev. Mol. Cell Biol.* **2020**, *21*, 384–397. [CrossRef] [PubMed]
3. Soria, J.-C.; Gauthier, L.R.; Raymond, E.; Granotier, C.; Morat, L.; Armand, J.-P.; Boussin, F.o.D.; Sabatier, L. Molecular Detection of Telomerase-positive Circulating Epithelial Cells in Metastatic Breast Cancer Patients. *Clin. Cancer Res.* **1999**, *5*, 971–975.
4. Guterres, A.N.; Villanueva, J. Targeting telomerase for cancer therapy. *Oncogene* **2020**, *39*, 5811–5824. [CrossRef] [PubMed]
5. Harley, C.B. Telomerase and cancer therapeutics. *Nat. Rev. Cancer* **2008**, *8*, 167–179. [CrossRef]
6. Dai, J.; Liu, Z.; Wang, L.; Huang, G.; Song, S.; Chen, C.; Wu, T.; Xu, X.; Hao, C.; Bian, Y.; et al. A Telomerase-Activated Magnetic Resonance Imaging Probe for Consecutively Monitoring Tumor Growth Kinetics and In Situ Screening Inhibitors. *J. Am. Chem. Soc.* **2023**, *145*, 1108–1117. [CrossRef] [PubMed]
7. Krupp, G.; Kühne, K.; Tamm, S.; Klapper, W.; Heidorn, K.; Rott, A.; Parwaresch, R. Molecular basis of artifacts in the detection of telomerase activity and a modified primer for a more robust 'TRAP' assay. *Nucleic Acids Res.* **1997**, *25*, 919–921. [CrossRef] [PubMed]
8. Herbert, B.-S.; Hochreiter, A.E.; Wright, W.E.; Shay, J.W. Nonradioactive detection of telomerase activity using the telomeric repeat amplification protocol. *Nat. Protoc.* **2006**, *1*, 1583–1590. [CrossRef]
9. Wu, L.; Wang, J.; Ren, J.; Qu, X. Ultrasensitive Telomerase Activity Detection in Circulating Tumor Cells Based on DNA Metallization and Sharp Solid-State Electrochemical Techniques. *Adv. Funct. Mater.* **2014**, *24*, 2727–2733. [CrossRef]
10. Ling, P.; Lei, J.; Ju, H. Nanoscaled Porphyrinic Metal–Organic Frameworks for Electrochemical Detection of Telomerase Activity via Telomerase Triggered Conformation Switch. *Anal. Chem.* **2016**, *88*, 10680–10686. [CrossRef]
11. Zheng, G.; Daniel, W.L.; Mirkin, C.A. A New Approach to Amplified Telomerase Detection with Polyvalent Oligonucleotide Nanoparticle Conjugates. *J. Am. Chem. Soc.* **2008**, *130*, 9644–9645. [CrossRef] [PubMed]
12. Hong, M.; Xu, L.; Xue, Q.; Li, L.; Tang, B. Fluorescence Imaging of Intracellular Telomerase Activity Using Enzyme-Free Signal Amplification. *Anal. Chem.* **2016**, *88*, 12177–12182. [CrossRef]
13. Ma, W.; Fu, P.; Sun, M.; Xu, L.; Kuang, H.; Xu, C. Dual Quantification of MicroRNAs and Telomerase in Living Cells. *J. Am. Chem. Soc.* **2017**, *139*, 11752–11759. [CrossRef] [PubMed]
14. Wang, W.; Huang, S.; Li, J.; Rui, K.; Bi, S.; Zhang, J.-R.; Zhu, J.-J. Evaluation of intracellular telomerase activity through cascade DNA logic gates. *Chem. Sci.* **2017**, *8*, 174–180. [CrossRef] [PubMed]
15. Liu, Z.; Zhao, J.; Zhang, R.; Han, G.; Zhang, C.; Liu, B.; Zhang, Z.; Han, M.-Y.; Gao, X. Cross-Platform Cancer Cell Identification Using Telomerase-Specific Spherical Nucleic Acids. *ACS Nano* **2018**, *12*, 3629–3637. [CrossRef] [PubMed]
16. Ran, X.; Wang, Z.; Pu, F.; Ju, E.; Ren, J.; Qu, X. Nucleic acid-driven aggregation-induced emission of Au nanoclusters for visualizing telomerase activity in living cells and in vivo. *Mater. Horiz.* **2021**, *8*, 1769–1775. [CrossRef]
17. Frangioni, J.V. In vivo near-infrared fluorescence imaging. *Curr. Opin. Chem. Biol.* **2003**, *7*, 626–634. [CrossRef]
18. Liang, H.; Hong, Z.; Li, S.; Song, X.; Zhang, D.; Chen, Q.; Li, J.; Yang, H. An Activatable X-Ray Scintillating Luminescent Nanoprobe for Early Diagnosis and Progression Monitoring of Thrombosis in Live Rat. *Adv. Funct. Mater.* **2021**, *31*, 2006353. [CrossRef]
19. Chen, X.; Song, J.; Chen, X.; Yang, H. X-ray-activated nanosystems for theranostic applications. *Chem. Soc. Rev.* **2019**, *48*, 3073–3101. [CrossRef]
20. Pei, P.; Chen, Y.; Sun, C.; Fan, Y.; Yang, Y.; Liu, X.; Lu, L.; Zhao, M.; Zhang, H.; Zhao, D.; et al. X-ray-activated persistent luminescence nanomaterials for NIR-II imaging. *Nat. Nanotechnol.* **2021**, *16*, 1011–1018. [CrossRef] [PubMed]
21. Hong, Z.; Chen, Z.; Chen, Q.; Yang, H. Advancing X-ray Luminescence for Imaging, Biosensing, and Theragnostics. *Acc. Chem. Res.* **2023**, *56*, 37–51. [CrossRef]

22. Chen, Z.; Tsytsarev, V.; Finfrock, Y.Z.; Antipova, O.A.; Cai, Z.; Arakawa, H.; Lischka, F.W.; Hooks, B.M.; Wilton, R.; Wang, D.; et al. Wireless Optogenetic Modulation of Cortical Neurons Enabled by Radioluminescent Nanoparticles. *ACS Nano* **2021**, *15*, 5201–5208. [CrossRef]
23. Ou, X.; Qin, X.; Huang, B.; Zan, J.; Wu, Q.; Hong, Z.; Xie, L.; Bian, H.; Yi, Z.; Chen, X.; et al. High-resolution X-ray luminescence extension imaging. *Nature* **2021**, *590*, 410–415. [CrossRef]
24. Huang, J.; Zhu, L.; Ju, H.; Lei, J. Telomerase Triggered DNA Walker with a Superhairpin Structure for Human Telomerase Activity Sensing. *Anal. Chem.* **2019**, *91*, 6981–6985. [CrossRef] [PubMed]
25. Wang, J.; Wu, L.; Ren, J.; Qu, X. Visual detection of telomerase activity with a tunable dynamic range by using a gold nanoparticle probe-based hybridization protection strategy. *Nanoscale* **2014**, *6*, 1661–1666. [CrossRef] [PubMed]
26. Li, C.H.; Lv, W.Y.; Yan, Y.; Yang, F.F.; Zhen, S.J.; Huang, C.Z. Nucleolin-Targeted DNA Nanotube for Precise Cancer Therapy through Förster Resonance Energy Transfer-Indicated Telomerase Responsiveness. *Anal. Chem.* **2021**, *93*, 3526–3534. [CrossRef] [PubMed]
27. Lin, Y.; Huang, Y.; Yang, Y.; Jiang, L.; Xing, C.; Li, J.; Lu, C.; Yang, H. Functional Self-Assembled DNA Nanohydrogels for Specific Telomerase Activity Imaging and Telomerase-Activated Antitumor Gene Therapy. *Anal. Chem.* **2020**, *92*, 15179–15186. [CrossRef]
28. Verjans, E.-T.; Doijen, J.; Luyten, W.; Landuyt, B.; Schoofs, L. Three-dimensional cell culture models for anticancer drug screening: Worth the effort? *J. Cell. Physiol.* **2018**, *233*, 2993–3003. [CrossRef]
29. Ruparel, S.B.; Siddiqa, A.; Marciniak, R.A. Targeting Telomerase for Cancer Therapy. *Curr. Cancer Ther. Rev.* **2011**, *7*, 215–226. [CrossRef]
30. Negrini, S.; De Palma, R.; Filaci, G. Anti-Cancer Immunotherapies Targeting Telomerase. *Cancers* **2020**, *12*, 2260. [CrossRef]
31. Rha, S.Y.; Izbicka, E.; Lawrence, R.; Davidson, K.; Sun, D.; Moyer, M.P.; Roodman, G.D.; Hurley, L.; Von Hoff, D. Effect of Telomere and Telomerase Interactive Agents on Human Tumor and Normal Cell Lines1. *Clin. Cancer Res.* **2000**, *6*, 987–993. [PubMed]
32. Mittal, A.; Pate, M.S.; Wylie, R.C.; Tollefsbol, T.O.; Katiyar, S.K. EGCG down-regulates telomerase in human breast carcinoma MCF-7 cells, leading to suppression of cell viability and induction of apoptosis. *Int. J. Oncol.* **2004**, *24*, 703–710. [CrossRef] [PubMed]
33. Ramachandran, C.; Fonseca, H.B.; Jhabvala, P.; Escalon, E.A.; Melnick, S.J. Curcumin inhibits telomerase activity through human telomerase reverse transcritpase in MCF-7 breast cancer cell line. *Cancer Lett.* **2002**, *184*, 1–6. [CrossRef] [PubMed]
34. Pascolo, E.; Wenz, C.; Lingner, J.; Hauel, N.; Priepke, H.; Kauffmann, I.; Garin-Chesa, P.; Rettig, W.J.; Damm, K.; Schnapp, A. Mechanism of Human Telomerase Inhibition by BIBR1532, a Synthetic, Non-nucleosidic Drug Candidate*. *J. Biol. Chem.* **2002**, *277*, 15566–15572. [CrossRef] [PubMed]
35. Barbosa, M.A.G.; Xavier, C.P.R.; Pereira, R.F.; Petrikaitė, V.; Vasconcelos, M.H. 3D Cell Culture Models as Recapitulators of the Tumor Microenvironment for the Screening of Anti-Cancer Drugs. *Cancers* **2022**, *14*, 190. [CrossRef]
36. Giacomotto, J.; Ségalat, L. High-throughput screening and small animal models, where are we? *Br. J. Pharmacol.* **2010**, *160*, 204–216. [CrossRef]
37. HogenEsch, H.; Nikitin, A.Y. Challenges in pre-clinical testing of anti-cancer drugs in cell culture and in animal models. *J. Control. Release* **2012**, *164*, 183–186. [CrossRef]
38. Debnath, A. Drug discovery for primary amebic meningoencephalitis: From screen to identification of leads. *Expert Rev. Anti. Infect. Ther.* **2021**, *19*, 1099–1106. [CrossRef]

targets

MDPI

Review

Recent Progress in Small Molecule Fluorescent Probes for Imaging and Diagnosis of Liver Injury

Shuo Liu [1,2], Fei Huang [2], Xinyi Huang [1,3,*], Fuxin Zhang [2], Dong Pei [2], Jinlong Zhang [2] and Jun Hai [2,*]

1 College of Science, Gansu Agricultural University, Lanzhou 730070, China; 1073323120453@st.gsau.edu.cn
2 CAS Key Laboratory of Chemistry of Northwestern Plant Resources, Key Laboratory for Natural Medicine of Gansu Province, Lanzhou Institute of Chemical Physics, Chinese Academy of Sciences (CAS), Lanzhou 730070, China; h15628982782@163.com (F.H.); zhangfuxin@licp.cas.cn (F.Z.); dongpei@licp.cas.cn (D.P.); zhangjinlong@licp.cas.cn (J.Z.)
3 Key Laboratory of Chemistry in Ethnic Medicinal Resources, State Ethnic Affairs Commission & Ministry of Education, Yunnan Minzu University, Kunming 650500, China
* Correspondence: huangxy@ymu.edu.cn (X.H.); haijun@licp.cas.cn (J.H.)

Abstract: The liver is an essential metabolic organ that is involved in energy metabolism, protein synthesis, and detoxification. Many endogenous and exogenous factors can cause liver injury, a complex pathological condition. It poses a serious risk to human health due to its extremely varied clinical manifestations, which range from mild fatty liver to liver fibrosis, cirrhosis, and even hepatocellular carcinoma. Because of their low specificity, lack of real-time monitoring, and invasiveness, traditional diagnostic techniques for liver injury, such as histopathological examination and serological analysis, have inherent limitations. Fluorescent probe technology, which offers high sensitivity, non-invasiveness, and real-time imaging capabilities, has become a potent tool in liver injury research and early diagnosis in recent years. The pathophysiology of liver injuries caused by alcohol, chemicals, drugs, and the immune system is methodically covered in this review, along with new developments in fluorescent probe development for their detection. The focused imaging properties of various fluorescent probes are highlighted, along with their possible uses in drug screening and early liver injury detection. This review attempts to offer theoretical insights to support the optimization of precision diagnostic and therapeutic approaches by summarizing these findings.

Keywords: fluorescent probe; liver injury; small molecule; fluorescence imaging

1. Introduction

The liver is an extremely intricate and multipurpose organ that is essential to many physiological functions, such as protein secretion, detoxification, glycogen storage, anabolism, excretion, and the recycling of vital nutrients [1]. A serious risk to human health, liver injury is one of the most prevalent conditions seen in clinical practice. The liver is especially susceptible to a variety of endogenous and exogenous pathogenic factors, which can cause a variety of liver diseases, because of its abundant blood supply and vital role in metabolic processes [2]. A major challenge in current medical research is the prevention and treatment of liver injury, which is a common pathological outcome of many liver disorders [3]. It can be caused by multiple factors, including alcoholic fatty liver disease, viral hepatitis, autoimmune hepatitis, and chemically induced liver damage [4]. These conditions may lead to inflammation, abnormal wound healing responses, and impaired liver function.

Based on the underlying causes, liver injury can be classified into two main categories: exogenous (e.g., alcohol, chemical toxins) and endogenous (e.g., metabolic disorders, immune dysfunction). Liver injury often manifests as hepatocyte necrosis and apoptosis, liver dysfunction, inflammatory responses, and structural alterations such as fibrosis [5]. Clinically, mild liver injury is typically reflected by abnormal liver enzyme levels, including AST and ALT, whereas severe liver damage may lead to coagulopathy, hepatic encephalopathy, and even multiple organ failure [6,7]. Furthermore, based on disease progression, liver injury can be categorized as ALI or CLI, each exhibiting distinct pathophysiological mechanisms and clinical outcomes [8].

Early diagnosis is critical for timely intervention and the preservation of liver function. However, traditional diagnostic methods have inherent limitations in detecting liver injury at an early stage. Serological testing, the most commonly used approach to assess liver function, primarily measures serum enzyme activity but is susceptible to external interferences [9]. Additionally, histopathological examination provides valuable diagnostic information but has a lag in detection and cannot dynamically monitor liver function in real time. Therefore, there is an urgent need for efficient and accurate diagnostic techniques to enable the early detection of liver injury.

In recent years, fluorescence imaging technology has emerged as a promising diagnostic tool in modern medical research due to its high sensitivity, real-time imaging capabilities, and ease of use [10]. Direct optical imaging using fluorescent probes targeting disease-related biomarkers (such as ROS, CYP450, and HClO) offers a valuable complement to traditional imaging techniques [11]. Over the past decade, numerous fluorescent sensors have been developed for rapid detection and in situ visualization of key biomolecules in vivo [12]. Meanwhile, there has been some literature summarizing and exploring the application of fluorescent probes in a certain type of liver injury disease or a certain type of biomarker related to the disease [13,14].

Given this background, this review systematically explores the pathological mechanisms underlying alcoholic, drug-induced, chemical, and immune-mediated liver injury, with a particular focus on the application and imaging capabilities of small-molecule fluorescent probes in a variety of liver injury models, excluding systems based on other technologies such as quantum dots or chemiluminescence. By summarizing recent advancements, this study aims to provide a theoretical foundation for the future design of fluorescent probes and their potential role in the precise diagnosis and treatment of liver diseases.

2. Alcoholic Liver Injury

2.1. Establishment of Alcoholic Liver Injury Models

ALI is a dynamic pathological process resulting from alcohol-induced hepatic dysfunction. Based on disease duration, ALI can be classified into acute ALI and chronic ALI. Acute ALI refers to sudden liver damage caused by excessive alcohol consumption within a short period. It is characterized by direct alcohol toxicity, significant inflammatory activation, hepatocellular necrosis, inflammatory infiltration, and a sharp decline in liver function. In severe cases, acute liver failure may occur, posing a life-threatening risk to patients [15]. In contrast, chronic ALI results from long-term, continuous, or recurrent alcohol consumption. With prolonged alcohol metabolism, ROS accumulate, while hepatic antioxidant capacity declines, leading to fatty degeneration, chronic inflammation, and hepatic fibrosis. If left untreated, chronic ALI may progress to alcoholic hepatitis, cirrhosis, or even hepatocellular carcinoma.

Upon alcohol ingestion, most of it is absorbed in the intestines and transported to the liver for metabolism via the bloodstream. Alcohol dehydrogenase and acetaldehyde dehydrogenase sequentially convert ethanol into acetaldehyde and acetic acid, which are ultimately metabolized into water and carbon dioxide via the tricarboxylic acid cycle [16,17]. However, prolonged and excessive alcohol consumption inhibits adenosine monophosphate-activated protein kinase, leading to hepatic triglyceride and cholesterol accumulation, thereby contributing to alcoholic fatty liver. Furthermore, excessive ethanol metabolism generates acetaldehyde, a cytotoxic metabolite that damages hepatocyte membranes, activates immune responses, and induces the production of ROS, exacerbating oxidative stress injury [18–20]. Additionally, alcohol metabolites activate Kupffer cells, stimulating the secretion of pro-inflammatory cytokines such as TNF-α, which further promotes hepatocyte apoptosis and leads to hepatic necrosis [21–23].

To simulate acute ALI in animal models (Figure 1), Yang et al. administered 12 mL/kg per day of 52° liquor via intragastric gavage in mice. After 14 days, a preliminary acute ALI model was successfully established [24]. Similarly, Zheng et al. induced acute liver injury in C57BL/6 mice by a single injection of 12 mL/kg of 50° white wine [25].

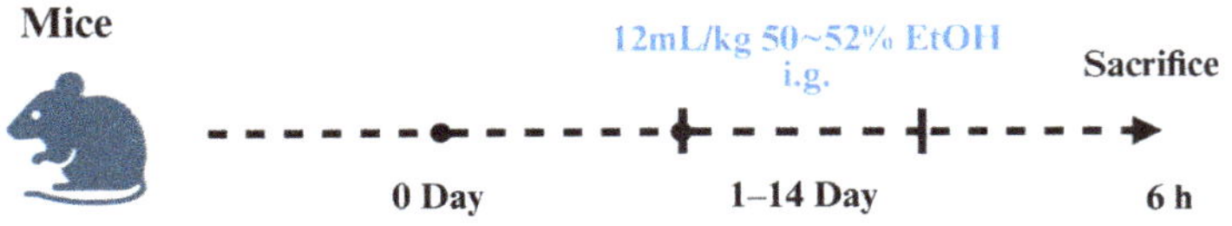

Figure 1. Schematic Representation of Animal Models for ALI.

Additionally, introducing dietary interventions such as high-fat and high-fructose diets during alcohol exposure significantly aggravates ALI severity, a phenomenon referred to as the "two-hit" model [26,27].

These experimental models are simple, cost-effective, and highly reproducible, with short induction periods, low mortality rates, and high stability. Furthermore, these models closely mimic alcohol-induced hepatic damage in humans, making them ideal tools for studying the pathogenesis and potential therapeutic strategies for acute ALI. Meanwhile a large number of fluorescent probes for the detection of ALI already exist (Table 1).

Table 1. Characterization of fluorescent probes in the ALI model.

Compounds	Trigger	λex (nm)	λem (nm)	Ref.
ICQ	Viscosity	630	740	[28]
MSOT	CYP450	710	750	[29]
DCI-Ac-Cys	Cys	540	705	[30]
BDP-ENE-Fur-HClO	HClO	620	700	[31]
HOTPy	viscosity	548	614	[32]
PPBI	Viscosity Polarity	575 400	750 497	[33]
WTH	HClO H_2O_2	440 360	500/605 450/605	[34]
BDP-NIR-ONOO	$ONOO^-$	920	700	[35]
Cy-ND	Viscosity	766	806	[36]
WZ-HOCl	HOCl	460	560	[37]

2.2. Application of Fluorescent Probes in Alcoholic Liver Injury

Feng et al. developed three mitochondria-targeted and viscosity-sensitive fluorescent probes (ICR, ICJ, and ICQ) for detecting AALI (Figure 2A). Among them, ICQ exhibits near-infrared fluorescence, a large Stokes shift, and high viscosity sensitivity. ICQ specifically targets mitochondria, enabling the sensitive tracking of mitochondrial viscosity changes. In vivo, ICQ localizes to the liver and facilitates noninvasive, real-time imaging of liver viscosity alterations. Using these properties, ICQ has been successfully applied to visualizing AALI progression, revealing a significant increase in viscosity during liver injury. This study provides a sensitive and noninvasive method for AALI detection and demonstrates the potential of mitochondrial viscosity as a biomarker for AALI [28].

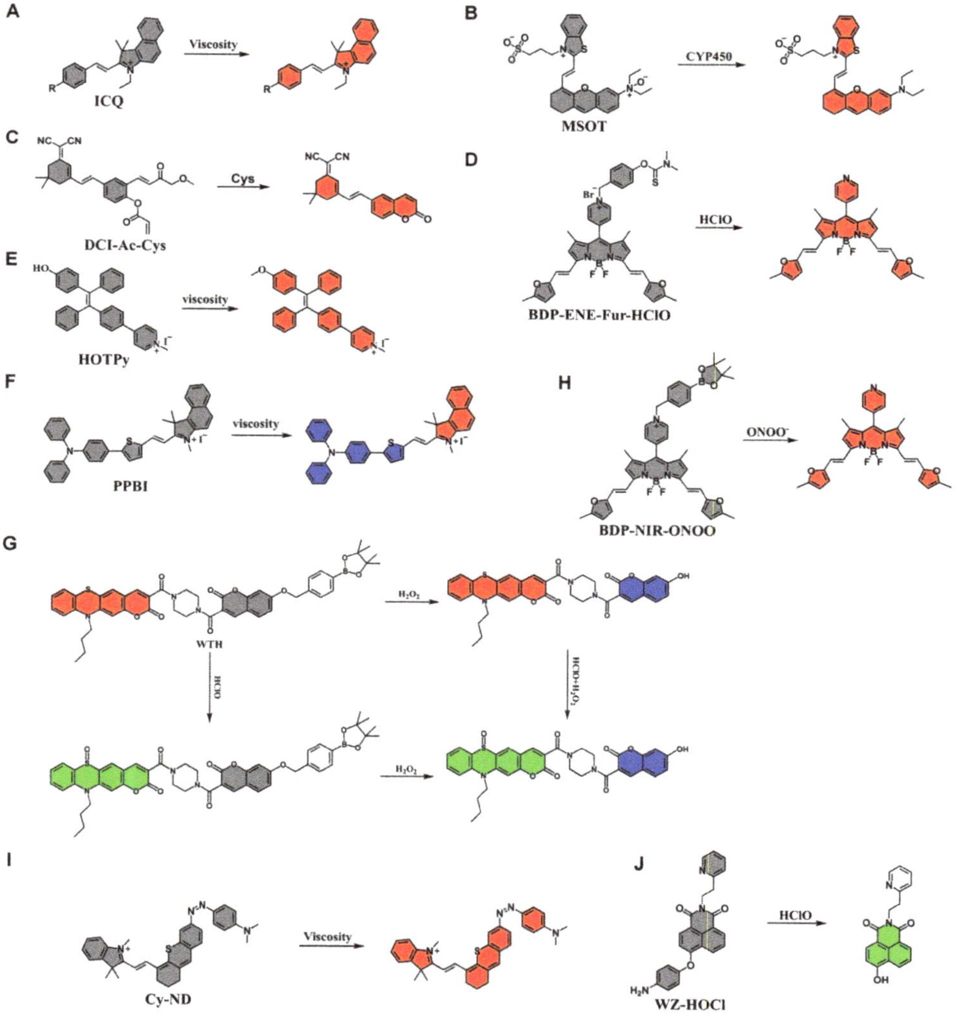

Figure 2. (**A–J**) Structure and detection mechanism of fluorescent probe in ALI models.

Wu et al. designed a water-soluble activatable fluorescent probe for imaging ALI using near-infrared fluorescence and multispectral optoacoustic tomography (MSOT) (Figure 2B). The probe is activated in situ in response to CYP450 reductase overexpression under hypoxic liver conditions, converting an N-oxide moiety into an aromatic tertiary amine, leading to enhanced fluorescence and red-shifted absorption. This activation generates both fluorescence and photoacoustic signals, enabling real-time imaging and monitoring of ALI progression and recovery in mouse models. Moreover, MSOT imaging precisely localizes liver damage, providing an essential tool for studying pathophysiological processes in ALI. This work highlights the potential of MSOT and fluorescence imaging in biomedical research [29].

Yang et al. developed a near-infrared fluorescent probe, DCI-Ac-Cys, which undergoes a Cys-triggered cascade reaction to form a coumarin fluorophore (Figure 2C). The acrylate-based DCI-Ac-Cys probe exhibits high selectivity for Cys, eliminating interference from other biothiols. Using DCI-Ac-Cys, researchers successfully achieved sensitive imaging of intracellular Cys and constructed cellular and mouse models of ALI. Additionally, the probe was applied to evaluate the therapeutic efficacy of antioxidant drugs (silybin and curcumin) in ALI treatment. These findings suggest that Cys may serve as a potential biomarker for ALD and introduce a novel method for ALD drug screening and diagnostic evaluation [30].

Zhang et al. developed a near-infrared fluorescent probe (BDP-ENE-Fur-HClO) capable of detecting HClO, enabling real-time imaging of ALD (Figure 2D). The probe demonstrated high sensitivity, selectivity, rapid detection, and near-infrared emission for in vitro HClO detection. It effectively tracked endogenous/exogenous HClO in living cells without interference from other ROS and was successfully applied in ALD cell models. Furthermore, BDP-ENE-Fur-HClO monitored exogenous HClO in normal mice and highly expressed HClO in peritonitis mice, enabling inflammation diagnosis and the evaluation of ALD drug treatments. This work provides a powerful tool for ALD diagnosis [31].

Chen et al. synthesized an AIE fluorescent probe (HOTPy) with pyridine and hydroxyl functional groups to enhance water solubility and chemical stability (Figure 2E). The probe exhibits high sensitivity and selectivity for viscosity detection, along with excellent biocompatibility and precise mitochondrial localization. Using live-cell imaging, HOTPy effectively sensed viscosity changes in ALI models. Additionally, HOTPy was successfully applied to in vivo imaging of AALI mice, allowing for real-time monitoring of liver viscosity fluctuations and tracking drug intervention effects. This study provides a novel AIE probe for the diagnosis and management of viscosity-related liver diseases [32].

Yu et al. have developed a novel dual-responsive fluorescent probe, PPBI, which simultaneously responds to polarity and viscosity (Figure 2F). By incorporating a positively charged half-carbonyl main chain into triphenylamine derivatives, PPBI exhibits deep red fluorescence along with AIE properties (Figure 3). The probe demonstrates excellent biocompatibility and can be used to precisely localize mitochondria and lipid droplets in live cells. Furthermore, it enables the distinction between cancerous and normal cells through fluorescence changes in both green and red channels. PPBI is also capable of monitoring abnormal fluctuations in intracellular viscosity and polarity linked to inflammation and ferroptosis. Notably, in a mouse model, in vivo imaging amplified these changes in polarity and viscosity, allowing for the monitoring of AALI and tracking the healing process of the damaged liver under drug intervention. This provides a powerful visual tool for both diagnosing and treating AALI. Overall, PPBI holds great promise as a tool for studying diseases related to biological polarity and viscosity, as well as for bioimaging and medical diagnostics [33].

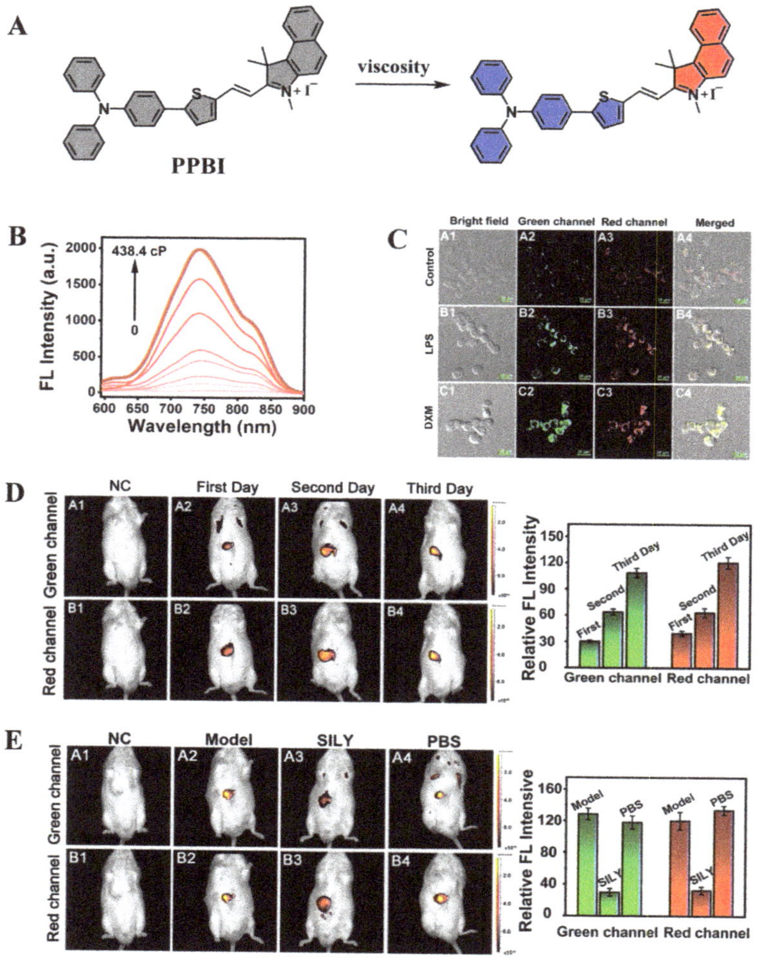

Figure 3. (**A**) Polarity/viscosity response mechanism of the fluorescent probe PPBI. (**B**) Fluorescent spectra of probe PPBI (10 μM) in PBS-glycerol mixture solution of different viscosities (0.89–438.4 cP). (**C**) Confocal laser scanning images of HepG2 cells, Control: the cells were pretreated with PPBI (10 μM) only LPS: the cells were pretreated with 20 μM LPS for 30 min and then incubated with PPBI (10 μM) for another 60 min, DXM: the cells were pretreated with 20 μM DXM for 30 min and then incubated with PPBI (10 μM) for another 60 min. (**D**) Time-dependent in vivo fluorescent imaging and relative fluorescent intensity of mice after intrahepatic injection of PPBI (10 μM) and continuous treatment with alcohol. (**E**) In vivo fluorescent imaging and relative fluorescence intensity of PPBI (10 μM) in normal contrast mice, model mice, SILY- and PBS-treated mice, respectively. Scale bar: 50 μm. Derived from Ref. [33] with authorization from Copyright 2023, Elsevier.

Yang et al. designed a dual-ratiometric fluorescent probe (WTH) based on phenothiazine-coumarin and hydroxycoumarin derivatives, capable of simultaneously detecting HClO and H_2O_2 (Figure 2G). The probe exhibited high sensitivity, selectivity, and stability across a broad pH range. Using confocal microscopy, researchers successfully imaged HClO and H_2O_2 levels in living cells and zebrafish, confirming the probe's ability to monitor oxidative stress changes during ALI (Figure 4). This study provides a valuable tool for understanding ALI-related inflammation and oxidative stress mechanisms [34].

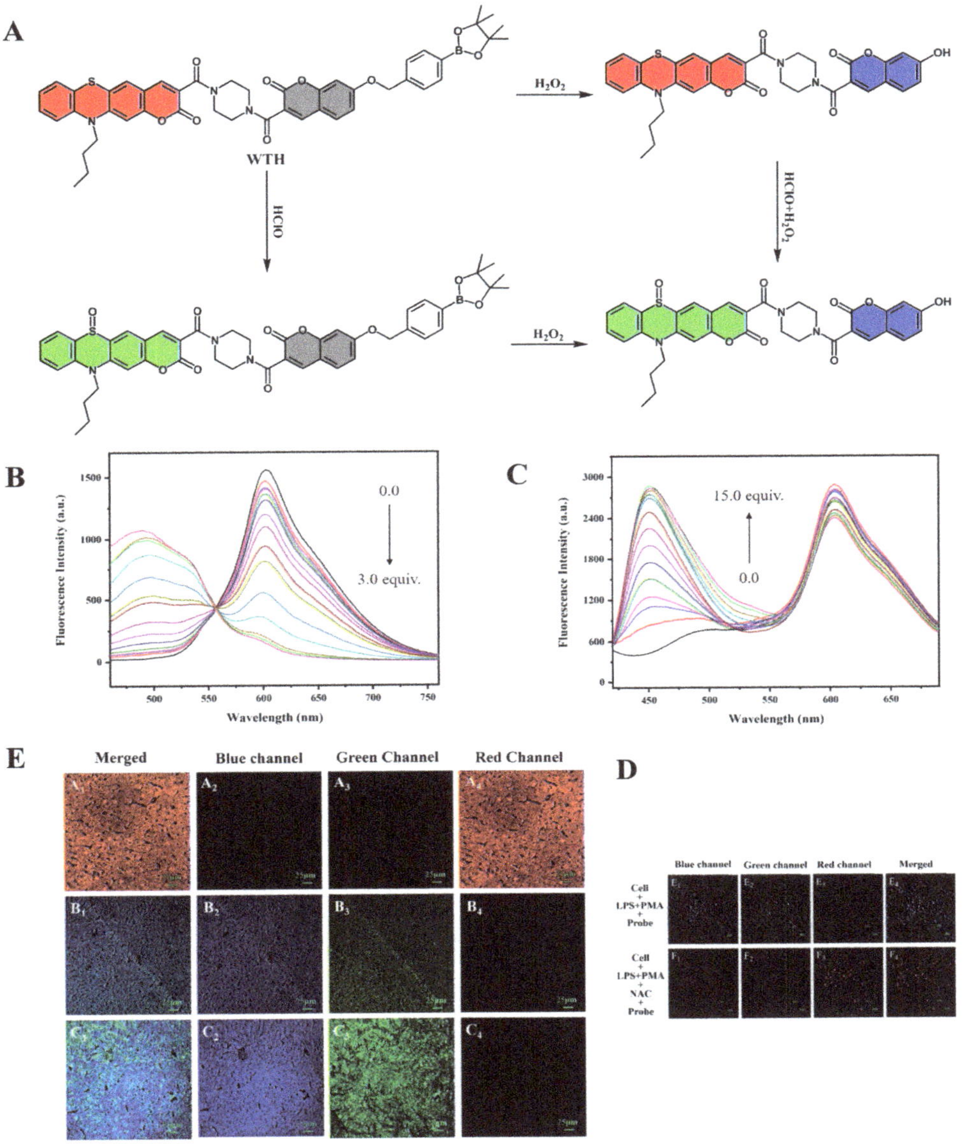

Figure 4. (**A**) H_2O_2/HClO response mechanism of the fluorescent probe WTH. (**B**,**C**) Fluorescence spectral changes of WTH in response to varying amounts of HClO and H_2O_2. (**D**) Imaging of endogenous HClO and H_2O_2 in HepG2 cells. (**E1–E4**) Cells cultivated with LPS (1 μg/mL) and PMA (1 μg/mL) for 1 h and then cultivated with probe 1 for 20 min. (**F1–F4**) Cells cultivated with LPS (1 μg/mL) and PMA (1 μg/mL) for 1 h, NAC (1.0 mM) for 1 h, and probe 1 for 20 min. (**E**) Imaging of alcoholic liver injury in mice. Fluorescence imaging of liver tissue of control mice (**A1–A4**), liver tissue of group 1 mice (**B1–B4**), and liver tissue sections of group 2 mice (**C1–C4**). Blue channel, Ex = 405 nm, Em = 450 ± 25 nm; green channel, Ex = 488 nm, Em = 525 ± 25 nm; red channel, Ex = 488 nm, Em = 595 ± 25 nm. Derived from Ref. [34] with authorization from Copyright 2022, American Chemical Society.

Zhang et al. synthesized a near-infrared fluorescent probe (BDP-NIR-ONOO) featuring a BODIPY dye core and an aryl boronic acid detection moiety, designed for real-time visualization of ALI (Figure 2H). The probe demonstrated high specificity, rapid response, and a high signal-to-noise ratio in vitro. BDP-NIR-ONOO successfully detected $ONOO^-$ fluctuations in ALD mouse models, providing a valuable tool for ALI diagnosis and inflammation research [35].

Zhou et al. designed and synthesized the Cy-ND probe, which exhibits significant sensitivity to viscosity changes in biological environments (Figure 2I). This probe also offers low cytotoxicity and minimal background interference, making it effective for identifying variations in biological viscosity. The Cy-ND probe successfully detected viscosity changes in models of acute alcohol-induced liver injury and liver ischemia-reperfusion injury, providing a valuable tool for further research on liver injury in animal models [36].

Hai et al. designed and synthesized the WZ-HOCl fluorescent probe, which exhibits rapid response and excellent selectivity for HOCl detection (Figure 2J). Furthermore, WZ-HOCl can distinguish between HepG2 and L02 cells, providing a novel approach for liver cancer detection. The probe has also successfully detected HOCl in models of alcoholic hepatitis, non-alcoholic fatty liver disease, and various inflammation-related diseases. WZ-HOCl holds significant potential for early clinical diagnosis and disease screening [37].

3. Chemical Liver Injury

3.1. Establishment of Chemical Liver Injury Models

CLI refers to liver function impairment caused by exposure to exogenous chemicals, including drugs, toxic compounds, and industrial chemicals. Its pathological mechanisms are complex and are closely linked to the direct toxic effects of active metabolites generated during hepatic metabolism as well as the subsequent inflammatory response. Depending on exposure mode and duration, CLI can be classified into acute CLI and chronic CLI. Acute CLI typically results from short-term exposure to high doses of toxic chemicals, such as CCl_4 or industrial solvents, leading to hepatocellular necrosis, increased oxidative stress, and an amplified inflammatory response. In severe cases, acute liver failure may occur. Chronic CLI is mainly induced by long-term exposure to low-dose toxicants, such as prolonged drug abuse or environmental toxin exposure, which gradually leads to hepatic fibrosis, fatty degeneration, and even cirrhosis.

CCl_4 is a widely studied hepatotoxin in liver toxicology due to its strong hepatotoxic effects, which can induce hepatic degeneration, fibrosis, carcinogenesis, and necrosis. Therefore, it is commonly used to establish acute CLI models. The hepatotoxicity of CCl_4 is dose- and time-dependent [38]. Upon metabolic activation by cytochrome P450 enzymes, CCl_4 generates trichloromethyl radicals and trichloromethyl peroxide radicals. These highly reactive metabolites interact with sulfhydryl-containing proteins, altering cell membrane fluidity and permeability, thereby inducing lipid peroxidation, damaging organelles (especially mitochondria), and ultimately triggering hepatocyte death [39,40].

Additionally, CCl_4-mediated liver damage is further exacerbated by the production of ROS and RNS, such as NO. At low concentrations, NO improves hepatic microcirculation, whereas at high concentrations, it accelerates liver damage [41]. Moreover, CCl_4 activates hepatic macrophages and Kupffer cells, stimulating the secretion of pro-inflammatory cytokines such as TNF-α, IL-1, and IL-6, further recruiting neutrophils, amplifying oxidative stress, and aggravating hepatocellular damage [42,43].

CCl_4-induced liver injury is also closely associated with apoptosis pathways. TNF-α can activate apoptotic proteins via the p55TNF receptor, leading to programmed cell death. Moreover, excess ROS disrupts mitochondrial function, promoting cytochrome C release and activating the mitochondrial apoptosis pathway [44,45].

To establish a CCl_4-induced liver injury model, rats or mice are commonly used. Intraperitoneal injection or oral gavage is the preferred administration route (Figure 5). Typically, CCl_4 is diluted in vegetable oil at a ratio of 1:1 to 1:3, and adult rats receive intraperitoneal or oral doses of 0.8–1.6 mL/kg body weight [46]. Since mice are more sensitive to CCl_4 toxicity, a lower concentration (0.1–1% CCl_4 in vegetable oil) is used, with doses of 10–20 mL/kg body weight administered via oral gavage or intraperitoneal injection [47,48]. After 12–24 h, hepatic function, lipid peroxidation markers, and histopathological changes are evaluated.

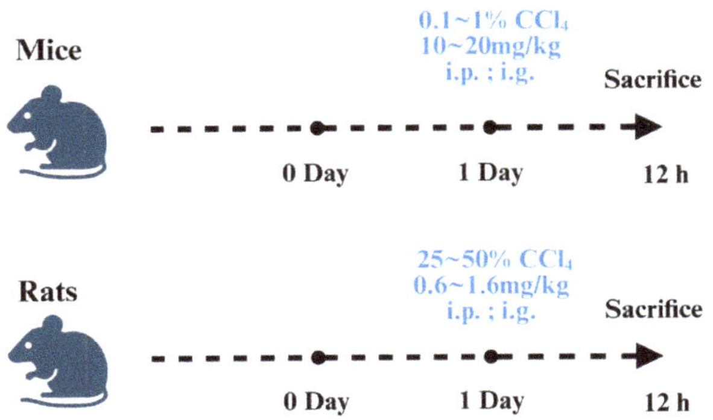

Figure 5. Schematic representation of CCl_4-induced liver injury in animal models.

The CCl_4-induced liver injury model accurately reflects hepatic functional, metabolic, and morphological alterations. This model is widely used due to its low technical requirements, reproducibility, and easy implementation. However, its effects are influenced by factors such as sex, dosage, and exposure duration, making precise experimental control essential. Meanwhile a large number of fluorescent probes for the detection of CLI already exist (Table 2).

Table 2. Characterization of fluorescent probes in the CLI model.

Compounds	Trigger	λex (nm)	λem (nm)	Ref.
YG-H_2S	H_2S	375	520	[49]
NRh-NNBA	GSTs	850	832	[50]
PX-1	$ONOO^-$	405	553	[51]
Lyso-ONOO	$ONOO^-$	450	555	[52]
NIR-II Cy3-988	HClO	980	1048	[53]
HZY	SO_3^{2-}	380	470	[54]
HM-ASPH-PF	MPO	400	586	[55]
Golgi-PER	$ONOO^-$	395	520/415	[56]
PC-Py	HOCl	420	570	[57]

3.2. Application of Fluorescent Probes in Chemical Liver Injury

Yang et al. developed YG-H_2S, a novel H_2S-responsive fluorescent probe, for imaging H_2S dynamics in CLI models (Figure 6A). Compared to traditional probes, YG-H_2S exhibits faster response kinetics, higher selectivity, and stability across a broad pH range, minimizing background noise from complex hepatic environments. Designed based on a deep understanding of H_2S metabolism dynamics, YG-H_2S enables precise localization of pathological sites while also elucidating the role of H_2S in free radical regulation. In CCl_4-induced acute liver injury models, YG-H_2S successfully visualized H_2S fluctuations, correlating its levels with injury severity (Figure 7). Moreover, its low cytotoxicity makes it well-suited for deep tissue imaging in live organisms [49].

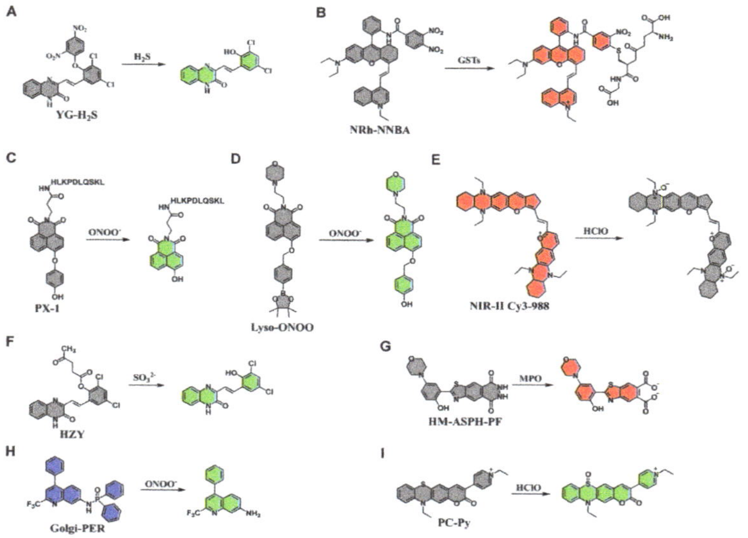

Figure 6. (**A–I**) Structure and detection mechanism of fluorescent probe in CLI models.

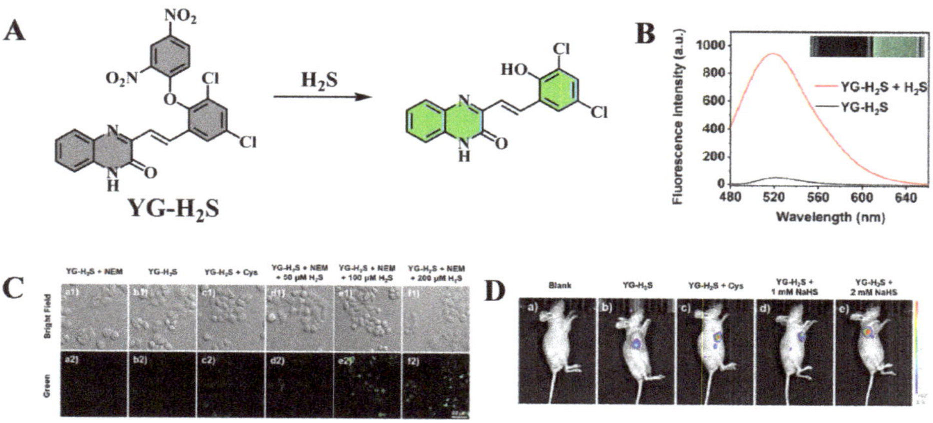

Figure 7. (**A**) H_2S response mechanism of the fluorescent probe YG-H_2S. (**B**) Fluorescence spectra of YG-H_2S (10 μM) in the absence and presence of H_2S in PBS buffer (10 mM, 1% DMSO, pH 7.4). Insert: the UV light of YG-H_2S (10 μM) without and with H_2S. (**C**) Confocal images of living HeLa cells with YG-H_2S (20 μM) under different treating conditions. (**D**) The fluorescence imaging of YG-H_2S (100 μM) in the 4T1-xenograft mice under various conditions. (**a**) Blank mice (**b**) Intravenous injection with YG-H_2S (100 μM); (**c**) Pre-treatment with Cys (2 mM) and then intravenous injection with YG-H_2S (100 μM); (**d,e**) Pre-treatment with NaHS and then intravenous injection with YG-H_2S (100 μM). Derived from Ref. [49] with authorization from Copyright 2023, Elsevier.

Wang et al. developed NRh-NNBA, a rhodamine-based upconversion luminescence probe, for the detection of GSTs in CLI models (Figure 6B). In CCl_4-induced CLI, NRh-

NNBA exhibited rapid and specific responsiveness to GSTs. The probe successfully quantified endogenous GSTs in human liver L02 cells, liver tissue sections, and mouse serum samples, offering a high signal-to-noise ratio. Furthermore, its upconversion luminescence properties effectively reduce autofluorescence interference in biological samples. These findings suggest that NRh-NNBA is a promising sensor for GST-based CLI diagnosis and monitoring [50].

Tang et al. developed PX-1, a non-peptide-based two-photon fluorescent probe, designed for $ONOO^-$ detection in CLI models (Figure 6C). PX-1 features high sensitivity, rapid response, and precise targeting capabilities. In vitro, PX-1 demonstrated excellent specificity and sensitivity for $ONOO^-$. Furthermore, in vivo imaging of CCl_4-induced CLI models revealed significantly upregulated $ONOO^-$ levels in injured livers. This work provides a new strategy for elucidating the physiological role of $ONOO^-$ in acute liver injury [51].

Yang et al. developed Lyso-ONOO, a fluorescent probe for $ONOO^-$ tracking in CLI mouse models (Figure 6D). Lyso-ONOO exhibited high selectivity, sensitivity, and resistance to interference, making it effective for monitoring $ONOO^-$ fluctuations in LX-2 cells. Moreover, its strong lysosomal targeting ability allowed for accurate visualization of $ONOO^-$ dynamics during CLI progression and therapeutic interventions. Both in vivo fluorescence imaging and deep-tissue imaging confirmed Lyso-ONOO's capability to correlate $ONOO^-$ levels with liver injury severity and therapeutic efficacy, providing a promising tool for assessing different treatment strategies [52].

Yuan et al. developed NIR-II Cy3-988, a NIR-II fluorescent probe, capable of selective HClO detection (Figure 6E). Notably, NIR-II Cy3-988 is reversible, allowing dynamic monitoring of HClO/RSS-mediated redox fluctuations. The probe was successfully applied to detect oxidative microenvironment alterations in acute inflammation and CLI repair models, as well as evaluate drug efficacy during acute inflammation. These findings suggest that NIR-II Cy3-988 is a valuable tool for precise real-time imaging of redox dynamics [53].

Hu et al. designed HZY, a fluorescent probe for SO_3^{2-} detection, which exhibits high specificity, rapid response, and a large Stokes shift (Figure 6F). HZY successfully visualized SO_3^{2-} levels in vitro and in vivo, tracking dynamic changes during CLI progression and treatment in CCl_4-induced models (Figure 8). These findings highlight HZY's potential for in situ SO_3^{2-} detection, making it a promising tool for CLI preclinical diagnosis and clinical applications [54].

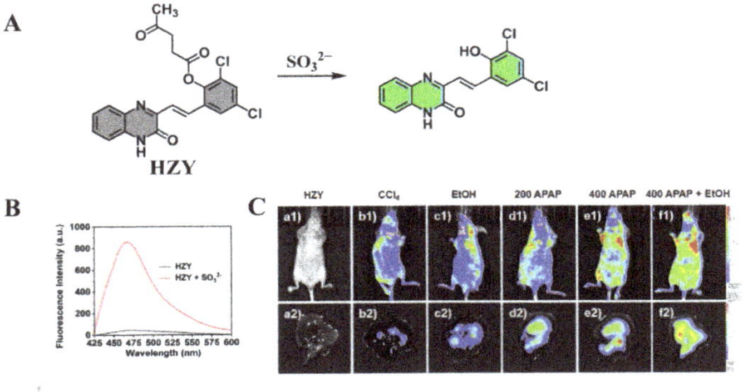

Figure 8. (**A**) SO_3^{2-} response mechanism of the fluorescent probe HZY. (**B**) The fluorescence of HZY with or without SO_3^{2-} in PBS buffer. (**C**) Intravenous injection of HZY in mice with normal and ALI. Liver tissues were dissected to perform in vivo imaging. Derived from Ref. [54] with authorization from Copyright 2023, Elsevier.

Duan et al. synthesized HM-ASPH-PF, a luminol-based fluorescent probe for MPO detection (Figure 6G). The probe, featuring benzothiazole and cyclic phthalic acid hydrazide moieties, exhibits high selectivity, sensitivity, stability, and excellent water solubility. HM-ASPH-PF successfully imaged MPO activity in ALI and acute liver failure mouse models, providing a novel approach for developing orange-emitting luminol derivatives and evaluating neutrophil-associated liver diseases [55].

Liu et al. designed Golgi-PER, a ratiometric fluorescent probe for $ONOO^-$ detection, which exhibited high sensitivity, a low detection limit, and excellent selectivity (Figure 6H). Golgi-PER efficiently tracked $ONOO^-$ levels in CCl_4-induced CLI models, correlating its fluctuations with disease progression and treatment response. These results suggest that Golgi-PER is a valuable tool for investigating liver disease pathogenesis and therapeutic strategies [56].

Wang et al. synthesized PC-Py, a highly selective HOCl-responsive fluorescent probe (Figure 6I). PC-Py exhibits fast response, high sensitivity, and excellent membrane permeability, allowing for accurate HOCl detection during CLI progression. Furthermore, PC-Py successfully differentiated liver fibrosis, acute liver injury, and cirrhosis by monitoring HOCl fluctuations. These findings highlight PC-Py's potential for HOCl-based liver disease diagnosis [57].

4. Drug-Induced Liver Injury

4.1. Establishment of Drug-Induced Liver Injury Models

DILI refers to hepatic damage caused by the toxic effects of drugs or their metabolites, as well as immune-mediated hypersensitivity reactions during pharmacological treatment [58]. The mechanisms underlying DILI are complex and can be broadly classified into two categories [59]. Idiosyncratic DILI arises from specific pharmacological effects of certain drugs, making its occurrence unpredictable by conventional screening methods. Although rare, idiosyncratic DILI is often severe and can lead to acute liver failure. In contrast, intrinsic DILI is associated with dose-dependent toxicity, typically resulting from drug overdose or the overloading of hepatic detoxification pathways, making its onset more predictable [60,61].

APAP is a widely used over-the-counter antipyretic and analgesic, and its hepatotoxicity is among the most common causes of DILI. Due to its well-characterized mechanisms, APAP-induced liver injury serves as a reliable experimental model for studying the pathogenesis, prevention, and treatment of DILI [62,63]. The hepatotoxicity of APAP is primarily mediated by its highly reactive metabolite, NAPQI, which targets mitochondrial proteins and forms covalent adducts with key hepatocellular components, including housekeeping proteins, glutathione peroxidase, and the α-subunit of ATP synthase [64,65]. The disruption of mitochondrial function by NAPQI leads to electron leakage in the electron transport chain, promoting the generation of superoxide radicals [66]. These radicals undergo further conversion into H_2O_2 and molecular oxygen through the action of manganese superoxide dismutase or react with endogenous NO to form $ONOO^-$ [67]. The accumulation of oxidative stress markers such as $ONOO^-$ contributes to mitochondrial damage, ultimately exacerbating hepatocyte injury.

The detoxification of H_2O_2 in hepatocytes primarily relies on GSH and enzymatic clearance by antioxidant systems, including glutathione peroxidase, catalase, and peroxiredoxin. However, during APAP-induced hepatotoxicity, mitochondrial cysteine reacts with GSH, further depleting antioxidant reserves and amplifying hepatic damage. The excessive accumulation of ROS and peroxynitrite leads to the depletion of GSH, the formation of nitrotyrosine-protein adducts, and mitochondrial DNA damage, collectively resulting in extensive liver cell necrosis and apoptosis. These molecular events make

APAP-induced hepatotoxicity a well-established model for investigating oxidative stress, lipid peroxidation, and mitochondrial dysfunction in DILI.

The APAP-induced acute liver injury model is widely used due to its ability to closely replicate human DILI pathogenesis. The most common experimental approach involves oral gavage or intraperitoneal injection of APAP in mice to induce hepatotoxicity (Figure 9). Sun et al. developed a DILI mouse model by fasting male C57BL/6J mice for 16 h, followed by intraperitoneal injection of 400 mg/kg APAP. After 12 h, liver and blood samples were collected for biochemical analysis [68]. Liang et al. established a similar model using Kunming mice, where animals were fasted for 24 h before receiving a 250 mg/kg intraperitoneal dose of APAP, with sampling conducted four hours post-injection [69]. Jiang et al. induced DILI by administering 300 mg/kg APAP intraperitoneally to male C57BL/6 mice following a seven-day adaptive feeding period, with liver samples collected 12 h later [70].

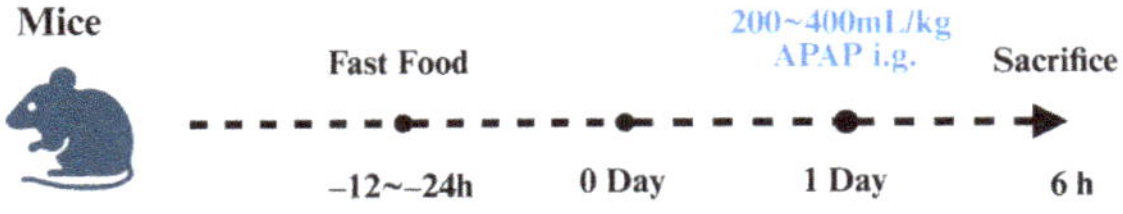

Figure 9. Schematic diagram of animal models for DILI.

The APAP-induced liver injury model offers high reproducibility and precise control over experimental variables, allowing researchers to closely examine oxidative stress, lipid peroxidation, DNA damage, and protein dysfunction. With a short modeling time and strong clinical relevance, this model serves as an optimal platform for studying the protective effects of natural compounds and pharmacological interventions against DILI. Meanwhile a large number of fluorescent probes for the detection of DILI already exist (Table 3).

Table 3. Characterization of fluorescent probes in the DILI model.

Compounds	Trigger	λex (nm)	λem (nm)	Ref.
LW-OTf	$O_2^{\bullet-}$ $ONOO^-$	670 360	715 461	[71]
ATP-LW	$ONOO^-$ ATP	450, 488 520	562, 568 587	[72]
VIS-HNO	HNO Viscosity	427 405	637 475	[73]
NOP	$ONOO^-$	580	656	[74]
IR-990	H_2O_2	808	990	[75]
RPC-1	H_2S HClO	360 545	445 580	[76]
RNH-X	HClO	442	500/710	[77]
Fluo-HNO	HNO	480	520	[78]
Mito-HNO	HNO	695	725	[78]
DOP-CO	CO	630	785/720	[79]
P-GST	GSH	420	550	[80]

4.2. Application of Fluorescent Probes in Drug-Induced Liver Injury

James et al. developed a near-infrared fluorescent probe LW-OTf, which consists of a triflate functional group and a cyanine dye, designed to detect $O_2^{\bullet-}$ and $ONOO^-$ during

DILI (Figure 10A). The probe enables in situ imaging of RNS and ROS in HL-7702 cells with APAP-induced hepatotoxicity. In vivo experiments demonstrated an increase in $O_2^{\bullet-}$ and $ONOO^-$ levels in DILI mouse livers, which was confirmed by SC (H&E) staining, validating the formation of DILI [71].

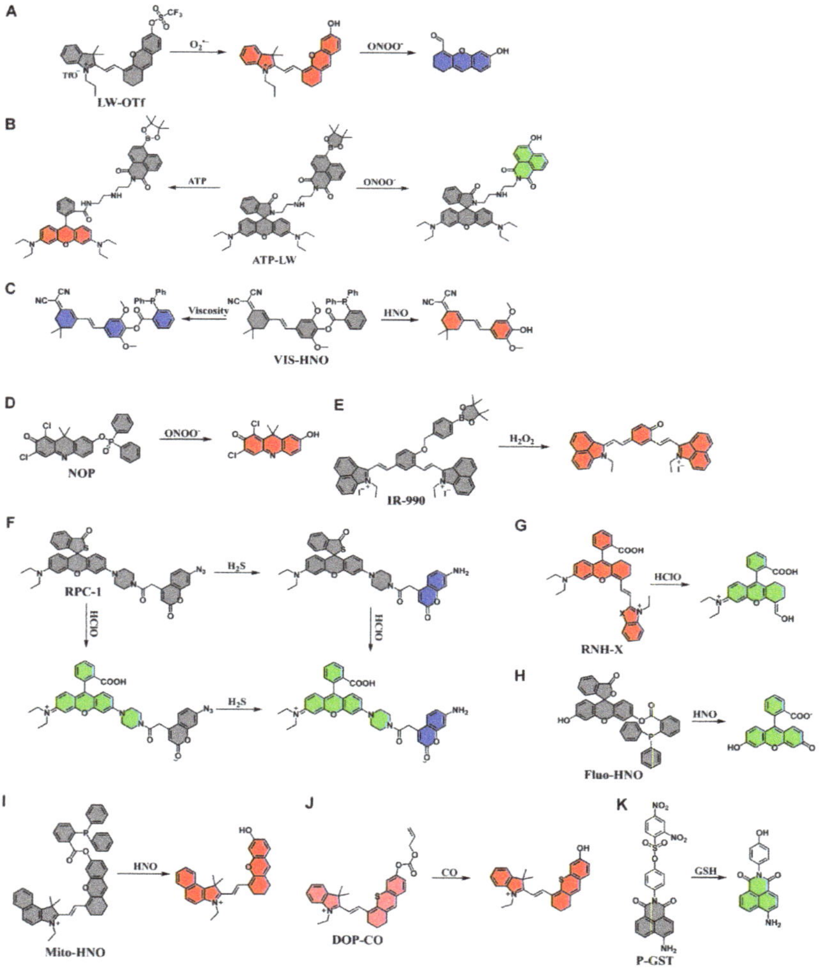

Figure 10. (**A–K**) Structure and detection mechanism of fluorescent probe in DILI models.

Sessler et al. reported ATP-LW, a hybrid fluorescent probe incorporating rhodamine lactam and 1,8-naphthalimide, capable of detecting $ONOO^-$ and ATP simultaneously (Figure 10B). When $ONOO^-$ was present, fluorescence emission was observed at 500–575 nm, whereas in the presence of ATP, rhodamine lactam ring opening resulted in fluorescence at 575–650 nm. Imaging of HL-7702 cells treated with oligomycin A revealed a decrease in red fluorescence and an increase in green fluorescence, indicating ATP depletion and $ONOO^-$ accumulation. Similar results were obtained when cells were treated with SIN-1. Further imaging of APAP-induced hepatotoxicity confirmed $ONOO^-$ elevation and ATP depletion, making ATP-LW a promising probe for simultaneous monitoring of oxidative stress and energy metabolism during DILI [72].

Ye et al. synthesized VIS-HNO, a dual-response fluorescent probe for HNO and viscosity, with high sensitivity, selectivity, and biocompatibility (Figure 10C). VIS-HNO enables independent evaluation of HNO and viscosity without cross-interference and was successfully applied to bioimaging in living cells and zebrafish, revealing the role of HNO in DILI. By assessing HNO and viscosity levels, VIS-HNO provides a valuable tool for DILI diagnosis, ferroptosis research, and therapeutic intervention studies [73].

Li et al. designed NOP, a fluorescent probe with high stability, broad pH adaptability, and superior specificity and sensitivity for $ONOO^-$ detection (Figure 10D). In the APAP-induced liver injury model, NOP successfully monitored $ONOO^-$ level fluctuations, with fluorescence signals correlating with histopathological staining and serum biomarker analysis (Figure 11). These results demonstrate that NOP is an effective tool for early DILI diagnosis, mechanistic studies, and therapeutic screening [74].

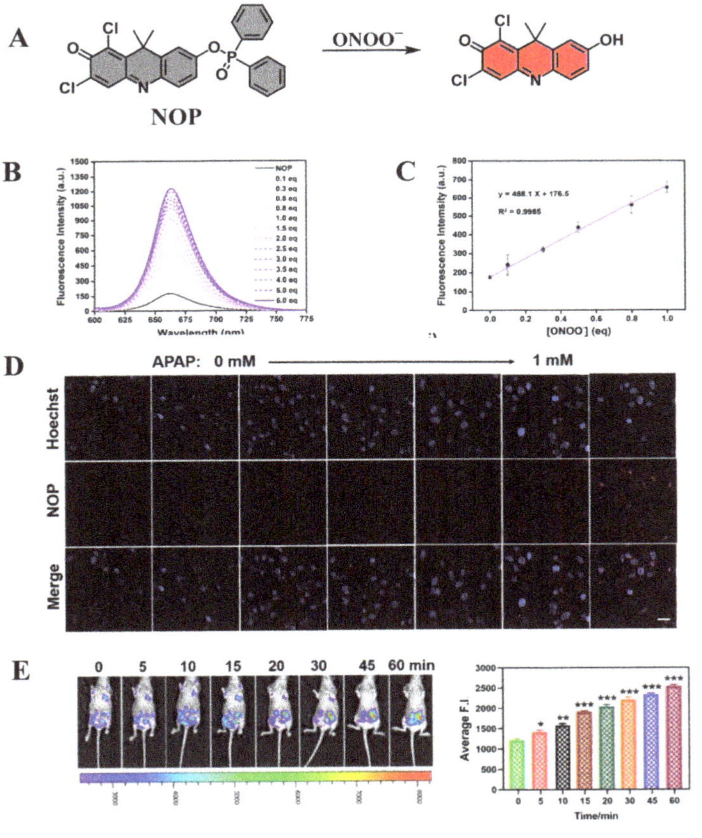

Figure 11. (**A**) $ONOO^-$ response mechanism of the fluorescent probe NOP. (**B**) Fluorescence spectra of 10 μM NOP and 0–60 μM $ONOO^-$ for 5 min. (**C**) The linear relationship between NOP (10 μM) and $ONOO^-$ (from 0 to 10 μM). (**D**) HepG2 cells were coincubated with APAP (0, 0.1, 0.2, 0.4, 0.6, 0.8, 1.0 mM) for 12 h, then the cells were washed with PBS three times. After added the NOP (10 μM) for 30 min. Finally, commercial Hoechst 33,342 dye was added and stained for 15min. Scale bar: 25 μM. (**E**) The nude mice were intraperitoneally injected with 100 μM NOP (200 μL in normal saline (NS)), and then intraperitoneally injected with 500 μM $ONOO^-$ at the same location. After fluorescence images and fluorescence intensity of mice were collected at different time points: 0, 5, 10, 15, 20, 30, 45, 60 min. ∗ p-value < 0.05, ∗∗ p-value < 0.01, ∗∗∗ p-value < 0.001. Derived from Ref. [74] with authorization from Copyright 2023, Elsevier.

Xiong et al. synthesized IR-990, a NIR-II fluorescent probe designed for H_2O_2 detection (Figure 10E). The receptor-π-receptor framework of IR-990 ensures high sensitivity, large Stokes shift, and low detection limits. IR-990 successfully tracked endogenous H_2O_2 variations in HepG2 cells and DILI mouse models, providing real-time imaging capabilities. As a powerful diagnostic tool, IR-990 enables H_2O_2 visualization in vivo, offering great potential for elucidating DILI pathogenesis [75].

Tang et al. developed RPC-1, a two-photon fluorescent probe for simultaneous differentiation and imaging of H_2S and HClO (Figure 10F). The probe was used to evaluate DILI induced by the antidepressants duloxetine and fluoxetine, revealing significant HClO upregulation and severe liver damage after co-administration. The correlated changes in HClO and H_2S levels confirmed the protective role of endogenous H_2S. Histological and serological analyses further validated HClO as a reliable DILI marker and H_2S as a potential therapeutic target, demonstrating RPC-1's utility in DILI prediction and antidote research [76].

Liu et al. designed RNH-X, a mitochondria-targeted ratiometric fluorescent probe for HClO detection in APAP-induced liver and kidney injury (Figure 10G). RNH-O, the most reactive derivative, exhibited high sensitivity, rapid response, and excellent selectivity. The probe successfully tracked HClO fluctuations in living cells, detected APAP-induced HClO upregulation, and enabled fluorescence imaging of inflammatory tissues in DILI mouse models, making it a valuable tool for assessing liver and kidney damage (Figure 12) [77].

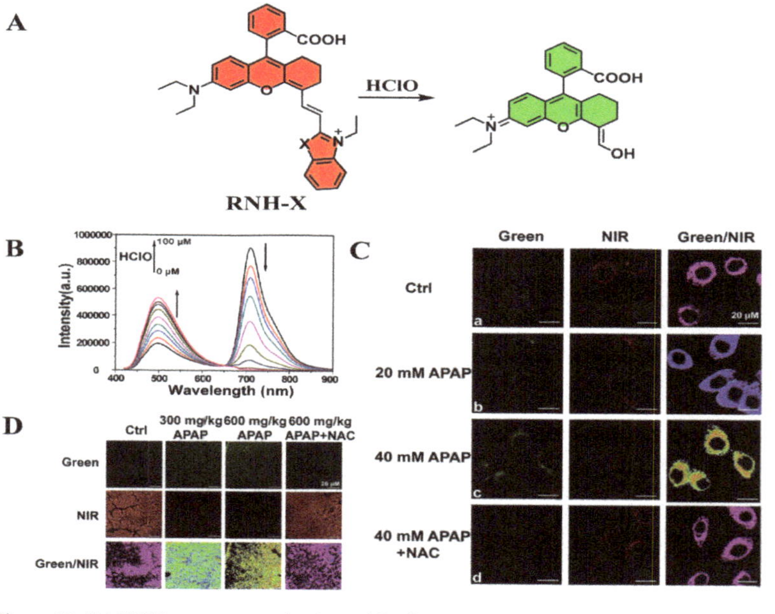

Figure 12. (**A**) HClO response mechanism of the fluorescent probe RNH-X. (**B**) Fluorescence value of the probe NRH-O (10 μM) at diverse concentrations of HClO (0–10 equiv.). (**C**) Confocal fluorescence imaging of human normal liver L02 cells at different fluorescence channels; scale bar is 20.0 μM. Cells were incubated with different contents of APAP (20, 40 mM), NAC (40 mM), and incubated with NRH-O (10 μM) for half an hour before imaging. (**D**) Fluorescence imaging of liver tissue slices from mice after intraperitoneal injection of NRH-O 10 μL (10 mM) after 1 h. Derived from Ref. [77] with authorization from Copyright 2022, American Chemical Society.

The production of RNS, such as HNO, is known to be an early signal of DILI (Figure 10H). Tang and colleagues developed Fluo-HNO as a specific fluorescent probe for in situ imaging of HNO in the Golgi apparatus. This work demonstrated for the first time that catalase plays a role in the production of intracellular HNO. The probe enables the detection of HNO concentration changes in the liver, providing a powerful tool for DILI diagnosis [78].

Tang's team also developed Mito-HNO, a specific fluorescent probe for in situ imaging of HNO in mitochondria (Figure 10I). Like Fluo-HNO, Mito-HNO enables multicolor imaging and provides detailed insights into HNO levels in the mitochondria of liver cells. Their study revealed that the concentration of HNO in the liver of mice induced by the anticancer drug bleomycin was significantly increased. These findings underscore the potential of HNO as a diagnostic marker for DILI, offering valuable information for liver injury assessment [78].

Chen et al. developed DOP-CO, a photoacoustic/fluorescence dual-mode imaging probe for CO detection, utilizing thione-semicyanine as the core structure (Figure 10J). The probe exhibited excellent biocompatibility and enabled quantitative CO detection in HepG2 cells. In vivo studies demonstrated CO fluctuations in liver injury and repair, providing insights into CO-mediated hepatoprotective mechanisms and making DOP-CO a promising tool for investigating CO-related physiological and pathological processes [79].

Feng et al. developed P-GST, a two-photon fluorescent probe for GST detection, utilizing a naphthalimide skeleton with a 2,4-dinitrobenzenesulfonyl recognition unit (Figure 10K). In the presence of GST and GSH, the recognition unit is cleaved, releasing fluorescence. P-GST was successfully applied to two-photon imaging of GST in DILI models, confirming its diagnostic potential for monitoring GST activity in complex biological systems [80].

5. Immune Liver Injury

5.1. Establishment of the LPS-Induced Liver Injury Models

ILI refers to hepatic damage triggered by abnormal immune responses, which involve autoimmune activation or responses to exogenous factors. The underlying pathological mechanisms primarily include immune cell-mediated attacks on hepatocytes or bile duct epithelial cells, excessive inflammatory cytokine release, and immune regulatory dysfunction [81,82]. Based on etiological factors, ILI can be classified into autoimmune liver injury and immune-mediated liver injury induced by drugs or infections. Autoimmune liver injury includes autoimmune hepatitis, primary biliary cholangitis, and primary sclerosing cholangitis, with pathogenesis closely linked to genetic predisposition, environmental factors (such as infections), and immune dysregulation. The main mechanism involves autoantigen exposure or abnormal antigen presentation, leading to T cell-mediated immune attacks [83,84]. Immune liver injury induced by drugs or infections is caused by immune checkpoint inhibitors or pathogenic infections (e.g., viral infections), which provoke an immune overreaction against liver tissues. This type of ILI is characterized by high individual variability, with drugs or infections acting as initiating triggers [85].

LPS is widely used in experimental models to mimic ILI due to its ability to elicit strong inflammatory responses via the TLR4 signaling pathway. LPS-induced liver injury effectively models exogenous infection-related immune activation and serves as a key tool for studying immune-mediated hepatic inflammation, oxidative stress, and apoptosis. Following entry into the bloodstream, LPS binds to lipopolysaccharide-binding protein and is transported via CD14 to the TLR4/myeloid differentiation protein-2 complex, leading to TLR4 dimerization and activation [86–88]. The activated TLR4 pathway stimulates NF-κB signaling through myeloid differentiation factor 88, thereby inducing inflammatory cytokine transcription and promoting acute inflammatory responses [89,90]. Upon NF-κB/p65 complex nuclear translocation, the expression of TNF-α, IL-1β, and IL-6 is upregulated, further exacerbating hepatic damage [91]. The severity of TLR4-mediated liver injury is dependent on NF-κB activation, which not only amplifies inflammatory responses but also enhances oxidative stress and apoptosis, leading to progressive liver pathology.

Studies have demonstrated that LPS activates Kupffer cells, triggering the release of ROS, proinflammatory cytokines, and chemokines, thereby promoting neutrophil infiltration and further aggravating hepatic damage [92]. The inflammatory cascade induced by LPS also activates anti-inflammatory cytokines such as IL-1RA, IL-4, IL-10, and IL-13 as part of a feedback regulation mechanism [93]. LPS-induced hepatic inflammation is characterized by increased TLR4 expression, excessive immune activation, and potential cellular damage, septic shock, or multi-organ failure. By administering intraperitoneal injections of LPS, researchers can simulate exogenous infections and construct experimental models of inflammatory liver injury. LPS-stimulated Kupffer cells generate ROS, proinflammatory mediators, and chemokines, which drive neutrophil infiltration, thereby reproducing key pathological features of immune-mediated hepatic injury.

Experimental models of LPS-induced liver injury have been successfully established in different mouse strains through intraperitoneal injection of LPS, leading to acute hepatic inflammation (Figure 13) [94]. Wang et al. developed an acute liver injury model using ICR mice, where 15 mg/kg of LPS was administered intraperitoneally, and hepatic damage was confirmed 12 h post-injection through histopathological and biochemical analyses [95]. Jia et al. established a similar model in Kunming mice, administering 5 mg/kg of LPS daily for three consecutive days, with one injection per day to induce progressive liver inflammation [96]. Miao et al. further optimized the model by injecting 3 mg/kg of LPS intraperitoneally in Kunming mice, and analyzing liver injury six hours post-injection [97].

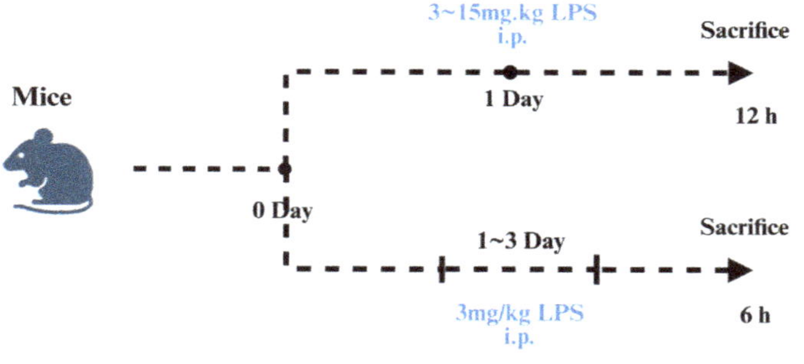

Figure 13. Schematic representation of LPS-induced liver injury in animal models.

This LPS-induced liver injury model offers high reproducibility and scalability, allowing researchers to replicate inflammatory hepatic damage across different animal groups. The model is simple to establish, with a short induction period and adjustable LPS dosages, making it a valuable platform for investigating immune-mediated hepatic inflammation, oxidative stress, and drug intervention strategies. The flexibility of this model also allows for combining LPS administration with additional treatments to simulate various pathological conditions in liver disease research. Meanwhile a large number of fluorescent probes for the detection of LPS-induced liver injury already exist (Table 4).

Table 4. Characterization of fluorescent probes in LPS-Induced Liver Injury.

Compounds	Trigger	λex (nm)	λem (nm)	Ref.
NIR-II-H_2S_2	H_2S_2	730, 808	840, 1000	[98]
AP-Sn	H_2Sn	660	715	[99]
Mito-ND	Viscosity	560	676	[100]
MNIP-Cu	NO	350	492	[101]
DHBP	Viscosity Polarity	450 480	630 660	[102]
KSQT	HOCl	405	475/589	[103]
SPN	$ONOO^-$	680	732	[104]
RDB-ClO	ClO^-	380	582/455	[105]
BDP-R-ClO	HOCl	560	661	[106]

5.2. Application of Fluorescent Probes in LPS-Induced Liver Injury

H_2S_2 plays a crucial role in redox biology and cellular signaling [107]. The ability to quantitatively visualize H_2S_2 in deep tissues is essential for understanding pathophysiological mechanisms related to immune liver injury. Chen et al. successfully developed NIR-II-H_2S_2, a dual-ratiometric NIR-II fluorescent probe designed for selective H_2S_2 detection, featuring a 1,3-dichloro recognition moiety and an NIR-II fluorophore scaffold (Figure 14A). The probe demonstrated high specificity, excellent sensitivity, and enhanced water solubility, as well as deep tissue penetration, making it suitable for real-time imaging of H_2S_2 levels in LPS-induced liver injury models. These findings highlight the potential of NIR-II-H_2S_2 as a promising tool for H_2S_2-related pathological research and future H_2S_2 probe development [98].

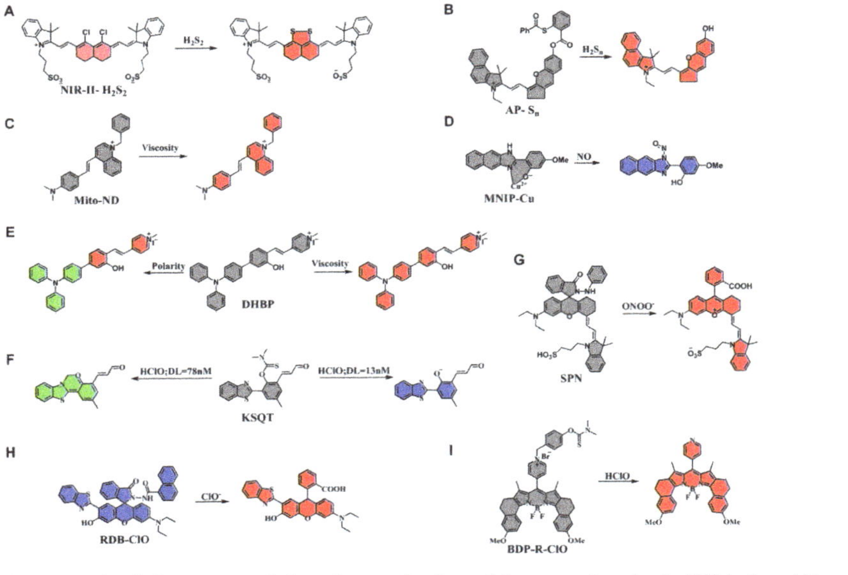

Figure 14. (**A–I**) Structure and detection mechanism of fluorescent probe in LPS-induced liver injury models.

Zhao et al. synthesized AP-Sn, an H_2Sn-activated photoacoustic probe, utilizing semi-anthocyanidin dye as the core structure and phenyl-2-(benzoylthio)benzoic acid as the reactive moiety (Figure 14B). H_2Sn, a key oxidation product of H_2S, plays a fundamental role in ion channel signaling and tumor suppression, while its dysregulation is associated with various inflammatory diseases. AP-Sn exhibits high stability, excellent biocompatibility, and high specificity for H_2S_4, making it a powerful tool for real-time monitoring of LPS-induced liver injury progression. This probe provides an advanced platform for in situ imaging and mechanistic studies of ALI [99].

Lin et al. designed Mito-ND, a viscosity-sensitive near-infrared fluorescent probe with mitochondrial localization, excellent chemical stability, and photostability (Figure 14C). Mito-ND enables fluorescence imaging of viscosity changes in vitro and successfully detected viscosity alterations in LPS-induced liver injury in mice, demonstrating its potential for clinical liver and kidney disease research (Figure 15) [100].

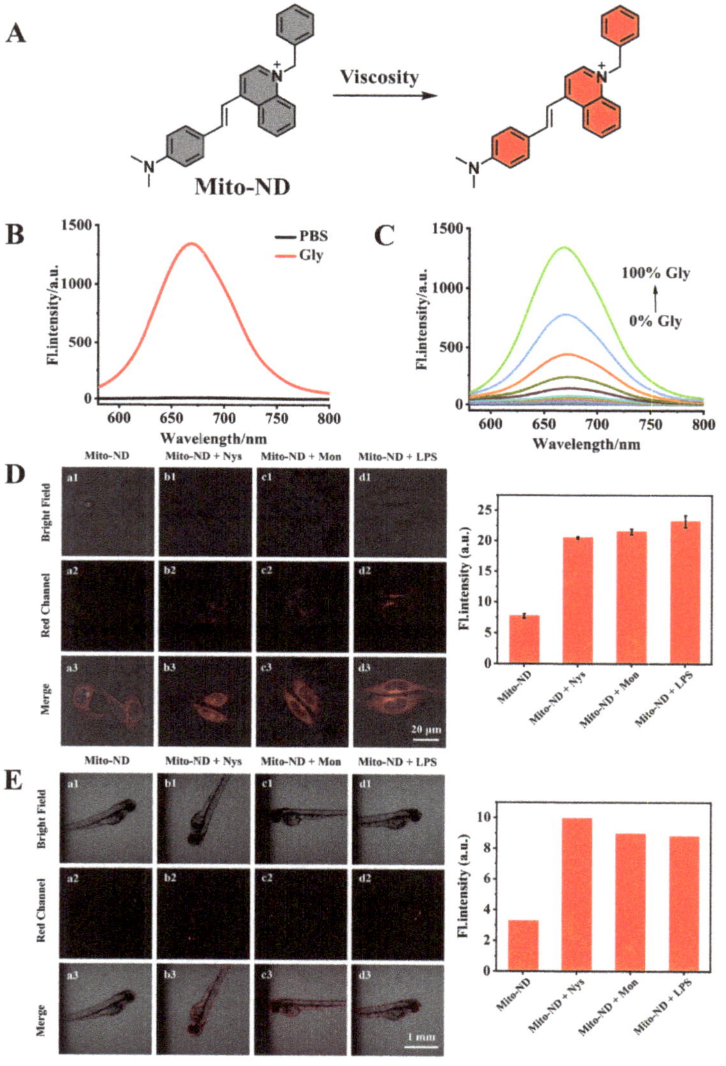

Figure 15. (**A**) Viscosity response mechanism of the fluorescent probe Mito-ND. (**B**) fluorescence spectra of Mito-ND (10 μM) in PBS and glycerol solution. (**C**) Fluorescence spectra of Mito-ND (10 μM)

in PBS-glycerol solution with different viscosity (1.5 cp-1099.5 cp). (**D**) Confocal fluorescence images and fluorescence intensities of HeLa cells incubated with Mito-ND (10 μM) only, and pretreated with different reagents (20 μM Nys, 20 μM Mon, or 20 μg/mL LPS) for 20 min and then incubated with Mito-ND (10 μM) for another 20 min (**a1,b1,c1,d1**) Bright field. (**a2,b2,c2,d2**) Fluorescence field. (**a3,b3,c3,d3**) Merge field. Scale bar: 20 μm (**E**) Confocal fluorescence images and fluorescence intensities of zebrafish incubated with Mito-ND (10 μM) only, and pretreated with different reagents (20 μM Nys, 20 μM Mon, or 20 μg/mL LPS) for 30 min and then incubated with Mito-ND (10 μM) for another 30 min (**a1,b1,c1,d1**) Bright field. (**a2,b2,c2,d2**) Fluorescence field. (**a3,b3,c3,d3**) Merge field. Scale bar: 1 mm. Derived from Ref. [100] with authorization from Copy-right 2022, Elsevier.

Zhang et al. developed MNIP-Cu, a fluorescent probe specifically designed for NO detection, which features rapid binding, high selectivity, and ease of synthesis (Figure 14D). Using MNIP-Cu, researchers successfully visualized and quantified NO distribution in LPS-induced acute liver injury mouse models, demonstrating its potential for biomedical applications in immune liver injury research [101].

Yu et al. synthesized DHBP, a dual-functional fluorescent probe capable of detecting microenvironmental viscosity and polarity fluctuations induced by LPS exposure (Figure 14E). DHBP was successfully applied to LPS-induced liver injury models, demonstrating its potential as a diagnostic tool for immune liver injury studies [102].

Guo et al. developed KSQT, a small-molecule fluorescent probe based on an HBT-derived scaffold, designed for real-time, dual-mode HOCl monitoring (Figure 14F). KSQT enables on/off fluorescence switching and ratiometric response for HOCl detection. Live-cell and zebrafish imaging demonstrated the probe's ability to track both exogenous and endogenous HOCl in real time, making it a powerful tool for high-resolution in vivo imaging of HOCl-related processes [103].

Xu et al. developed SPN, a phenylhydrazine-based near-infrared fluorescent probe, optimized for $ONOO^-$ detection (Figure 14G). SPN exhibits low detection limits, high sensitivity, and excellent water solubility, making it suitable for cellular and in vivo imaging. It successfully detected $ONOO^-$ in LPS-induced zebrafish liver injury models, confirming its utility in diagnosing immune liver injury and evaluating therapeutic interventions (Figure 16) [104].

Hou et al. developed RDB-ClO, a rhodamine-based fluorescent probe for ClO^- detection, featuring high selectivity, low detection limits, rapid response, and high sensitivity (Figure 14H). The probe effectively tracked endogenous and exogenous ClO^- in live cells and zebrafish models and was successfully applied to detect ClO^- fluctuations in E. coli and LPS-induced liver injury models. These findings highlight RDB-ClO's potential as a powerful tool for exploring disease pathogenesis and studying antimicrobial drug effects [105].

Zhao et al. synthesized BDP-R-ClO, a HOCl-responsive fluorescent probe incorporating benzothiazole coumarin and hydroxycoumarin via a piperazine linker (Figure 14I). The probe demonstrated high specificity, rapid response, and broad pH adaptability. BDP-R-ClO was effectively applied to cellular imaging and in vivo imaging of LPS-induced liver injury models, confirming its potential as a diagnostic tool for immune liver injury research [106].

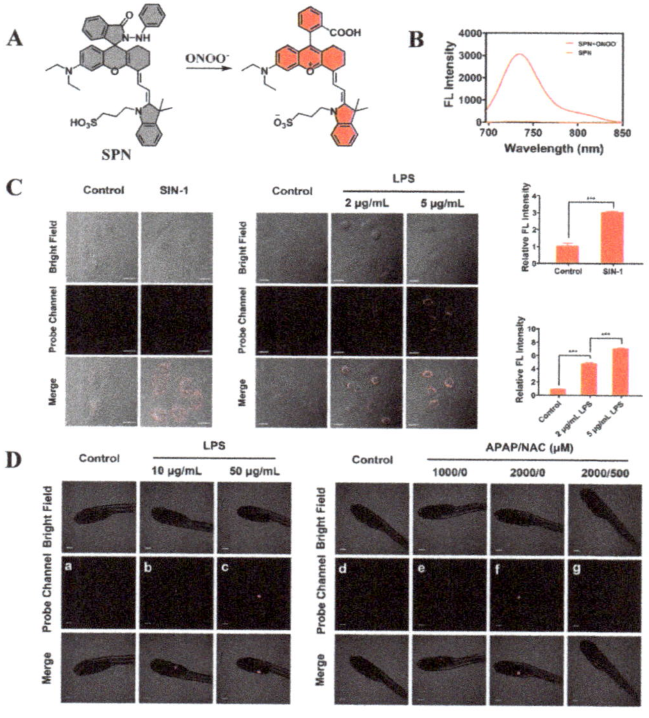

Figure 16. (**A**) ONOO⁻ response mechanism of the fluorescent probe SPN. (**B**) Fluorescence spectra of SPN with ONOO⁻. (**C**) Fluorescence imaging and relative fluorescence intensity of exogenous and endogenous ONOO⁻ in HepG2 cells. Scale bar: 20 μm. Error bars are ± SD (n = 3). Statistical significance levels: *** $p < 0.001$. (**D**) In vivo imaging of LPS/APAP-induced liver injury in zebrafish. Scale bar: 200 μm. Derived from Ref. [104] with authorization from Copyright 2023, Elsevier.

5.3. D-GalN/LPS-Induced Liver Injury Models

D-GalN is a well-established sensitizer of LPS that significantly amplifies the hepatotoxic effects of LPS [108,109]. The combined administration of D-GalN and LPS leads to severe hepatic necrosis and inflammation, making this model a widely used tool for studying ILI.

D-GalN exerts its hepatotoxic effects through multiple mechanisms. Upon entering the body, D-GalN reacts with UDP to form UDP-galactosamine, which leads to rapid depletion of UDP. Since UDP is essential for the biosynthesis of nucleic acids, glycoproteins, and lipopolysaccharides, prolonged UDP depletion severely disrupts cellular metabolism and detoxification processes, further aggravating hepatic damage. Moreover, D-GalN triggers ROS production, inducing severe lipid peroxidation in hepatocyte membranes and initiating oxidative stress-induced apoptosis.

Hepatocyte apoptosis occurs via three main pathways: the death receptor (extrinsic) pathway, the mitochondrial (intrinsic) pathway, and the endoplasmic reticulum stress-induced pathway. Under pathological conditions, these three pathways form a cascading apoptotic response, leading to DNA fragmentation and cytoskeletal protein degradation, ultimately resulting in cell death. Meanwhile, LPS activates hepatic macrophages (Kupffer cells), stimulating the secretion of proinflammatory cytokines such as TNF-α, IL-6, and IL-1β, thereby exacerbating hepatic necrosis and disrupting antioxidant enzyme homeostasis. The synergistic hepatotoxicity of D-GalN and LPS creates a highly specific liver-targeted immune response, which is widely utilized in experimental models of acute liver failure [110,111].

D-GalN/LPS co-administration induces specialized hepatic macrophage activation, which plays a pivotal role in host immune defense. Kupffer cells secrete proinflammatory cytokines (TNF-α, IL-6, IL-1β) and ROS, triggering NF-κB and p38 MAPK activation, leading to the upregulation of multiple cytokine synthesis pathways. Furthermore, this model induces organelle-specific stress, including endoplasmic reticulum stress, while impairing the mitochondrial ROS defense network, ultimately inhibiting antioxidant enzyme production and worsening hepatic damage [112–114].

Wan et al. used male mice, injecting them intraperitoneally with 10 μg/kg of LPS and 700 mg/kg of D-GalN, followed by sacrificing the animals 5 h later to collect serum, liver, and spleen samples [115]. Qu et al. developed a similar model using C57BL/6 mice, dissolving D-GalN/LPS in phosphate-buffered saline and injecting 30 μg/kg of LPS and 600 mg/kg of D-GalN intraperitoneally. Mice were euthanized 6 h post-injection using 2% sodium pentobarbital, and blood samples were collected via orbital puncture (Figure 17) [116]. Feng et al. used BALB/c mice, administering 10 μg/kg of LPS and 700 mg/kg of D-GalN intraperitoneally, establishing a liver injury model within 6 h [117].

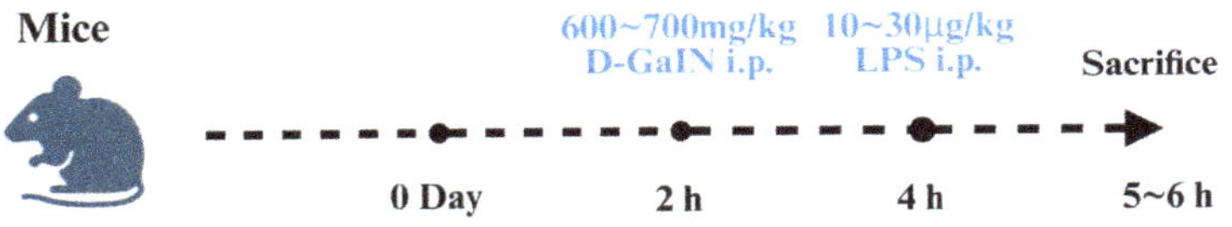

Figure 17. Schematic representation of D-GalN/LPS-induced liver injury in animal models.

The D-GalN/LPS model is highly efficient due to its ability to induce acute liver injury with low LPS doses. As a potent LPS sensitizer, D-GalN enhances LPS hepatotoxicity by several orders of magnitude, allowing researchers to induce acute liver failure with minimal LPS exposure. This model closely resembles human infectious liver injury, making it an ideal tool for studying the pathogenesis of acute liver injury, as well as for screening therapeutic agents with hepatoprotective effects against infection-induced liver damage. Meanwhile a large number of fluorescent probes for the detection of D-GalN/LPS-induced liver injury already exist (Table 5).

Table 5. Characterization of fluorescent probes in D-GalN/LPS-Induced Liver Injury.

Compounds	Trigger	λex (nm)	λem (nm)	Ref.
DMOPB	NO	574	622	[118]
MBTD	NO	421	625/550	[119]
Nil-ClO	HClO/ClO$^-$	560	650	[120]
PTZCy	ClO$^-$	625	700/820	[121]
S-BODIPY	ClO$^-$	540	587/619	[122]
TMSDNPOB	H_2S	574	592	[123]
Cou-dhz-Ph-NO_2	OCl$^-$	430	478	[124]
Hcy-Biot	$O_2^{\bullet-}$ H_2Sn	730 730	780 780	[125]
DMONPB	H_2S	576	627	[126]

5.4. Application of Fluorescent Probes in D-GalN/LPS-Induced Liver Injury

Wang et al. developed DMOPB, a BODIPY-based fluorescent probe designed for NO detection (Figure 18A). The probe exhibits high sensitivity, strong selectivity, low toxicity, and minimal background fluorescence, enabling precise visualization of NO in cells and tissues. By expanding the Stokes shift, DMOPB enhances NO detection capabilities and has been successfully applied to D-GalN/LPS-induced liver injury models. These findings demonstrate that BODIPY-based fluorescent probes provide a powerful tool for NO imaging in immune liver injury studies [118].

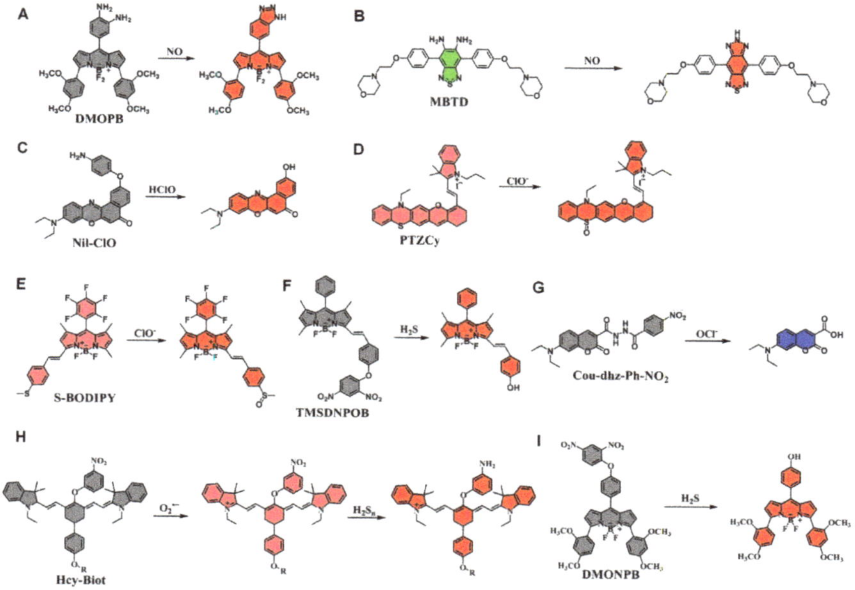

Figure 18. (**A–I**) Structure and detection mechanism of fluorescent probe in D-GalN/LPS-induced liverinjury models.

Ding et al. synthesized MBTD, a dual-intramolecular charge transfer fluorescent probe, which incorporates an OPD fragment into a donor-acceptor-donor fluorophore, allowing for far-red emission and NO detection (Figure 18B). MBTD exhibits a large fluorescence on-off ratio, long fluorescence lifetime, high photostability, and excellent specificity for NO. Notably, MBTD is lysosome-compatible, making it suitable for specific imaging of lysosomal NO levels. In D-GalN/LPS-induced liver injury models, MBTD effectively monitored NO fluctuations in lysosomes, revealing the key role of inducible nitric oxide synthase in NO production during hepatic inflammation. These findings suggest that NO serves as a crucial biomarker for acute liver injury, and MBTD represents a valuable tool for precise NO detection in lysosomal environments [119].

Kong et al. designed Nil-ClO, a two-photon near-infrared fluorescent probe derived from Nile Red, capable of selective $HClO/ClO^-$ detection (Figure 18C). Nil-ClO exhibits high stability, excellent selectivity, and deep tissue penetration, enabling in vivo imaging of $HClO/ClO^-$ in D-GalN/LPS-induced liver injury models. This probe serves as an effective molecular tool for investigating the role of $HClO/ClO^-$ in immune-mediated hepatic inflammation [120].

Xu et al. synthesized PTZCy, a benzothiazole-semicyanine-based ratiometric fluorescent probe for ClO^- detection. PTZCy responds rapidly and selectively to ClO^-, facilitating the real-time monitoring of endogenous and exogenous ClO^- fluctuations in D-GalN/LPS-induced liver injury models (Figures 18D and 19). These results highlight PTZCy as a promising tool for early-stage liver injury diagnosis and in vivo ClO^- imaging [121].

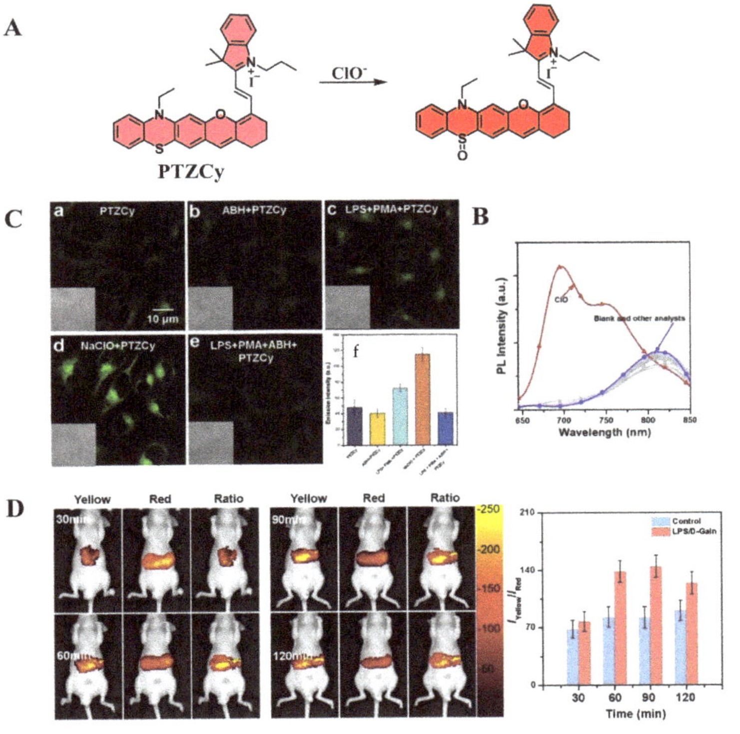

Figure 19. (**A**) OCl^- response mechanism of the fluorescent probe PTZCy. (**B**) Emission spectra of PTZCy (5 μM) with the increasing OCl^- concentrations (0–150 μM) in EtOH/PBS buffer. (**C**) Confocal fluorescence images and fluorescence intensity of the probe in RAW264.7 cells under different conditions. (**a**–**e**) PTZCy group: cells were incubated with PTZCy (5 μM, 30 min), then imaged; ABH+PTZCy group: cells were pretreated with ABH (250 μM, 4 h), next incubated with PTZCy (5 μM, 30 min), then imaged; LPS/PMA+PTZCy group: cells were treated with LPS (5 μg·mL^{-1}) and PMA (5 μg·mL^{-1}) for 12 h, then incubated with PTZCy (5 μM, 30 min) and imaged; NaClO+PTZCy group: cells were treated with NaClO (200 μM, 4 h), then incubated with PTZCy (5 μM, 30 min) and imaged; LPS/PMA+ABH+PTZCy group: cells were treated with LPS (5 μg·mL^{-1}) and PMA (5 μg·mL^{-1}) for 12 h, then treated with ABH (250 μM, 4 h), finally incubated with PTZCy (5 μM, 30 min) and imaged. Scale bar: 10 μm. (λex = 633 nm, λem = 650–780 nm). (**f**) Intensity of green fluorescence obtained in the above five groups, data = mean ± S.D., n = 3. (**D**) In vivo imaging and fluorescence intensity ratio of mice treated with LPS/D-GalN at different times, I.V. injected with PTZCy. (λex = 620 nm, yellow channel: λem = 650–780 nm, red channel: λem = 790–850 nm). Data = mean ± S.D., n = 3. Derived from Ref. [121] with authorization from Copyright 2024, Elsevier.

Kim et al. developed S-BODIPY, a BODIPY-based ratiometric fluorescent probe for $HClO/ClO^-$ detection, enabling real-time imaging of ClO^- both in vitro and in vivo (Figure 18E). S-BODIPY exhibits high fluorescence quantum yield, strong pH stability, good membrane permeability, and excellent biocompatibility, making it well-suited for tracking ClO^- fluctuations in D-GalN/LPS-induced liver injury models (Figure 20). These results demonstrate the potential of S-BODIPY as an efficient tool for ClO^- detection in hepatic inflammation research [122].

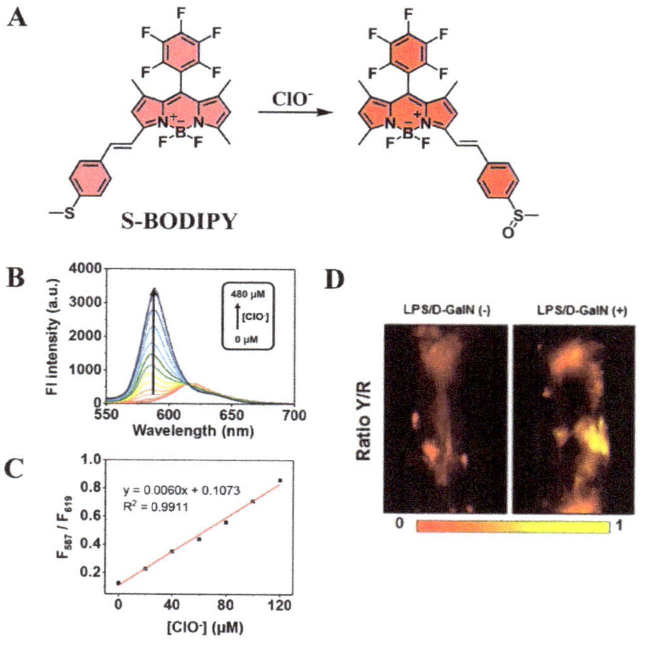

Figure 20. (**A**) ClO^- response mechanism of the fluorescent probe S-BODIPY. (**B**) Fluorescence spectral changes of S-BODIPY (5 μM) in DMF/PBS solution (v/v = 1/1, pH 7.4, 10 mM) upon addition of increasing amounts of NaClO (0–480 μM). Each spectrum was recorded after 1 min. (**C**) Linear relationships between fluorescence intensity ratios (F587/F619) of S-BODIPY (5 μM) versus concentrations of NaClO. (**D**) In vivo imaging of endogenous HClO production from the peritoneal cavity of the mice treated with LPS/D-GalN. Representative images of mice intravenously treated with S-BODIPY (1 mM) for 2 h, pretreated via intraperitoneal injection of PBS (left) and LPS/D-GalN (right). Ratiometric fluorescence images of mice (λex = 550 nm, yellow channel (λem = 550–610 nm) and red channel (λem = 610–670 nm)). Derived from Ref. [122] with authorization from Copyright 2019, American Chemical Society.

Wang et al. designed TMSDNPOB, a dinitrophenyl ether-based fluorescent probe capable of detecting H_2S via thiolysis reactions (Figure 18F). The probe exhibits long-wavelength fluorescence, minimal background interference, and low photodamage, making it suitable for in vivo H_2S imaging. TMSDNPOB enables real-time visualization of both exogenous and endogenous H_2S in D-GalN/LPS-induced liver injury models, confirming its high sensitivity, biocompatibility, and selectivity [123].

Mo et al. synthesized Cou-dhz-Ph-NO_2, a coumarin-based fluorescent probe incorporating aromatic dihydrazines as the linking unit (Figure 18G). The probe exhibits high specificity, rapid response kinetics, and low background fluorescence, facilitating sensitive OCl^- detection. Cou-dhz-Ph-NO_2 successfully visualized OCl^- dynamics in living cells and monitored OCl^- fluctuations in acute liver injury models, demonstrating its potential as a diagnostic tool for OCl^- related liver diseases [124].

Chen et al. developed Hcy-Mito and Hcy-Biot, two near-infrared fluorescent probes designed for simultaneous $O_2^{\bullet-}$ and H_2Sn detection (Figure 18H). These probes exhibit high sensitivity, deep tissue penetration, and strong selectivity, enabling real-time imaging of $O_2^{\bullet-}$ and H_2Sn in D-GalN/LPS-induced liver injury models. Their application in ex vivo and in vivo imaging highlights their potential as valuable tools for investigating oxidative stress and sulfur metabolism in immune liver injury [125].

Wang et al. synthesized DMONPB, a BODIPY-based fluorescent probe with a large Stokes shift, specifically designed for H_2S detection (Figure 18I). The probe exhibits strong photostability, pH insensitivity, low cytotoxicity, and high selectivity for H_2S, making it highly suitable for cellular and tissue imaging. DMONPB was successfully applied to D-GalN/LPS-induced liver injury models, providing an effective method for visualizing H_2S dynamics in immune liver injury research [126].

6. Conclusions

This review systematically summarizes the mechanisms and modeling methods of various liver injury models used in experimental studies, as well as the latest advances in fluorescent probe technology developed for liver injury detection. Liver injury is an intrinsically complex pathological process, and traditional diagnostic methods often suffer from limitations such as low sensitivity, time-consuming procedures, and invasive sample collection. In contrast, fluorescent probes have emerged as powerful analytical tools due to their rapid response, simplicity, and high sensitivity. Recent breakthroughs in fluorescent probe technology have significantly enhanced its applications in liver injury detection, particularly in target identification, localization of injury sites, and quantitative assessment of lesion severity. These advancements have not only provided valuable visualization tools for liver disease research but also established a foundation for early intervention and personalized therapeutic strategies.

Despite the distinct etiologies of alcoholic, chemical, drug-induced, and immune-mediated liver injuries, they share common pathological features such as oxidative stress, inflammatory cytokine release, and hepatocyte apoptosis. Key molecular markers, including ROS and TNF-α, play a central role in multiple liver injury mechanisms, making them universal targets for fluorescent probe development. However, the variability in oxidative stress intensity and biomarker profiles across different liver injury models highlights the need for highly specific and ultrasensitive fluorescent probes capable of dynamic detection and cross-model analysis.

While significant progress has been made in the field of fluorescent probe design, several challenges remain. Most existing probes are designed to detect a single biomarker, whereas liver injury is typically characterized by the interplay of multiple biomolecules and enzymes. This necessitates the development of dual- or multi-target fluorescent probes, which will likely become the future direction in the field. However, the potential for signal interference between multiple fluorophores in multiplexed probes must be carefully addressed to ensure detection accuracy and reliability. Therefore, future research should focus on optimizing key probe properties, including photostability, biocompatibility, low toxicity, high selectivity, and ultra-high sensitivity, while simultaneously advancing multifunctional, synergistic detection strategies.

Another critical challenge is the limited tissue penetration depth of fluorescence imaging, which currently restricts its application to small animal models or ex vivo human liver tissues. To enable non-invasive, real-time diagnosis in clinical settings, the development of NIR-II fluorescent probes has gained significant attention. These NIR-II probes offer deeper tissue penetration, reduced background interference, and enhanced imaging resolution, potentially overcoming the constraints of conventional fluorescence imaging.

Future advancements in fluorescent probe technology will also benefit from interdisciplinary integration with other cutting-edge approaches, further enhancing the precision and efficacy of liver disease diagnosis and therapy. For instance, combining fluorescent probes with nanomaterials could enable simultaneous imaging, quantitative assessment, and targeted drug delivery to the site of liver injury. Additionally, emerging therapeutic modalities such as PDT and PTT have demonstrated promising potential for liver disease treatment, yet their clinical applicability remains limited by poor tissue penetration and suboptimal cell targeting. The integration of fluorescent probes with PDT and PTT could significantly enhance their spatial precision and therapeutic efficiency, broadening their clinical applications.

In summary, fluorescent probes hold immense potential for advancing liver injury detection, early diagnosis, and personalized medicine. Their continued development will not only deepen our understanding of liver disease pathophysiology but also provide critical support for precision medicine and targeted therapeutic strategies. The pursuit of more efficient, multifunctional, and clinically translatable fluorescent probes will have a profound impact on elucidating liver injury mechanisms and accelerating innovations in liver disease diagnostics and therapeutics.

Author Contributions: Writing—original draft preparation, S.L.; software, F.H.; investigation, F.Z.; supervision, X.H.; visualization, J.Z.; supervision, D.P.; funding acquisition, J.H. All authors have read and agreed to the published version of the manuscript.

Funding: This work was financed by Science and Technology program of Gansu Province (24JRRA063), Yunnan Province Major Scientific and Technological Project (202302AE090007 and 202402AA310034), Key R&D Program of Yunnan Province (202203AD150003), Gansu Science Fund for Basic Creative Research Groups (25JRRA469), Gansu Province Intellectual Property Plan Project (23ZSCQG026), Key Research and Development Program of Sichuan Province (2023YFS0339).

Institutional Review Board Statement: Not Applicable.

Informed Consent Statement: Not Applicable.

Data Availability Statement: Not Applicable.

Conflicts of Interest: The authors declare no conflicts of interest.

Abbreviations

The following abbreviations are used in this manuscript:

ALI	Alcoholic liver injury
CLI	Chronic liver injury
AALI	Acute alcoholic liver injury
DILI	Drug-induced liver injury
ILI	Immune-mediated liver injury
ALD	Alcoholic liver disease
CCl_4	Carbon tetrachloride
APAP	Acetaminophen
D-GalN	D-galactosamine
LPS	Lipopolysaccharide
AST	Aspartate aminotransferase
ALT	Alanine aminotransferase
ROS	Reactive oxygen species
RNS	Reactive nitrogen species
CYP450	Cytochrome P450
HClO	Hypochlorous acid
$O_2^{\bullet-}$	Superoxide anions
Cys	Cysteine
NO	Nitric oxide
GSTs	Glutathione S-transferase
H_2S_2	Hydrogen persulfide
H_2O_2	Hydrogen peroxide
GSH	Glutathione

$ONOO^-$	Peroxynitrite
IL-1β	Interleukin-1β
IL-1	Interleukin-1
IL-6	Interleukin-6
TNF-α	Tumor necrosis factor-alpha
IL-1RA	Interleukin-1 receptor antagonist
NF-κB	Nuclear factor-kappa B
NAFLD	Non-alcoholic fatty liver disease
AIE	Aggregation-induced emission
NIR-II	Near-infrared second-window
UDP	Uridine diphosphate
NAPQI	N-acetyl-p-benzoquinoneimine
H&E	Hematoxylin and eosin
PDT	Photodynamic therapy
PTT	Photothermal therapy

References

1. Trefts, E.; Gannon, M.; Wasserman, D.H. The liver. *Curr. Biol.* **2017**, *27*, R1147–R1151. [CrossRef] [PubMed]
2. Damm, T.W.; Kramer, D.J. The Liver in Critical Illness. *Crit. Care Clin.* **2016**, *32*, 425–438. [CrossRef] [PubMed]
3. Crabb, D.W.; Im, G.Y.; Szabo, G.; Mellinger, J.L.; Lucey, M.R. Diagnosis and Treatment of Alcohol-Associated Liver Diseases: 2019 Practice Guidance From the American Association for the Study of Liver Diseases. *Hepatology* **2020**, *71*, 306–333. [CrossRef]
4. Zhang, Y.; Lu, Y.; Ji, H.; Li, Y. Anti-inflammatory, anti-oxidative stress and novel therapeutic targets for cholestatic liver injury. *Biosci. Trends* **2019**, *13*, 23–31. [CrossRef]
5. Duarte, S.; Baber, J.; Fujii, T.; Coito, A.J. Matrix metalloproteinases in liver injury, repair and fibrosis. *Matrix Biol.* **2015**, *44–46*, 147–156. [CrossRef]
6. Giordano, C.M.; Zervos, X.B. Clinical manifestations and treatment of drug-induced hepatotoxicity. *Clin. Liver Dis.* **2013**, *17*, 565–573. [CrossRef]
7. Thomson, J.; Hargrove, L.; Kennedy, L.; Demieville, J.; Francis, H. Cellular crosstalk during cholestatic liver injury. *Liver Res.* **2017**, *1*, 26–33. [CrossRef]
8. Guicciardi, M.E.; Gores, G.J. Apoptosis: A mechanism of acute and chronic liver injury. *Gut* **2005**, *54*, 1024–1033. [CrossRef]
9. Ozer, J.; Ratner, M.; Shaw, M.; Bailey, W.; Schomaker, S. The current state of serum biomarkers of hepatotoxicity. *Toxicology* **2008**, *245*, 194–205. [CrossRef]
10. Ueno, T.; Nagano, T. Fluorescent probes for sensing and imaging. *Nat. Methods* **2011**, *8*, 642–645. [CrossRef]
11. Huang, J.; Pu, K. Activatable Molecular Probes for Second Near-Infrared Fluorescence, Chemiluminescence, and Photoacoustic Imaging. *Angew. Chem. Int. Engl.* **2020**, *59*, 11717–11731. [CrossRef] [PubMed]
12. Shuhendler, A.J.; Pu, K.; Cui, L.; Uetrecht, J.P.; Rao, J. Real-time imaging of oxidative and nitrosative stress in the liver of live animals for drug-toxicity testing. *Nat. Biotechnol.* **2014**, *32*, 373–380. [CrossRef] [PubMed]
13. Zhou, Y.; Yang, X.; Wei, X.; Zhang, S.-S.; Yan, M. Recent progress in small molecule fluorescent probes for imaging and diagnosis of non-alcoholic fatty liver disease. *Coord. Chem. Rev.* **2024**, *513*, 215864. [CrossRef]
14. Li, Z.-J.; Wang, C.-Y.; Xu, L.; Zhang, Z.-Y.; Tang, Y.-H.; Qin, T.-Y.; Wang, Y.-L. Recent Progress of Activity-Based Fluorescent Probes for Imaging Leucine Aminopeptidase. *Biosensors* **2023**, *13*, 752. [CrossRef]
15. Liu, X.; Yu, S.; Zhang, Y.; Zhang, W.; Zhong, H.; Lu, X.; Guan, R. A review on the protective effect of active components in Antrodia camphorata against alcoholic liver injury. *J. Ethnopharmacol.* **2023**, *300*, 115740. [CrossRef]
16. Meier, P.; Seitz, H.K. Age, alcohol metabolism and liver disease. *Curr. Opin. Clin. Nutr. Metab. Care* **2008**, *11*, 21–26. [CrossRef]
17. Jones, B.E.; Liu, H.; Lo, C.R.; Koop, D.R.; Czaja, M.J. Cytochrome P450 2E1 expression induces hepatocyte resistance to cell death from oxidative stress. *Antioxid. Redox Signal.* **2002**, *4*, 701–709. [CrossRef]
18. Yu, S.; Rao, S.; Reddy, J.K. Peroxisome proliferator-activated receptors, fatty acid oxidation, steatohepatitis and hepatocarcinogenesis. *Curr. Mol. Med.* **2003**, *3*, 561–572. [CrossRef]
19. Lieber, C.S. Alcoholic fatty liver: Its pathogenesis and mechanism of progression to inflammation and fibrosis. *Alcohol* **2004**, *34*, 9–19. [CrossRef]
20. Tuma, D.J.; Casey, C.A. Dangerous byproducts of alcohol breakdown--focus on adducts. *Alcohol Res. Health* **2003**, *27*, 285–290.
21. Miranda-Mendez, A.; Lugo-Baruqui, A.; Armendariz-Borunda, J. Molecular basis and current treatment for alcoholic liver disease. *Int. J. Environ. Res. Public Health* **2010**, *7*, 1872–1888. [CrossRef] [PubMed]
22. Ma, Y.; Wang, D.; Xu, X.; Yang, X.; Wang, X.; Zhu, Z.; Zhao, Y.; Chen, M.; Xu, F.; Fu, L.; et al. Dynamic changes of ROS, MDA and SOD during arsenic-induced neoplastic transformation in human keratinocytes. *J. Hyg. Res.* **2015**, *44*, 456–461.

23. Zhao, X.; Xue, X.; Wang, C.; Wang, J.; Peng, C.; Li, Y. Emerging roles of Sirtuins in alleviating alcoholic liver Disease: A comprehensive review. *Int. Immunopharmacol.* **2022**, *108*, 108712. [CrossRef] [PubMed]
24. Chen, J.; Zhang, Q.; Wang, R.; Yang, Y.; Wang, Y.; Liu, X.; Zhang, X.; Qiao, X.; Zhong, G.; Wei, J.; et al. Preliminary study on the effective site and mechanism of action of Meconopsis quintuplinervia Regel in alleviating acute alcoholic liver injury in mice. *Int. Immunopharmacol.* **2023**, *308*, 116230. [CrossRef]
25. Liu, X.; Hou, R.; Yan, J.; Xu, K.; Wu, X.; Lin, W.; Zheng, M.; Fu, J. Purification and characterization of Inonotus hispidus exopolysaccharide and its protective effect on acute alcoholic liver injury in mice. *Int. J. Biol. Macromol.* **2019**, *129*, 41–49. [CrossRef]
26. Wang, J.W.; Chen, X.Y.; Hu, P.Y.; Tan, M.M.; Tang, X.G.; Huang, M.C.; Lou, Z.H. Effects of Linderae radix extracts on a rat model of alcoholic liver injury. *Exp. Ther. Med.* **2016**, *11*, 2185–2192. [CrossRef]
27. Song, M.; Chen, T.; Prough, R.A.; Cave, M.C.; McClain, C.J. Chronic Alcohol Consumption Causes Liver Injury in High-Fructose-Fed Male Mice Through Enhanced Hepatic Inflammatory Response. *Alcohol. Clin. Exp. Res.* **2016**, *40*, 518–528. [CrossRef]
28. Tan, X.; Hong, J.; Jiang, S.; Zhang, S.; Chen, Y.; Feng, G. Near-Infrared Fluorescent Probe Reveals Elevated Mitochondrial Viscosity during Acute Alcoholic Liver Injury. *Anal. Chem.* **2024**, *96*, 14860–14866. [CrossRef]
29. Chen, J.; Fang, Y.; Sun, L.; Zeng, F.; Wu, S. An activatable probe for detecting alcoholic liver injury via multispectral optoacoustic tomography and fluorescence imaging. *Chem. Commun.* **2020**, *56*, 11102–11105. [CrossRef]
30. Shen, L.; Ma, M.; Zhou, K.; Jin, M.; Wang, S.; Liu, H.; Yang, Y. Cysteine triggered cascade reaction forming coumarin: Visualization of cysteine fluctuation in alcoholic liver disease by a NIR fluorescent probe. *Spectrochim. Acta Part A Mol. Biomol. Spectrosc.* **2025**, *324*, 124974. [CrossRef]
31. Ma, X.; Han, R.; Wang, J.; Zhang, B.; Ruan, M.; Zhao, W.; Zhang, J. Novel NIR fluorescent probe based on BODIPY for diagnosis and treatment evaluation of alcoholic liver disease via visualizing HClO fluctuation. *Spectrochim. Acta Part A Mol. Biomol. Spectrosc.* **2025**, *328*, 125497. [CrossRef] [PubMed]
32. Feng, Y.; Nie, G.; Liang, W.; Li, W.; Zhang, Y.; Wang, K.; Chen, D. Real-time imaging of acute alcoholic liver injury in vivo via a robust viscosity probe with aggregation-induced emission nature. *Sens. Actuators B Chem.* **2022**, *355*, 131285. [CrossRef]
33. Wang, S.; Zhao, X.; Liu, M.; Yang, L.; Yu, M.; Li, Z. A dual-responsive crimson fluorescent probe for real-time diagnosis of alcoholic acute liver injury. *Biosens. Bioelectron.* **2023**, *239*, 115596. [CrossRef]
34. Niu, P.; Liu, J.; Xu, F.; Yang, L.; Li, Y.; Sun, A.; Wei, L.; Liu, X.; Song, X. Dual-Ratiometric Fluorescent Probe for H2O2 and HClO in Living Cells and Zebrafish and Application in Alcoholic Liver Injury Monitoring. *ACS Appl. Bio Mater.* **2022**, *5*, 1683–1691. [CrossRef]
35. Zhang, B.; Qin, S.; Wang, J.; Ma, X.; Han, R.; Ruan, M.; Zhao, W.; Zhang, J. A near-infrared $ONOO^-$-activated fluorescent probe for real-time visualizing of alcoholic liver disease. *J. Mol. Struct.* **2025**, *1323*, 140805. [CrossRef]
36. Wang, R.; Zhou, X.; Li, B.; Ding, R.; Han, J.; Wang, Y.; Meng, A.; Zhou, J. A novel dual near-infrared fluorescent probe for bioimaging and visualization of viscosity in acute alcoholic liver injury. *Chem. Commun.* **2024**, *60*, 5804–5807. [CrossRef]
37. Zhang, P.; Wei, L.; Jiang, Q.; Gai, S.; Zhou, Z.; Bian, J.; Zhang, Y.; Han, W.; Shu, W.; He, Y. A turn-on fluorescent probe for detecting and bioimaging of HOCl in inflammatory and liver disease models. *Bioorg. Chem.* **2024**, *143*, 107051. [CrossRef]
38. Lee, Y.S.; Cho, I.J.; Kim, J.W.; Lee, M.K.; Ku, S.K.; Choi, J.S.; Lee, H.J. Hepatoprotective effects of blue honeysuckle on CCl(4)-induced acute liver damaged mice. *Food Sci. Nutr.* **2019**, *7*, 322–338. [CrossRef]
39. Unsal, V.; Cicek, M.; Sabancilar, İ. Toxicity of carbon tetrachloride, free radicals and role of antioxidants. *Rev. Environ. Health* **2021**, *36*, 279–295. [CrossRef]
40. Wang, W.; Jiang, L.; Ren, Y.; Shen, M.; Xie, J. Characterizations and hepatoprotective effect of polysaccharides from Mesona blumes against tetrachloride-induced acute liver injury in mice. *Int. J. Biol. Macromol.* **2019**, *124*, 788–795. [CrossRef]
41. Iwakiri, Y. Nitric oxide in liver fibrosis: The role of inducible nitric oxide synthase. *Clin. Mol. Hepatol.* **2015**, *21*, 319–325. [CrossRef] [PubMed]
42. Li, S.; Liu, J.; Zhang, M.; Chen, Y.; Zhu, T.; Wang, J. Protective Effect of Eckol against Acute Hepatic Injury Induced by Carbon Tetrachloride in Mice. *Mar. Drugs* **2018**, *16*, 300. [CrossRef] [PubMed]
43. Gu, X.; Manautou, J.E. Molecular mechanisms underlying chemical liver injury. *Expert Rev. Mol. Med.* **2012**, *14*, e4. [CrossRef] [PubMed]
44. Badr, G.; Sayed, E.A.; Waly, H.; Hassan, K.A.; Mahmoud, M.H.; Selamoglu, Z. The Therapeutic Mechanisms of Propolis Against CCl(4) -Mediated Liver Injury by Mediating Apoptosis of Activated Hepatic Stellate Cells and Improving the Hepatic Architecture through PI3K/AKT/mTOR, TGF-β/Smad2, Bcl2/BAX/P53 and iNOS Signaling Pathways. *Cell. Physiol. Biochem.* **2019**, *53*, 301–322. [CrossRef]
45. Choi, S.Y.; Song, P.; Hwang, J.S.; Lee, Y.K.; Shin, M.S.; Son, H.J.; Kim, Y.J.; Kim, W.; Lee, K.M. Cereblon deficiency ameliorates carbon tetrachloride-induced acute hepatotoxicity in HepG2 cells by suppressing MAPK-mediated apoptosis. *Front. Immunol.* **2024**, *15*, 1457636. [CrossRef]

46. Salim, N.S.; Abo El-Maati, M.F.; Abdelnour, S.A.; Abdel-Alim, M.E. Hepatoprotective activity of taraxacum officinale extract against CCl4-induced liver injury in rats. *Food Biosci.* **2025**, *68*, 106708. [CrossRef]
47. Lu, Y.; Hu, D.; Ma, S.; Zhao, X.; Wang, S.; Wei, G.; Wang, X.; Wen, A.; Wang, J. Protective effect of wedelolactone against CCl4-induced acute liver injury in mice. *Int. Immunopharmacol.* **2016**, *34*, 44–52. [CrossRef]
48. Lee, M.-M.; Kim, H.-G.; Lee, J.-S.; Lee, S.-B.; Cho, J.-H.; Lee, D.-S.; Lee, N.-H.; Lee, H.-D.; Moon, S.-O.; Son, C.-G. Synergistic hepatoprotective effects of CGplus on CCl4-induced acute injury. *J. Ethnopharmacol.* **2020**, *249*, 112441. [CrossRef]
49. Wang, K.; Guo, R.; Chen, X.-Y.; Li, X.-L.; Hu, Z.-G.; Wang, X.; Wang, C.-Y.; Qin, Y.-J.; Yao, K.; Yang, Y.-S. A novel fluorescent probe for imaging endogenous hydrogen sulfide in living cells and mice models of acute liver injury. *Chem. Eng. J.* **2023**, *468*, 143611. [CrossRef]
50. Zhu, W.; Qian, X.; Wang, J.; Yu, H.; Liu, W.; Wang, H.-Y.; Liu, Y. Near-infrared frequency upconversion probes for monitoring glutathione S-transferase to evaluate acute liver injury. *Sens. Actuators B Chem.* **2021**, *347*, 130640. [CrossRef]
51. Zhou, Y.; Li, P.; Fan, N.; Wang, X.; Liu, X.; Wu, L.; Zhang, W.; Zhang, W.; Ma, C.; Tang, B. In situ visualization of peroxisomal peroxynitrite in the livers of mice with acute liver injury induced by carbon tetrachloride using a new two-photon fluorescent probe. *Chem. Commun.* **2019**, *55*, 6767–6770. [CrossRef] [PubMed]
52. Wang, K.; Guo, R.; Chen, X.-Y.; Yang, Y.-S.; Qiao, L.-Q.; Wang, M.-L. Multifunctional lysosome-targetable fluorescent probe for imaging peroxynitrite in acute liver injury model. *Chem. Eng. J.* **2023**, *455*, 140491. [CrossRef]
53. He, L.; He, L.H.; Xu, S.; Ren, T.B.; Zhang, X.X.; Qin, Z.J.; Zhang, X.B.; Yuan, L. Engineering of Reversible NIR-II Redox-Responsive Fluorescent Probes for Imaging of Inflammation in vivo. *Angew. Chem. Int. Engl.* **2022**, *61*, e202211409. [CrossRef]
54. Hu, Z.-Y.; Chen, X.-Y.; Yang, X.; Li, T.; Yang, Y.-S.; Wang, S.-J.; Wang, K.; Hu, Z.-G. Imaging and detection of sulfite in acute liver injury with a novel quinoxaline-based fluorescent probe. *Anal. Chim. Acta* **2023**, *1261*, 341177. [CrossRef]
55. Meng, T.; Zhang, X.; Tang, W.; Liu, C.; Duan, X. A Small Molecule Chemiluminophore with near 600 nm Emission for In Vivo Imaging of Myeloperoxidase and Inflammatory Diseases. *Chem. Biomed. Imaging* **2024**, *2*, 205–212. [CrossRef]
56. Zhang, T.; Li, Z.; Qin, M.; Zhang, J.; Sun, Y.; Liu, C. Visulization of peroxynitrite variation for accurate diagnosis and assessing treatment response of hepatic fibrosis using a Golgi-targetable ratiometric fluorescent probe. *J. Photochem. Photobiol. B Biol.* **2024**, *257*, 112950. [CrossRef]
57. Yuan, F.; Wang, B.; Hou, J.-T.; Li, J.; Shen, J.; Duan, Y.; Ren, W.X.; Wang, S. Demonstrating HOCl as a potential biomarker for liver fibrosis using a highly sensitive fluorescent probe. *Sens. Actuators B Chem.* **2023**, *378*, 133219. [CrossRef]
58. Garcia-Cortes, M.; Robles-Diaz, M.; Stephens, C.; Ortega-Alonso, A.; Lucena, M.I.; Andrade, R.J. Drug induced liver injury: An update. *Arch. Toxicol.* **2020**, *94*, 3381–3407. [CrossRef]
59. Andrade, R.J.; Chalasani, N.; Björnsson, E.S.; Suzuki, A.; Kullak-Ublick, G.A.; Watkins, P.B.; Devarbhavi, H.; Merz, M.; Lucena, M.I.; Kaplowitz, N.; et al. Drug-induced liver injury. *Nat. Rev. Dis. Primers* **2019**, *5*, 58. [CrossRef]
60. Katarey, D.; Verma, S. Drug-induced liver injury. *Clin. Med.* **2016**, *16*, s104–s109. [CrossRef]
61. Björnsson, H.K.; Björnsson, E.S. Drug-induced liver injury: Pathogenesis, epidemiology, clinical features, and practical management. *Eur. J. Intern. Med.* **2022**, *97*, 26–31. [CrossRef] [PubMed]
62. Wan, M.; Gao, H.; Liu, X.; Zhang, Y. Rutaecarpine Aggravates Acetaminophen-Induced Acute Liver Injury by Inducing CYP1A2. *Toxics* **2024**, *12*, 515. [CrossRef] [PubMed]
63. Hossen, M.S.; Akter, A.; Azmal, M.; Rayhan, M.; Islam, K.S.; Islam, M.M.; Ahmed, S.; Abdullah-Al-Shoeb, M. Unveiling the molecular basis of paracetamol-induced hepatotoxicity: Interaction of N-acetyl-p-benzoquinone imine with mitochondrial succinate dehydrogenase. *Biochem. Biophys. Rep.* **2024**, *38*, 101727. [CrossRef]
64. Hanawa, N.; Shinohara, M.; Saberi, B.; Gaarde, W.A.; Han, D.; Kaplowitz, N. Role of JNK translocation to mitochondria leading to inhibition of mitochondria bioenergetics in acetaminophen-induced liver injury. *J. Biol. Chem.* **2008**, *283*, 13565–13577. [CrossRef]
65. Du, K.; Ramachandran, A.; Weemhoff, J.L.; Chavan, H.; Xie, Y.; Krishnamurthy, P.; Jaeschke, H. Editor's Highlight: Metformin Protects Against Acetaminophen Hepatotoxicity by Attenuation of Mitochondrial Oxidant Stress and Dysfunction. *Toxicol. Sci.* **2016**, *154*, 214–226. [CrossRef]
66. Lee, K.K.; Imaizumi, N.; Chamberland, S.R.; Alder, N.N.; Boelsterli, U.A. Targeting mitochondria with methylene blue protects mice against acetaminophen-induced liver injury. *Hepatology* **2015**, *61*, 326–336. [CrossRef]
67. Jaeschke, H.; Ramachandran, A. Central Mechanisms of Acetaminophen Hepatotoxicity: Mitochondrial Dysfunction by Protein Adducts and Oxidant Stress. *Drug Metab. Dispos.* **2024**, *52*, 712–721. [CrossRef]
68. Wu, Y.; Li, W.; Zhang, J.; Lin, J.; You, L.; Su, J.; Zheng, C.; Gao, Y.; Kong, X.; Sun, X. Shaoyao-Gancao Decoction, a famous Chinese medicine formula, protects against APAP-induced liver injury by promoting autophagy/mitophagy. *Phytomedicine* **2024**, *135*, 156053. [CrossRef]
69. Zhong, Y.H.; Wu, X.W.; Zhang, X.Y.; Zhang, S.W.; Feng, Y.; Zhang, X.M.; Xu, B.B.; Zhong, G.Y.; Huang, H.L.; He, J.W.; et al. Intestinal microbiota-mediated serum pharmacochemistry reveals hepatoprotective metabolites of Platycodonis Radix against APAP-induced liver injury. *J. Chromatogr. B* **2025**, *1251*, 124395. [CrossRef]

70. Lei, P.; Li, X.; Jiang, L.; Yu, H.; Zhang, P.; Han, L.; Jiang, M. Alisma plantago-aquatica polysaccharides ameliorate acetaminophen-induced acute liver injury by regulating hepatic metabolic profiles and modulating gut microbiota. *Int. J. Biol. Macromol.* **2025**, *285*, 138345. [CrossRef]
71. Wu, L.; Liu, J.; Tian, X.; Groleau, R.R.; Bull, S.D.; Li, P.; Tang, B.; James, T.D. Fluorescent probe for the imaging of superoxide and peroxynitrite during drug-induced liver injury. *Chem. Sci.* **2021**, *12*, 3921–3928. [CrossRef] [PubMed]
72. Wu, L.; Liu, J.; Tian, X.; Groleau, R.R.; Feng, B.; Yang, Y.; Sedgwick, A.C.; Han, H.-H.; Wang, Y.; Wang, H.-M.; et al. Dual-Channel Fluorescent Probe for the Simultaneous Monitoring of Peroxynitrite and Adenosine-5′-triphosphate in Cellular Applications. *J. Am. Chem. Soc.* **2022**, *144*, 174–183. [CrossRef] [PubMed]
73. Liu, J.; Li, C.; Sun, J.; Zhang, D.; Yang, X.; Ye, Y.; Zhao, Y. A dual-response fluorescent probe to reveal the role of ferroptosis in drug-induced liver injury. *Chem. Eng. J.* **2024**, *495*, 153592. [CrossRef]
74. Dong, J.; Yang, Y.; Fan, X.; Zhu, H.-L.; Li, Z. Accurate imaging in the processes of formation and inhibition of drug-induced liver injury by an activable fluorescent probe for $ONOO^-$. *Mater. Today Bio* **2023**, *21*, 100689. [CrossRef]
75. Tian, Y.; Liu, S.; Cao, W.; Wu, P.; Chen, Z.; Xiong, H. H_2O_2-Activated NIR-II Fluorescent Probe with a Large Stokes Shift for High-Contrast Imaging in Drug-Induced Liver Injury Mice. *Anal. Chem.* **2022**, *94*, 11321–11328. [CrossRef]
76. Jiao, X.; Xiao, Y.; Li, Y.; Liang, M.; Xie, X.; Wang, X.; Tang, B. Evaluating Drug-Induced Liver Injury and Its Remission via Discrimination and Imaging of HClO and H_2S with a Two-Photon Fluorescent Probe. *Anal. Chem.* **2018**, *90*, 7510–7516. [CrossRef]
77. Shangguan, L.; Wang, J.; Qian, X.; Wu, Y.; Liu, Y. Mitochondria-Targeted Ratiometric Chemdosimeter to Detect Hypochlorite Acid for Monitoring the Drug-Damaged Liver and Kidney. *Anal. Chem.* **2022**, *94*, 11881–11888. [CrossRef]
78. Wang, H.; Liu, C.; He, Z.; Li, P.; Zhang, W.; Zhang, W.; Tang, B. Dual-Colored Fluorescence Imaging of Mitochondrial HNO and Golgi-HNO in Mice with DILI. *Anal. Chem.* **2021**, *93*, 6551–6558. [CrossRef]
79. Liu, L.; Xu, J.; Zhang, S.; Chen, H.; Wang, L.; Shen, X.; Chen, H. Rational design of dual ratiometric photoacoustic and fluorescent probe for reliable imaging and quantitative detection of endogenous CO during drug-induced liver injury and repair. *Sens. Actuators B Chem.* **2022**, *367*, 132171. [CrossRef]
80. Zhang, J.; Jin, Z.; Hu, X.-X.; Meng, H.-M.; Li, J.; Zhang, X.-B.; Liu, H.-W.; Deng, T.; Yao, S.; Feng, L. Efficient Two-Photon Fluorescent Probe for Glutathione S-Transferase Detection and Imaging in Drug-Induced Liver Injury Sample. *Anal. Chem.* **2017**, *89*, 8097–8103. [CrossRef]
81. Kubes, P.; Jenne, C. Immune Responses in the Liver. *Annu. Rev. Immunol.* **2018**, *36*, 247–277. [CrossRef] [PubMed]
82. Polakos, N.K.; Cornejo, J.C.; Murray, D.A.; Wright, K.O.; Treanor, J.J.; Crispe, I.N.; Topham, D.J.; Pierce, R.H. Kupffer cell-dependent hepatitis occurs during influenza infection. *Am. J. Pathol.* **2006**, *168*, 1169–1178. [CrossRef] [PubMed]
83. Santodomingo-Garzon, T.; Swain, M.G. Role of NKT cells in autoimmune liver disease. *Autoimmun. Rev.* **2011**, *10*, 793–800. [CrossRef] [PubMed]
84. Eilat, D. Introduction: Mechanisms of tissue injury in autoimmune diseases. *Semin. Immunopathol.* **2014**, *36*, 491–493. [CrossRef]
85. Gerussi, A.; Natalini, A.; Antonangeli, F.; Mancuso, C.; Agostinetto, E.; Barisani, D.; Di Rosa, F.; Andrade, R.; Invernizzi, P. Immune-Mediated Drug-Induced Liver Injury: Immunogenetics and Experimental Models. *Int. J. Mol. Sci.* **2021**, *22*, 4557. [CrossRef]
86. Xuan, S.; Ma, Y.; Zhou, H.; Gu, S.; Yao, X.; Zeng, X. The implication of dendritic cells in lung diseases: Immunological role of toll-like receptor 4. *Genes Dis.* **2024**, *11*, 101007. [CrossRef]
87. Ciesielska, A.; Matyjek, M.; Kwiatkowska, K. TLR4 and CD14 trafficking and its influence on LPS-induced pro-inflammatory signaling. *Cell. Mol. Life Sci.* **2021**, *78*, 1233–1261. [CrossRef]
88. Giordano, N.P.; Cian, M.B.; Dalebroux, Z.D. Outer Membrane Lipid Secretion and the Innate Immune Response to Gram-Negative Bacteria. *Infect. Immun.* **2020**, *88*, 7. [CrossRef]
89. Wang, J.; Sun, Z.; Xie, J.; Ji, W.; Cui, Y.; Ai, Z.; Liang, G. Inflammasome and pyroptosis in autoimmune liver diseases. *Front. Immunol.* **2023**, *14*, 1150879. [CrossRef]
90. Płóciennikowska, A.; Hromada-Judycka, A.; Borzęcka, K.; Kwiatkowska, K. Co-operation of TLR4 and raft proteins in LPS-induced pro-inflammatory signaling. *Cell. Mol. Life Sci.* **2015**, *72*, 557–581. [CrossRef]
91. Chae, B.S. Pretreatment of Low-Dose and Super-Low-Dose LPS on the Production of In Vitro LPS-Induced Inflammatory Mediators. *Toxicol. Res.* **2018**, *34*, 65–73. [CrossRef] [PubMed]
92. Kawai, T.; Akira, S. Signaling to NF-kappaB by Toll-like receptors. *Trends Mol. Med.* **2007**, *13*, 460–469. [CrossRef] [PubMed]
93. Ntoufa, S.; Vilia, M.G.; Stamatopoulos, K.; Ghia, P.; Muzio, M. Toll-like receptors signaling: A complex network for NF-κB activation in B-cell lymphoid malignancies. *Semin. Cancer Biol.* **2016**, *39*, 15–25. [CrossRef]
94. Wang, X.; Wang, L.; Dong, R.; Huang, K.; Wang, C.; Gu, J.; Luo, H.; Liu, K.; Wu, J.; Sun, H.; et al. Luteolin ameliorates LPS-induced acute liver injury by inhibiting TXNIP-NLRP3 inflammasome in mice. *Phytomedicine* **2021**, *87*, 153586. [CrossRef]
95. Wang, D.; Gao, Q.; Zhao, G.; Kan, Z.; Wang, X.; Wang, H.; Huang, J.; Wang, T.; Qian, F.; Ho, C.T.; et al. Protective Effect and Mechanism of Theanine on Lipopolysaccharide-Induced Inflammation and Acute Liver Injury in Mice. *J. Agric. Food Chem.* **2018**, *66*, 7674–7683. [CrossRef]

96. Gao, Z.; Zhang, C.; Tian, C.; Ren, Z.; Song, X.; Wang, X.; Xu, N.; Jing, H.; Li, S.; Liu, W.; et al. Characterization, antioxidation, anti-inflammation and renoprotection effects of selenized mycelia polysaccharides from Oudemansiella radicata. *Carbohydr. Polym.* **2018**, *181*, 1224–1234. [CrossRef]
97. Miao, F.; Geng, S.; Ning, D. Hydroxytyrosol ameliorates LPS-induced acute liver injury (ALI) in mice by modulating the balance between M1/M2 phenotype macrophage and inhibiting TLR4/NF-κB activation. *J. Funct. Foods* **2023**, *102*, 105455. [CrossRef]
98. Hu, B.; Liu, Q.; Jiang, Y.; Huang, Y.; Ji, H.; Zhang, J.; Wang, X.; Shen, X.C.; Chen, H. NIR-II Fluorescence/Photoacoustic Dual Ratiometric Probes with Unique Recognition Site for Quantitatively Visualizing H_2S_2 in Vivo. *Angew. Chem. Int. Engl.* **2025**, *64*, e202418378. [CrossRef]
99. He, Z.; Qiu, Z.; Liao, L.; Zhang, C.; Hu, S.; Zhao, S. Design and synthesis of photoacoustic probe for in situ imaging of hydrogen polysulfides in organs in vivo. *Sens. Actuators B Chem.* **2023**, *381*, 133423. [CrossRef]
100. Tang, Y.; Peng, J.; Zhang, Q.; Song, S.; Lin, W. A new NIR emission mitochondrial targetable fluorescent probe and its application in detecting viscosity changes in mouse liver and kidney injury. *Talanta* **2022**, *249*, 123647. [CrossRef]
101. Ouyang, J.; Hong, H.; Shen, C.; Zhao, Y.; Ouyang, C.; Dong, L.; Zhu, J.; Guo, Z.; Zeng, K.; Chen, J.; et al. A novel fluorescent probe for the detection of nitric oxide in vitro and in vivo. *Free Radic. Biol. Med.* **2008**, *45*, 1426–1436. [CrossRef] [PubMed]
102. Bian, C.; Liu, M.; Cheng, J.; Yang, L.; Li, Z.; Yu, M. Dual-Functional Fluorescent Probe in the Diagnosis of Liver Injury and the Evaluation of Drug Therapy with Double Signal Amplification. *Chem. Biomed. Imaging* **2024**, *2*, 156–164. [CrossRef] [PubMed]
103. Tang, H.; Qiang, X.; Gao, Y.; Teng, H.; Chen, X.; Zhang, Y.; Tian, J.; Qin, B.; Guo, Y. Real-time tracking and dual-mode imaging of hypochlorous acid in vivo by a small-sized fluorescence probe. *Dye. Pigment.* **2021**, *188*, 109219. [CrossRef]
104. Pan, Y.; Yang, Q.; Xu, H.; Yuan, Z.; Xu, H. Screening and optimization of a water-soluble near-infrared fluorescent probe for drug-induced liver injury monitoring. *Anal. Chim. Acta* **2023**, *1276*, 341654. [CrossRef]
105. He, C.; Liu, Q.; Zhang, X.; Wang, L.; Fu, S.; Zhang, H.; Li, S.; Li, Q.; Chen, S.; Hou, P. Visualizing ClO- fluctuations in drug-induced liver injury and bacterium via a robust ratiometric fluorescent probe. *Spectrochim. Acta Part A Mol. Biomol. Spectrosc.* **2024**, *323*, 124944. [CrossRef]
106. Fan, G.; Wang, N.; Zhang, J.; Ji, X.; Qin, S.; Tao, Y.; Zhao, W. BODIPY-based near-infrared fluorescent probe for diagnosis drug-induced liver injury via imaging of HClO in cells and in vivo. *Dye. Pigment.* **2022**, *199*, 110073. [CrossRef]
107. Roy, B.; Shieh, M.; Takata, T.; Jung, M.; Das, E.; Xu, S.; Akaike, T.; Xian, M. Phototriggered Hydrogen Persulfide Donors via Hydrosulfide Radical Formation Enhancing the Reactive Sulfur Metabolome in Cells. *J. Am. Chem. Soc.* **2024**, *146*, 30502–30509. [CrossRef]
108. Jia, R.; Zhang, H.; Zhang, W.; Zhao, H.; Zha, C.; Liu, Y. Protective effects of tenuigenin on lipopolysaccharide and d-galactosamine-induced acute liver injury. *Microb. Pathog.* **2017**, *112*, 83–88. [CrossRef]
109. Galanos, C.; Freudenberg, M.A.; Reutter, W. Galactosamine-induced sensitization to the lethal effects of endotoxin. *Proc. Natl. Acad. Sci. USA* **1979**, *76*, 5939–5943. [CrossRef]
110. Ansari, M.A.; Raish, M.; Bin Jardan, Y.A.; Ahmad, A.; Shahid, M.; Ahmad, S.F.; Haq, N.; Khan, M.R.; Bakheet, S.A. Sinapic acid ameliorates D-galactosamine/lipopolysaccharide-induced fulminant hepatitis in rats: Role of nuclear factor erythroid-related factor 2/heme oxygenase-1 pathways. *World J. Gastroenterol.* **2021**, *27*, 592–608. [CrossRef]
111. Wang, W.; Zhang, Y.; Li, H.; Zhao, Y.; Cai, E.; Zhu, H.; Li, P.; Liu, J. Protective Effects of Sesquiterpenoids from the Root of Panax ginseng on Fulminant Liver Injury Induced by Lipopolysaccharide/d-Galactosamine. *J. Agric. Food Chem.* **2018**, *66*, 7758–7763. [CrossRef] [PubMed]
112. Zheng, X.F.; Hu, X.Y.; Ma, B.; Fang, H.; Zhang, F.; Mao, Y.F.; Yang, F.Y.; Xiao, S.C.; Xia, Z.F. Interleukin-35 Attenuates D-Galactosamine/Lipopolysaccharide-Induced Liver Injury via Enhancing Interleukin-10 Production in Kupffer Cells. *Front. Pharmacol.* **2018**, *9*, 959. [CrossRef] [PubMed]
113. Wang, X.; Wu, L.; Zhang, Q.; Li, L.; Xie, Y.; Wan, X.; Wu, H.; Xiang, Y. Methyl 3,4-dihydroxybenzoate protects against d-galN/LPS-induced acute liver injury by inhibiting inflammation and apoptosis in mice. *J. Pharm. Pharmacol.* **2019**, *71*, 1082–1088. [CrossRef] [PubMed]
114. Wen, J.; Lin, H.; Zhao, M.; Tao, L.; Yang, Y.; Xu, X.; Jia, A.; Zhang, J.; Weng, D. Piceatannol attenuates D-GalN/LPS-induced hepatoxicity in mice: Involvement of ER stress, inflammation and oxidative stress. *Int. Immunopharmacol.* **2018**, *64*, 131–139. [CrossRef]
115. Yang, S.; Kuang, G.; Zhang, L.; Wu, S.; Zhao, Z.; Wang, B.; Yin, X.; Gong, X.; Wan, J. Mangiferin Attenuates LPS/D-GalN-Induced Acute Liver Injury by Promoting HO-1 in Kupffer Cells. *Front. Immunol.* **2020**, *11*, 285. [CrossRef]
116. Ai, G.; Wu, X.; Dou, Y.; Huang, R.; Zhong, L.; Liu, Y.; Xian, Y.; Lin, Z.; Li, Y.; Su, Z.; et al. Oxyberberine, a novel HO-1 agonist, effectively ameliorates oxidative stress and inflammatory response in LPS/D-GalN induced acute liver injury mice via coactivating erythrocyte metabolism and Nrf2 signaling pathway. *Food Chem. Toxicol.* **2022**, *166*, 113215. [CrossRef]
117. Guo, P.; Zeng, M.; Liu, M.; Zhang, Y.; Jia, J.; Zhang, Z.; Liang, S.; Zheng, X.; Feng, W. Isolation of Calenduloside E from Achyranthes bidentata Blume and its effects on LPS/D-GalN-induced acute liver injury in mice by regulating the AMPK-SIRT3 signaling pathway. *Phytomedicine* **2024**, *125*, 155353. [CrossRef]

118. Zhu, X.-Y.; Yao, H.-W.; Fu, Y.-J.; Guo, X.-F.; Wang, H. Effect of substituents on Stokes shift of BODIPY and its application in designing bioimaging probes. *Anal. Chim. Acta* **2019**, *1048*, 194–203. [CrossRef]
119. Wang, F.; Yu, S.; Xu, Z.; Li, L.; Dang, Y.; Xu, X.; Luo, Y.; Cheng, Z.; Yu, H.; Zhang, W.; et al. Acid-Promoted D-A-D Type Far-Red Fluorescent Probe with High Photostability for Lysosomal Nitric Oxide Imaging. *Anal. Chem.* **2018**, *90*, 7953–7962. [CrossRef]
120. Lin, X.; Chen, Y.; Bao, L.; Wang, S.; Liu, K.; Qin, W.; Kong, F. A two-photon near-infrared fluorescent probe for imaging endogenous hypochlorite in cells, tissue and living mouse. *Dye. Pigment.* **2020**, *174*, 108113. [CrossRef]
121. Wang, X.; Wang, H.; Duan, J.; Sun, Q.; Zhang, C.; Xu, L.; Liu, Z. Phenothiazine-hemicyanine hybrid as a near-infrared fluorescent probe for ratiometric imaging of hypochlorite in vivo. *Sens. Actuators B Chem.* **2024**, *407*, 135453. [CrossRef]
122. Duan, C.; Won, M.; Verwilst, P.; Xu, J.; Kim, H.S.; Zeng, L.; Kim, J.S. In Vivo Imaging of Endogenously Produced HClO in Zebrafish and Mice Using a Bright, Photostable Ratiometric Fluorescent Probe. *Anal. Chem.* **2019**, *91*, 4172–4178. [CrossRef] [PubMed]
123. Ji, Y.; Xia, L.-J.; Chen, L.; Guo, X.-F.; Wang, H.; Zhang, H.-J. A novel BODIPY-based fluorescent probe for selective detection of hydrogen sulfide in living cells and tissues. *Talanta* **2018**, *181*, 104–111. [CrossRef] [PubMed]
124. Li, T.; Wang, L.; Lin, S.; Xu, X.; Liu, M.; Shen, S.; Yan, Z.; Mo, R. Rational Design and Bioimaging Applications of Highly Specific "Turn-On" Fluorescent Probe for Hypochlorite. *Bioconjugate Chem.* **2018**, *29*, 2838–2845. [CrossRef]
125. Huang, Y.; Yu, F.; Wang, J.; Chen, L. Near-Infrared Fluorescence Probe for in Situ Detection of Superoxide Anion and Hydrogen Polysulfides in Mitochondrial Oxidative Stress. *Anal. Chem.* **2016**, *88*, 4122–4129. [CrossRef]
126. Zhu, X.-Y.; Wu, H.; Guo, X.-F.; Wang, H. Novel BODIPY-based fluorescent probes with large Stokes shift for imaging hydrogen sulfide. *Dye. Pigment.* **2019**, *165*, 400–407. [CrossRef]

targets

MDPI

Review

The Design and Prospects of Influenza Virus Vaccines Based on Conserved Epitopes and Adjuvant Optimization

Meng-Qian Zhang [1,†], Jin-Wei Bu [2,†], Zhi-Gang Wang [1] and Shu-Lin Liu [1,2,*]

1 State Key Laboratory of Medicinal Chemical Biology, Tianjin Key Laboratory of Biosensing and Molecular Recognition, Frontiers Science Centre for New Organic Matter, Research Centre for Analytical Sciences, College of Chemistry, School of Medicine and Frontiers Science Center for Cell Responses, Nankai University, Tianjin 300071, China
2 Engineering Research Center of Nano-Geomaterials of Ministry of Education, Faculty of Materials Science and Chemistry, China University of Geosciences, Wuhan 430074, China
* Correspondence: shulin.liu@nankai.edu.cn
† These authors contributed equally to this work.

Abstract: Influenza viruses pose a significant threat to human health, and vaccination remains the most cost-effective and efficient strategy for controlling outbreaks. This review first introduces the molecular characteristics of influenza A virus (IAV) and examines how conserved epitopes contribute to overcoming its high variability, laying the foundation for broadly protective vaccine design. Different vaccine platforms are then categorized and analyzed through representative examples to highlight their research significance and application potential. The discussion further extends to the role of adjuvants in modulating immune responses, with a focus on how their optimization enhances vaccine efficacy. We explore future directions in vaccine design, highlighting the synergistic potential of conserved epitope targeting and adjuvant improvement in advancing the next generation of influenza vaccines.

Keywords: influenza A virus; conserved epitopes; vaccine platforms; universal vaccine; adjuvants

1. Introduction

Influenza is a persistent global health threat, with seasonal epidemics causing substantial morbidity and mortality worldwide [1–4]. Influenza viruses are classified into four types—A (IAV), B (IBV), C (ICV), and D (IDV)—based on antigenic differences in two internal structural proteins: nucleoprotein (NP) and matrix protein (M1). Among them, types A and B pose the greatest threat to human health. IAV, in particular, is responsible for most pandemics and severe seasonal outbreaks, making it the primary focus of universal vaccine development. The influenza virus genome consists of eight segmented RNA strands, which are prone to mutation and reassortment, leading to high genetic diversity. This variation is driven by two major mechanisms: antigenic drift, which refers to the accumulation of point mutations in the genes encoding hemagglutinin (HA) and neuraminidase (NA), causing seasonal strain changes; and antigenic shift, which involves the reassortment of gene segments between different viral subtypes in a shared host, potentially generating novel HA or NA subtypes that can trigger pandemics [5]. The ongoing antigenic variability presents significant challenges for vaccine design and necessitates frequent updates to match circulating strains. While traditional influenza vaccines provide strain-specific protection, their limited cross-protection underscores the urgent need for next-generation vaccines capable of inducing broader and more durable immunity [6–9].

 https://doi.org/10.3390/targets3020016

A promising approach to addressing this challenge is the targeting of conserved epitopes—relatively stable regions of viral proteins that exhibit minimal variation across diverse influenza strains [10–13]. Vaccines designed to elicit immune responses against these conserved elements hold the potential to circumvent strain-specific limitations and provide broad-spectrum protection against influenza [14–19]. However, the effectiveness of such vaccines hinges not only on the strategic selection of epitopes but also on the choice of vaccine platforms and the integration of adjuvants to enhance immunogenicity and durability of the immune response [20–24].

Recent advancements in vaccine platforms, such as recombinant protein-based vaccines, virus-like particles, and nanoparticle-based formulations, offer promising new avenues for enhancing vaccine efficacy [25–30]. These platforms provide improved antigen presentation, stability, and scalability, positioning them as strong candidates for next-generation influenza vaccines. Beyond the choice of vaccine platform, adjuvants play a crucial role in modulating immune responses, enhancing antigen processing, and promoting long-lasting immunity [31–36]. The synergistic combination of conserved epitope targeting, advanced vaccine platforms, and optimized adjuvants represents a powerful strategy for improving the effectiveness of influenza vaccines [37–39].

This review examines the molecular characteristics of IAV and the role of conserved epitopes in addressing antigenic variability. We then discuss various vaccine platforms, evaluating their potential through representative examples, and explore the role of adjuvants in optimizing immune responses. Finally, we conducted an in-depth discussion on the development of universal influenza vaccines and the practical challenges in their clinical translation.

2. Molecular Characterization and Conserved Epitopes of Influenza A Virus

The high variability of influenza viruses poses a significant challenge to the effectiveness of conventional vaccines, prompting the search for universal influenza vaccines targeting conserved antigenic epitopes. Here, we revisit the structural features of influenza viruses and highlight recent advances in recognizing conserved antigenic epitopes, including hemagglutinin (HA), neuraminidase (NA), and the matrix protein 2 extracellular structural domain (M2e). In addition, we discuss the immunological roles of these epitopes and their potential applications in the design of next-generation influenza vaccines.

2.1. Structure of the Influenza Virus

IAV belongs to the Orthomyxoviridae family and is an enveloped, negative-sense, single-stranded RNA virus. IAV exhibits pleomorphism, with the most common form being roughly spherical, approximately 120 nm in diameter (Figure 1) [40–43]. Its envelope is derived from the host cell membrane and is embedded with three transmembrane proteins: the HA, NA, and M2 proteins [44]. Among them, HA is the most abundant and mediates viral entry into host cells, NA facilitates viral release, and M2 functions as an ion channel involved in viral uncoating [45–48]. Beneath the envelope, the matrix protein M1 surrounds the viral ribonucleoprotein complex (vRNP), which consists of genomic RNA, nucleoprotein (NP), and an RNA polymerase complex composed of polymerase acidic protein (PA) and polymerase basic proteins (PB1 and PB2) [48]. The IAV genome comprises eight segments encoding 12 proteins, including the aforementioned structural proteins and four non-structural proteins (NS1, NEP/NS2, PB1-F2, and N40), which play crucial roles in viral replication, host immune regulation, and adaptive evolution [49–51].

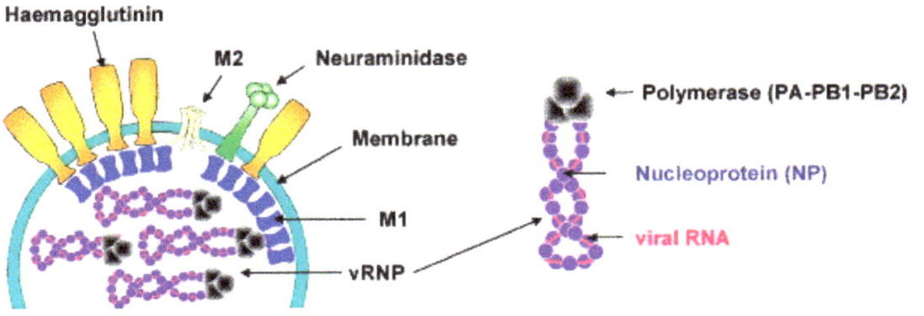

Figure 1. Influenza virus particles contain HA, NA, M1, and M2. Schematic diagram of viral RNA, nucleoprotein, and RNA polymerase complex. Reproduced with permission from ref. [40], copyright (2006) Elsevier B.V. All rights reserved.

2.2. Conserved Epitopes

HA, a transmembrane glycoprotein on the influenza viral envelope, comprises two distinct structural domains: the globular head and the helical stalk regions [52–54]. The HA head domain demonstrates potent immunogenicity capable of eliciting antibody responses [55–57]. However, frequent genetic mutations and reassortment in influenza viruses induce antigenic drift in the HA head, thereby compromising the broad protective efficacy of conventional vaccines [58]. In contrast, the HA stalk region exhibits relative structural conservation with cross-subtype similarity, establishing it as a promising target for universal influenza vaccine development [53,54,59–61]. The principal challenge lies in overcoming its weak immunodominance in natural immune responses. Recent advancements involve engineered headless HA immunogens and recombinant stalk-based vaccines that significantly enhance stalk-directed immunity (Figure 2) [8,59,62–64]. Notably, Yassine et al. implemented six rounds of iterative optimization (Gen1–Gen6) using HA from H1N1 A/New Caledonia/20/1999 as a template [65]. Through the sequential removal of immunodominant head domains and incorporation of the HIV-1 gp41 trimerization domain with ferritin self-assembling nanoparticle platform (H1-SS-np), they achieved enhanced conformational stability of stalk epitopes. This vaccine induced cross-reactive antibodies conferring complete protection in murine models and partial resistance in ferrets against H5N1 challenge, validating the protective mechanism mediated by Fc effector functions of non-neutralizing antibodies. These findings underscore the translational potential of stalk-focused strategies despite the continued dominance of head-targeting approaches in current vaccine design.

NA, a crucial influenza surface glycoprotein, facilitates viral release through the cleavage of sialic acid residues anchoring nascent virions to host cells [45,66]. While NA exhibits inter-subtype variability, the catalytic site residues remain highly conserved across N1 to N9 subtypes, making it a viable target for universal vaccine development (Figure 3) [62,67]. Recent studies demonstrate that NA-specific monoclonal antibodies (mAbs) confer substantial protective efficacy in animal models through dual mechanisms, namely, the direct inhibition of NA enzymatic activity and Fc-mediated effector mechanisms [68,69]. These findings reinforce NA's potential as a complementary component in broad-spectrum influenza vaccine formulations.

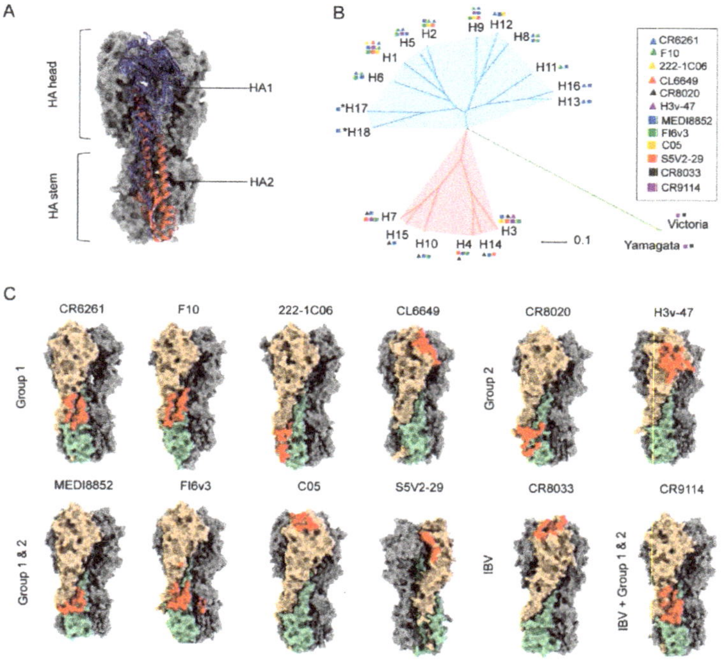

Figure 2. Structural basis for the induction of broadly neutralizing antibodies against HA: (**A**) Structure of Influenza hemagglutinin (HA).). Ha1 subunit is blue, H2 subunit is red (**B**) Phylogenetic tree of influenza A and influenza B HA subtypes. (**C**) The epitopes of anti-HA head and stem antibodies, including IAV group 1 neutralizing mAbs CR6261 (PDB 3GBN), F10 (PDB 3FKU), 222-1C06 (PDB 7T3D), CL6649 (PDB 5W6G), group 2 neutralizing mAbs CR8020 (PDB 3SDY), H3v-47 (PDB 5W42), IAV cross-group neutralizing mAbs FI6v3 (PDB 3ZTJ), MEDI8852 (PDB 5JW4), C05 (PDB 4FQR), S5V2-29 (PDB 6E4X), IBV neutralizing mAbs CR8033 (PDB 4FQM), and both IAV and IBV reactive mAb CR9114 (PDB 4FQI) on their cognate HA ligands. HA1 and HA2 monomer molecules are shown in light brown and dark sea green, respectively, and the other two monomer molecules of the HA trimer are shown in dark gray. Antibody epitopes are shown in red The epitope of S5V2-29 is hidden between the two HA head interfaces of the trimeric HA. Reproduced with permission from ref. [62], copyright (2023) Elsevier B.V.

M2e is the exposed portion of the M2 protein in influenza A virus, consisting of 24 amino acid residues. M2, an ion channel protein, plays a critical role in the viral uncoating process [70]. M2e is relatively conserved across various influenza A strains, and it can induce broad inhibitory effects against multiple subtypes, making it a promising target for universal influenza vaccines. A highly conserved epitope, SLLTEVET, is found within the 2–9 amino acid region of M2e [71]. Monoclonal antibodies targeting this epitope have been shown to effectively inhibit H1 and H3 subtype influenza A viruses in vitro. Two key amino acids in this epitope—threonine at position 5 and glutamic acid at position 6—have been identified as associated with antibody escape variants [72]. Despite some variability, the specific region of M2e continues to be considered a potential conserved epitope for universal influenza vaccines, offering promising prospects for broad-spectrum protection [73].

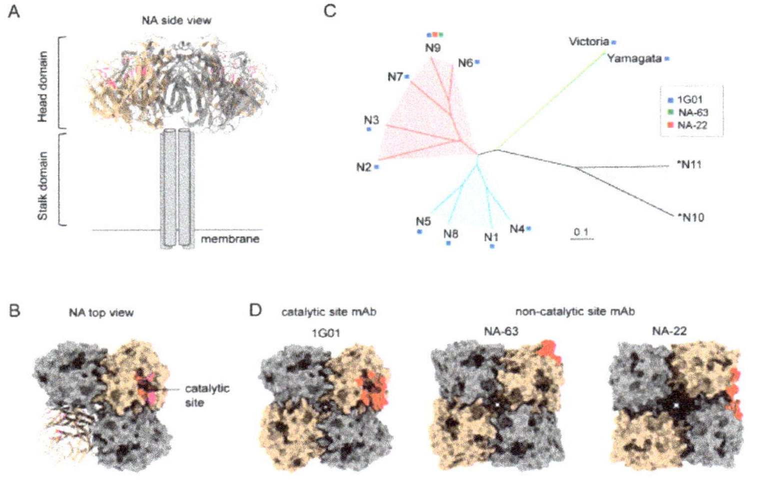

Figure 3. Structural basis for the induction of broadly neutralizing antibodies against NA: (**A**) The side view structure of N1 NA (PDB 6Q23). (**B**) The top view structure of N1 NA (PDB 6Q23). (**C**) Phylogenetic tree of influenza A and influenza B NA subtypes. Based on sequence variations of NA, IAV NA can be divided into three genetically distinct subgroups: group 1 consists of N1, N4, N5, and N8; group 2 consists of N2, N3, N6, N7, and N9; and group 3 consists of two bats-derived *N10 and *N11 subtypes. (**D**) Anti-NA antibody epitopes include catalytic (1G01, 6Q23) and non-catalytic binders (NA-63, 6U02; NA-22, 6PZW). The protomers of tetramer NA are shown as surface in tan and dark gray, respectively. The epitopes of antibodies are shown as red.

2.3. Computational Design and Epitope Screening Strategies

Traditional vaccine development often spans several years, making it difficult to respond swiftly to emerging infectious diseases or rapidly mutating viruses. In recent years, advances in structural biology, artificial intelligence, and big data have accelerated the rise of computational vaccinology, which offers a powerful means to shorten vaccine development timelines [74,75]. For example, during the COVID-19 pandemic, candidate vaccines entered clinical trials within months of the virus's identification—an achievement enabled in part by computational strategies.

In the design of broadly protective influenza vaccines, computational methods are extensively used to identify highly conserved T or B cell epitopes that possess strong immunogenic potential across diverse viral subtypes. A representative platform is the Integrated Vaccine Design Toolkit (iVAX), developed by a team based in Rhode Island, USA [76,77]. iVAX provides a comprehensive suite of immunoinformatic tools, including algorithms for scoring and ranking candidate antigens, selecting conserved and immunogenic T cell epitopes, redesigning or removing regulatory T cell epitopes, and engineering antigens to enhance immunogenicity and confer protection in both humans and animals [76,78]. The platform is accessible online and has been applied to a range of human and zoonotic pathogens.

In collaboration with Saint Louis University, iVAX has been used to identify conserved $CD4^+$ and $CD8^+$ T cell epitopes across various IAV strains [79]. These epitopes were shown to be immunogenic in individuals with diverse MHC genotypes and were further validated in HLA-transgenic mouse models. The resulting vaccines elicited robust T cell responses and provided cross-strain protection, supporting the feasibility of T cell–targeted universal influenza vaccines.

Furthermore, antigens identified through computational screening can be structurally optimized using emerging protein structure technologies. Methods such as X-ray crystallography, cryo-electron microscopy (cryo-EM), cryo-electron tomography (cryo-ET), and antibody-binding assays allow the precise characterization of antigen conformation [80,81]. These approaches help eliminate misfolded or non-functional antigen designs, reducing the risk of vaccine failure. Structure-based vaccine design is poised to play a central role in accelerating the development of next-generation influenza vaccines.

3. Research Progress of Vaccine Platform

Influenza vaccines have long been a critical tool in the fight against seasonal flu outbreaks, and they continue to be a major focus of public health research. However, as the influenza virus rapidly evolves and new strains emerge, traditional vaccines face challenges in terms of limited adaptability and immune protection. In response to these challenges, recent scientific and technological advancements have led to the development of novel influenza vaccine platforms, including inactivated vaccines, subunit vaccines, viral vector vaccines, and mRNA vaccines. These innovative platforms offer greater options and possibilities for the development and production of influenza vaccines. They not only enhance vaccine efficacy but also provide new solutions to address viral antigen variation and improve global vaccine accessibility. This review will explore the research advancements of various influenza vaccine platforms, analyzing their contributions to improving vaccine effectiveness, combating antigenic drift, addressing the challenges in vaccine development, including the hurdles encountered in translating animal model successes to human applications.

3.1. *Characteristics of Inactivated Influenza Vaccines*

Inactivated influenza vaccines (IIVs) are among the most widely used flu vaccines worldwide (Table 1). They are produced by chemically or physically inactivating the virus while preserving its main antigenic properties to induce immune protection [82,83]. The production process includes virus cultivation (e.g., in embryonated chicken eggs or MDCK cells), purification, and inactivation. Common inactivation agents include formaldehyde, which cross-links viral genomes to render them non-infectious, and beta-propiolactone (BPL), which chemically modifies nucleic acids and membrane proteins to inhibit viral activity [84].

To provide broader coverage against co-circulating strains, current influenza vaccines typically include two influenza A strains (H1N1 and H3N2) and one or two influenza B strains, formulated as trivalent (TIV) or quadrivalent (QIV) vaccines based on annual global surveillance. Their primary mechanism of protection relies on neutralizing IgG antibodies targeting the hemagglutinin (HA) head, preventing viral attachment to host cells. A hemagglutination inhibition (HI) antibody titer of $\geq$40 is generally considered protective [85].

The limitations of inactivated influenza vaccines primarily stem from their reliance on HA head-specific immunity and their mode of administration. Due to the high mutation rate of the HA head, frequent strain updates are required, and vaccine effectiveness can be significantly reduced when mismatches occur, as seen in the 2014–2015 H3N2 season, where protection dropped to 7% [86]. Additionally, these vaccines provide limited cross-protection since they induce primarily HA head-specific antibodies while eliciting weak immune responses against conserved antigens such as the HA stalk, NA, and NP. Inactivated influenza vaccines primarily elicit humoral immune responses dominated by strain-specific antibodies; however, they induce limited cellular immunity, including the weak activation of $CD8^+$ T cells and poor engagement of innate immune pathways, which together restrict their long-term protective efficacy [87]. Overall, despite being a mainstay of influenza prevention, inactivated vaccines are limited by their narrow strain-specific protection and weak cellular immune activation, highlighting the need for next-generation vaccines with broader and longer-lasting efficacy.

Table 1. Comparative analysis of influenza vaccine platforms.

Vaccine Platform	Antigens	Immune Response Strength	Broad-Spectrum Protection	Production Difficulty	Scalability	Supply Chain	Other Considerations
Inactivated Influenza Vaccines	HA	Moderate; primarily humoral; limited cellular immunity	Low (vaccine-matched strains only)	Medium—Requires virus propagation, inactivation, and purification in eggs	Mature process; scalable but slow to adapt	2–8 °C refrigeration	Needs yearly updates; struggles to scale rapidly during epidemics.
Subunit Vaccine	HA, NA, NP, Conserved epitopes	Weak; adjuvant-dependent	Moderate (multivalent design improves broad spectrum)	Low-Medium—Recombinant protein production and purification	High scalability using expression systems	2–8 °C refrigeration	-
Peptide Vaccine (Chemically Synthesized)	Conserved epitopes	Moderate; strong $CD8^+$ T cell responses (rational sequence design enhances MHC binding)	High (targets conserved epitopes)	Low—Chemically synthesized or recombinant production	Very high; synthesis-based and rapid	Room-temperature preservation	-
Viral Vector Vaccines	HA, NA, NP, Conserved epitopes	Strong; humoral & robust $CD8^+$ T-cell responses (via MHC-I) and TRM induction	High (adaptable antigen cassette)	High—Requires vector design, virus construction, and safety validation	Moderate; depends on vector type and cell-based production	2–8 °C refrigeration	Pre-existing immunity to Ad5 may reduce efficacy; HVT vectors may persist latently and reactivate
mRNA vaccines	HA, NA, NP, Conserved epitopes	Strong; elicits specific antibody and $CD4^+$ T-cell responses	High (rapid antigen updates)	Very high—Requires RNA synthesis, purification, LNP formulation, stringent	High scalability via synthetic, cell-free production	−70 °C ultra-cold	LNP delivery may cause reactogenicity

3.2. Characteristics of Subunit Influenza Vaccines

Subunit vaccines are produced by extracting or synthesizing key antigenic components of pathogens [88]. These antigens can be derived from purified natural viruses, recombinant expression systems, or chemical synthesis (Table 1). Lacking viral nucleic acids, subunit vaccines are inherently safer, with well-defined compositions and controlled production processes, making them suitable for standardized and large-scale manufacturing [89,90]. Due to their inherently low immunogenicity, subunit vaccines usually require adjuvants to enhance immune responses. Adjuvants function by stimulating and amplifying the immune system, thereby improving vaccine immunogenicity and promoting stronger antibody and cellular responses [91,92]. In addition, the incorporation of multivalent antigen design can broaden the protective scope of the vaccine. For example, the Cui research team developed a multivalent antigen nanoparticle, termed HMNF, by displaying conserved epitopes from influenza surface proteins—HA A α-Helix (H), M2e (M), and NA HCA-2 (N)—on self-assembling human ferritin (F) nanoparticles [93]. This vaccine effectively induced both antibody and cellular responses and provided broad protection in a mouse model.

As a specialized form of subunit vaccines, peptide vaccines use short synthetic peptide epitopes (typically 8–30 amino acids) from pathogen proteins to mimic natural antigen structures and elicit both humoral and cellular immunity [94,95]. Their design usually relies on structural biology and bioinformatics to precisely identify conserved epitopes such as those in the HA stalk region and nucleoprotein (NP). Compared with protein-based vaccines, peptide vaccines can be produced via recombinant systems or entirely through chemical synthesis, avoiding potential expression system contaminants and offering greater batch-to-batch consistency and safety. FLU-v, developed by PepTcell (now SEEK), is a synthetic peptide vaccine composed of four peptides derived from conserved internal antigens of influenza viruses [96,97], namely, M1 (32 amino acids), M2 (24 amino acids), the NP of influenza A (20 amino acids), and the NP of influenza B (19 amino acids). These peptides are engineered to be recognized by human leukocyte antigen (HLA) molecules, immune proteins that present antigens and activate T cells, thus inducing virus-specific T cell responses [97–99]. In both preclinical and clinical studies, FLU-v has been shown to induce interferon-γ (IFN-γ) and granzyme B-secreting cells and to confer protection in challenge models. In a randomized, double-blind, placebo-controlled phase IIb clinical trial, FLU-v demonstrated favorable safety and immunogenicity and provided protection against strains such as H3N2.

In summary, subunit vaccines, particularly peptide-based ones, are emerging as a promising avenue toward broad-spectrum influenza protection by leveraging the precise design of conserved antigenic sites in combination with adjuvant enhancement strategies.

3.3. Characteristics of Viral Vector Influenza Vaccines

Viral vector vaccines are a type of vaccine that uses viruses as "delivery vehicles" to introduce specific pathogen genes—typically those encoding antigenic proteins—into host cells to elicit an immune response (Table 1). Viral vector vaccines employ adenoviruses (Ad5s), Newcastle disease virus (NDV), herpesviruses (HVTs), or other vectors to deliver influenza antigen genes (HA, NA) via genetic recombination. For example, adenoviral vectors delete replication-essential genes (e.g., E1/E3 regions) to insert HA genes, forming replication-defective recombinant viruses [100].

After entering host cells via natural infection routes (e.g., respiratory tract or intramuscular injection), recombinant vectors utilize host transcription/translation machinery to express influenza antigens. Adenoviral double-stranded DNA enters the nucleus, where host RNA polymerase II transcribes HA mRNA for cytoplasmic ribosome translation, ultimately presenting HA proteins on cell surfaces or secreting them extracellularly [101,102]. Vectors like Pichinde virus (PICV) naturally target dendritic cells (DCs) and macrophages, enhancing antigen presentation efficiency [103]. Experimental studies have shown that PICV vectors induce robust $CD8^+$ T-cell responses in murine lungs and spleens via MHC-I antigen presentation [104]. HVT vectors sustain low-level antigen expression, eliciting long-term tissue-resident memory T cells (TRM). For instance, HVT-H9 vaccines maintain high IFN-γ^+/$CD8^+$ T-cell levels in chickens 35 days post-vaccination [105]. Limitations involve pre-existing Ad5 antibodies in populations that may neutralize vectors (requiring high-dose compensation) [106,107] and risks of HVT reactivation due to latent persistence [108].

3.4. Characteristics of mRNA Influenza Vaccines

mRNA vaccines encapsulate influenza antigen-encoding mRNA (HA, NA) within lipid nanoparticles (LNPs) for delivery to the host cytoplasm (Table 1). Since the mRNA sequence is modular, it can be updated with new sequences encoding antigens from evolving viral strains. This enables the rapid development of variant-specific vaccines. In addition, mRNA's simplicity as a small molecule (~2–3 kb) and highly specific immune responses due to s encode only one or a few target proteins [109].

Current mRNA vaccine platforms include both conventional non-replicating mRNA and self-amplifying mRNA (saRNA). Conventional non-replicating mRNA contains antigen-coding sequences, 5′ cap, and 3′ poly-A tail, relying on host ribosomes for direct antigen translation [110,111]. Self-amplifying mRNA (saRNA) uses alphavirus or flavivirus-derived genomes (Figure 4), replacing nonstructural (NS) genes with antigen-coding sequences. After cytoplasmic replication, subgenomic RNA fragments express antigens without producing infectious particles, enabling full protection with 0.1–1.0 μg doses (vs. 10 μg for conventional vaccines) and reduced reactogenicity [112,113].

Key challenges include balancing interferon (IFN) signaling: Moderate IFN enhances DC maturation and antigen presentation; however, excessive activation inhibits translation and causes inflammation. Double-stranded RNA (dsRNA) contaminants in in vitro-transcribed mRNA trigger TLR3/RIG-I sensing, inducing rapid IFN-α/β secretion and the degradation of cellular/exogenous RNA [114,115]. Even dsRNA-free single-stranded mRNA activates TLR7/8 as pathogen-associated molecular patterns (PAMPs), suppressing antigen translation via IFN [116]. Current strategies employ naturally modified nucleotides to alter RNA secondary structures or block TLR7/8 binding [117]. To address these challenges, current strategies include the use of chemically modified nucleotides (e.g., pseudouridine) and purification methods that reduce dsRNA contaminants, as well as sequence design that minimizes immune stimulation [108].

However, mRNA vaccines typically require ultra-cold storage conditions (e.g., −70 °C), due to the temperature sensitivity of both the mRNA and the LNP carriers. This presents logistical challenges for global distribution and long-term stockpiling.

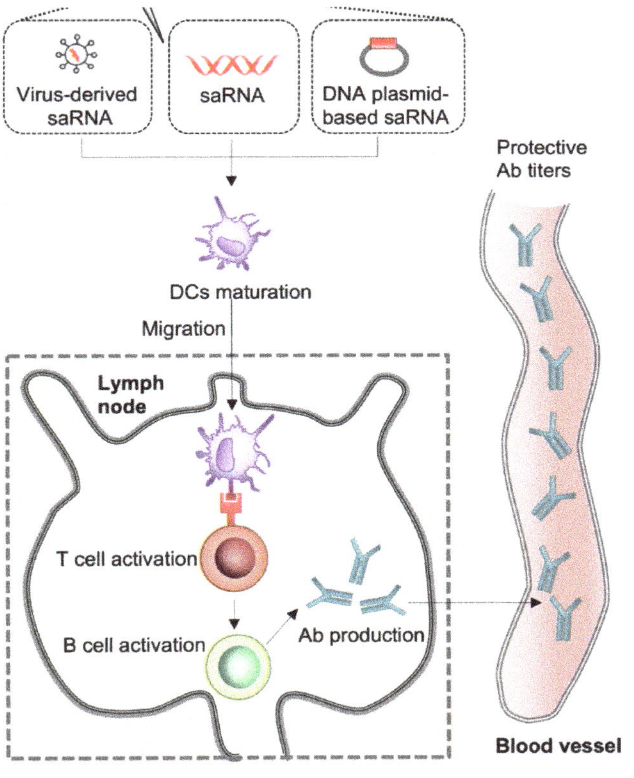

Figure 4. Self-amplified RNA (Sarna) vaccine can be delivered in the form of virus-like RNA particles, in vitro transcriptional RNA, and plasmid DNA. DCs recognize Sarna in muscle and cells differentiate. Differentiated DCs act as antigen-presenting cells and migrate to lymph nodes, resulting in the activation of T cells and B cells and the production of antibodies. Reproduced with permission from ref. [118], copyright (2021) Lee and Ryu.

3.5. Key Challenges in Influenza Vaccine Development

A wide range of influenza vaccines is being developed globally. Table 2 summarizes both licensed vaccines and those currently in clinical trials. However, compared to the numerous candidates being studied, only a small fraction ever make it to market. This gap between research and real-world applicability underscores the high attrition rate in influenza vaccine development. Failures occur for multiple reasons, often unrelated to a single technical flaw. A major issue is the insufficient immune response in humans. For example, the MVA-NP+M1 vaccine developed by Vaccitech showed strong immune cell activation in animal models but failed to reduce clinical symptoms of influenza in a Phase II trial. Another common reason is safety or reactogenicity concerns: the adjuvanted vaccine made by Acambis (ACAM-FLU-A) was halted after reports of adverse events during early human testing. Manufacturing and stability issues have also contributed to setbacks—for example, certain VLP-based vaccines faced scalability problems and antigen degradation during storage. Lastly, regulatory or commercial viability can lead to discontinuation; BiondVax's universal vaccine candidate (M-001) progressed to Phase III but was terminated after failing to show efficacy, despite extensive investment and early-phase success.

Table 2. Marketed and clinical development of influenza vaccines by platform type.

Platform Type	Vaccine Name	Developer	Development Status	Dosage Form
Inactivated Influenza Vaccines	Afluria Quadrivalent	Seqirus	Complete	Injection
Inactivated Influenza Vaccines	Fluarix Quadrivalent	GSK	Complete	Injection
Inactivated Influenza Vaccines	FluLaval Quadrivalent	GSK	Complete	Injection
Inactivated Influenza Vaccines	Fluzone Quadrivalent	Sanofi Pasteur	Complete	Injection
Inactivated Influenza Vaccines	Flucelvax Quadrivalent	Seqirus	Complete	Injection
Subunit Vaccine	Flublok Quadrivalent	Sanofi Pasteur	Complete	Injection
Subunit Vaccine	Influvac Tetra	Viatris	Complete	Injection
Peptide Vaccine	KD-295	Valneva	Complete	Injection
Peptide Vaccine	FLU-v	SEEK	Phase II/III	Injection
Viral Vector Vaccines	FluMist Quadrivalent	AstraZeneca	Complete	Intranasal
Viral Vector Vaccines	VXA-A1.1	Vaxart	Phase I	Oral
mRNA vaccines	VAL-506440	Moderna	Phase I	Injection
mRNA vaccines	VAL-339851	Moderna	Phase I	Injection
mRNA vaccines	NCT04956575	Moderna	Phase II/III	Injection

One of the most persistent challenges is the gap between animal models and human responses. While mice and ferrets are commonly used in preclinical studies, they differ significantly from humans in terms of immune receptor distribution and pre-existing immunity. Vaccines that show promise in these animal models, particularly in animals with no prior exposure to the virus, often fail to provide the same protective effect in humans with more complex and diverse immune systems. This issue is particularly pronounced for newer vaccine platforms, such as viral vectors or mRNA vaccines, which rely on precise control of antigen expression and immune modulation. To improve translation to humans, there is an urgent need for more relevant preclinical models, including humanized mice, organoids, or human lymphoid tissue cultures, as well as early-phase clinical trials that focus on immune biomarkers, not just antibody titers.

4. Advances in Vaccine Adjuvant Research

Adjuvants are substances that enhance the immune response to vaccines, typically by promoting the immunogenicity of the antigen or boosting the immune system's response to the antigen [119]. Adjuvants do not directly trigger an immune response; rather, they assist in the effectiveness of the vaccine by modulating various mechanisms of the immune system, which is particularly important in the development of vaccines against emerging or variant pathogens. Traditional adjuvants, such as aluminum salts (e.g., aluminum hydroxide) and oil-in-water emulsions (e.g., MF59), have been used in vaccines for many years and have been shown to significantly enhance immune responses. However, as vaccine development progresses, the limitations of traditional adjuvants are becoming more apparent. The effectiveness of traditional adjuvants may be constrained by the specific antigen type or the characteristics of the pathogen, creating a need to develop novel adjuvants to overcome these challenges. New adjuvants can not only improve the immunogenicity of vaccines but also optimize immune responses through multiple pathways, thereby enhancing the broad-spectrum and long-term protective effects of vaccines.

4.1. Conventional Adjuvants

Aluminum-based adjuvants (e.g., aluminum hydroxide, aluminum oxyhydroxide) function as immunostimulants through their unique antigen adsorption capacity [120]. Aluminum salts form localized depots by binding antigens, enabling sustained release at injection sites for days to weeks (Figure 5). This mechanism was first validated in diphtheria toxin experiments: the removal of aluminum-containing injection sites abolished immunity in animals, while transplanting these sites transferred specific immunity to recipients [121]. This physical retention prolongs antigen exposure, creating a window for immune recognition and processing. Aluminum adjuvants also activate the NLRP3 inflammasome to trigger innate immunity [122], recruiting neutrophils, inflammatory monocytes, and other immune cells to injection sites [123]. They induce cellular damage to release endogenous danger signals (e.g., DNA) [124], stimulating pro-inflammatory cytokines (IL-5, IL-6) and fostering a Th2-biased immune microenvironment [125]. This depot-inflammation synergy ultimately generates high-affinity IgG antibody responses [126]. However, their Th2 bias limits utility in vaccines requiring robust cellular immunity, and formulation differences (e.g., Alhydrogel® vs. Imject® alum) yield variable immunogenicity [127,128].

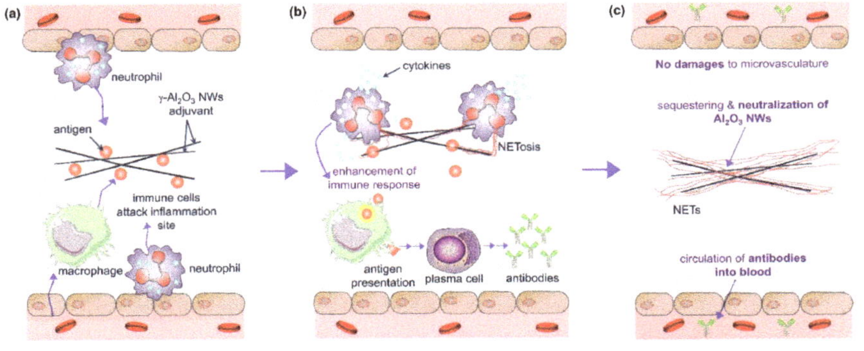

Figure 5. The enhancement mechanism of γ-Al_2O_3 nanowires (NWS) aluminum adjuvant+antigen: (**a**) γ-Al_2O_3 nanoparticles adsorb antigen and release antigen after injection, resulting in the activation of local macrophages and neutrophils. Macrophages process antigen and stimulate antibody production through plasma cells; (**b**) In the early stage of neutrophils contacting with γ-Al_2O_3 NWs, cytokines were released to enhance antibody production; (**c**) γ-Al_2O_3 NWs were isolated by neutrophil extracellular traps, which limited the inflammation caused by the presence of exogenous γ-Al_2O_3 NW particles; γ-Al_2O_3 NWs locally exist in the trap and will not damage the microvessels or cause local inflammation. Reproduced with permission from ref. [120], copyright (2018) Elsevier Ltd. All rights reserved.

MF59, an oil-in-water emulsion adjuvant, comprises 4.3% squalene, surfactants Tween 80 and Span 85 in citrate buffer, forming stable ~160 nm oil droplets. Licensed in over 30 countries for influenza vaccines (e.g., Fluad®), it is particularly effective in children, pregnant women, and the elderly [129,130]. Upon intramuscular injection, MF59 activates immunity via the following three mechanisms: Inducing a local chemokine storm (CCL2, IL-8) to recruit neutrophils and monocytes, which capture antigens and migrate to lymph nodes [131,132]; Promoting monocyte differentiation into dendritic cells (DCs) within lymph nodes [133,134]; Driving IgG class-switching and $CD8^{+}$ T cell responses via double-negative T cells ($CD4^{-}CD8^{-}$) and apoptotic cell-derived uric acid signals, even in $CD4^{+}$ T cell-deficient mice [135,136]. Clinical data show that MF59-adjuvanted H1N1 vaccines elevate hemagglutination inhibition (HI) antibodies two-to-five-fold, confer cross-protection against variants, and exhibit maternal-fetal safety profiles comparable to non-adjuvanted vaccines [137,138]. Compared to alum, MF59 recruits significantly more neutrophils, monocytes, eosinophils, macrophages, and DCs. Its "low antigen, high response" property enables protective thresholds with 3.75 μg antigen in elderly subjects, making it critical for pandemic preparedness [139,140].

AS03, another oil-in-water emulsion, contains 4.3% squalene, 11.86 mg/dose α-tocopherol, and surfactants Tween 80/Span 85 in phosphate buffer, forming ~160 nm droplets (Figure 6) [141]. Approved in >30 countries for seasonal/pandemic influenza vaccines (e.g., Pandemrix®, Arepanrix®) [142–144], it shares MF59's chemokine-driven recruitment mechanisms but differs via α-tocopherol's MyD88-dependent, TLR-independent signaling. This enhances antigen-specific antibodies (IgG1, IgG2a) and $CD8^{+}$ T cell responses [145]. AS03 induces apoptosis in DCs and macrophages, releasing endogenous danger signals (e.g., uric acid) to amplify immunity [146]. In H5N1/H1N1 vaccines, AS03 reduces antigen doses to 3.75 μg HA/dose (vs. 15–30 μg in conventional vaccines) while maintaining high HI titers [147]. In $CD4^{+}$ T cell-deficient models, it drives IgG isotype switching and $CD8^{+}$ T cell activation via double-negative T cells and MHC II upregulation [145]. However, the 2009 Pandemrix® formulation was linked to increased narcolepsy risk in children, potentially due to H1N1 HA-derived $CD4^{+}$ T cell-mimicking epitopes [148]. Reformulated versions mitigate this risk, and current research explores combining AS03 with TLR agonists to balance Th1/Th2 responses [149].

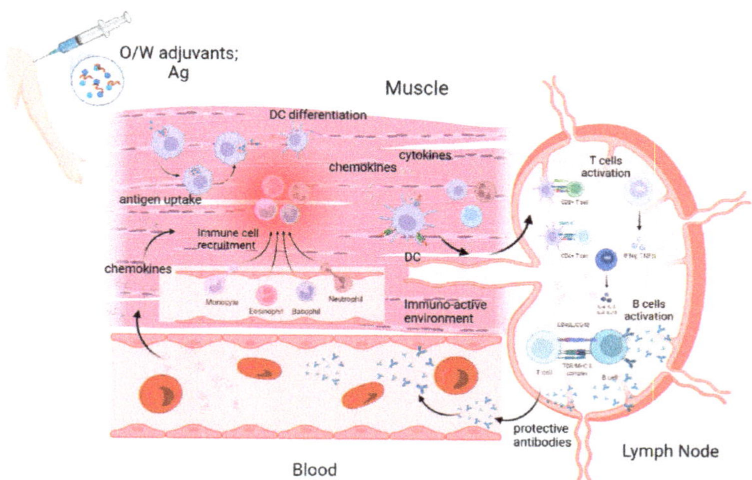

Figure 6. Possible mechanism of action of oil in water emulsion (MF59, AS03). Reproduced with permission from ref. [141], copyright (2024) Elsevier Ltd. All rights reserved.

4.2. Novel Adjuvants Based on Pattern Recognition Receptor Agonists

Pattern recognition receptors (PRRs) constitute essential proteins that enable the immune recognition of pathogen-associated molecular patterns (PAMPs) and damage-associated molecular patterns (DAMPs) [150–152]. Recent advancements highlight the emerging potential of PRR-targeting strategies in adjuvant development, with growing attention to their application in enhancing vaccine efficacy [153]. PRR activation potentiates immune responses by improving immunogenicity and critically bridging innate and adaptive immunity. Major PRR families include Toll-like receptors (TLRs), RIG-I-like receptors (RLRs), and NOD-like receptors (NLRs), as well as the cGAS–STING pathway, which senses cytosolic DNA and triggers potent type I interferon responses [154]. Figure 7 show some key PRR agonists currently being researched or used in influenza vaccines.

PRR Types	RIG-I	TLR4	TLR5	TLR3	TLR9	Sting
Adjuvant Names	IVT-RNA	Glucopyranosyl Lipid A (GLA)	Vax128	Poly(I:C)	CpG	cGAMP
Adjuvant Types/Mechanisms	Simulating Viral Genome RNA to Activate RIG-I/MDA5	Synthesis of Monophosphoryl Lipid A Analogues as TLR4 Agonists	Flagellin-based TLR5 Agonist Vaccines	Mimicking Viral dsRNA to Activate TLR3	CpG Motif-containing Oligonucleotides to Activate TLR9	Secondary Signaling Molecules Activating the STING Pathway
Representative Vaccines/Uses	CureVac's mRNA-based Influenza Vaccine CVSQIV	Synthetic Adjuvant GLA-SE for Seasonal Influenza Vaccines	Adjuvant Activity in H5N1 Influenza Vaccines	-	Use in Conjunction with HA for H5N1 Vaccine Research	Enhancing Anti-Influenza Immunity
Clinical Stage Entry	Yes	Yes	Yes	Yes	Yes	No

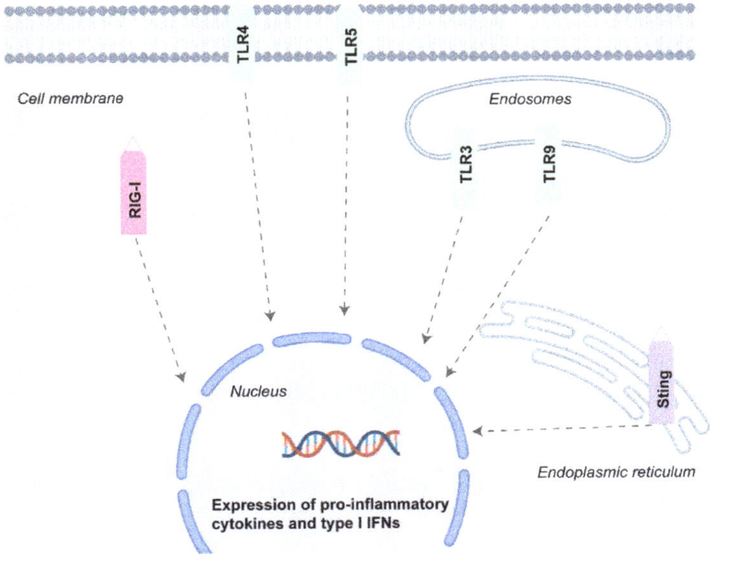

Figure 7. Representative PRR-targeting adjuvants in influenza vaccine development.

Glucopyranosyl Lipid A–Stable Emulsion (GLA-SE) is a synthetic TLR4 agonist that has been tested in seasonal influenza vaccines [155]. By activating TLR4, GLA-SE induces a potent immune response that promotes both innate and adaptive immunity. This activation results in a strong Th1-biased immune response, which is particularly beneficial for enhancing the immune system's ability to recognize and fight the influenza virus [156].

GLA-SE has shown promise in enhancing immune protection and is being investigated for its potential to improve the overall efficacy of influenza vaccines.

Vax128 is a TLR5 agonist based on bacterial flagellin, which is the major component of bacterial flagella [157]. Flagellin activates TLR5, triggering a strong innate immune response. When fused with influenza hemagglutinin (HA), Vax128 enhances the immunogenicity of the vaccine. This combination boosts the body's ability to recognize and respond to the H5N1 influenza virus. Vax128 has been used as an adjuvant in H5N1 vaccines, demonstrating its ability to increase the effectiveness of influenza vaccines.

CpG 1018 is a TLR9 agonist that has been licensed for use in the hepatitis B vaccine Heplisav-B [158,159]. This synthetic oligonucleotide contains CpG motifs that activate TLR9, stimulating B cells and promoting the production of antibodies. In the context of influenza vaccines, CpG 1018 has been explored in combination with HA antigens. It has shown promise in rapidly inducing protective antibodies, enhancing the immune response, and providing quicker protection against influenza viruses. Its potential use in influenza vaccines offers a powerful tool for improving vaccine efficacy.

Poly(I:C) is a synthetic TLR3 ligand that mimics viral double-stranded RNA (dsRNA), an intermediate produced by viruses during replication [160,161]. By activating TLR3, Poly(I:C) triggers antiviral responses, preparing the immune system to recognize and fight viral infections. Poly(I:C) has been used as an adjuvant in influenza vaccines, enhancing immune activation and improving the body's defense against influenza viruses. Its application highlights the potential of using TLR3 agonists to boost immune responses in vaccine development.

IVT-RNA is a RIG-I agonist used in mRNA-based influenza vaccines. RIG-I detects RNA viruses, including influenza, by recognizing viral RNA patterns [162]. IVT-RNA mimics the viral genome, activating RIG-I to stimulate an immune response. By including IVT-RNA in mRNA vaccines, researchers aim to enhance the immune system's recognition of the virus and improve the effectiveness of the vaccine [163]. This strategy is being explored in the development of next-generation influenza vaccines that leverage the power of RNA-based platforms.

The STING pathway is increasingly recognized as a promising target to enhance vaccine efficacy [164]. Although influenza is an RNA virus and lacks DNA, it can indirectly activate the cGAS-STING pathway through cellular stress during infection, such as mitochondrial damage or nuclear DNA leakage [165]. This activation triggers type I interferon responses, which play a crucial role in boosting both innate and adaptive immune responses [166]. The activation of STING not only enhances the breadth and specificity of immune responses but also facilitates antigen presentation and immune cell activation. Upon recognizing cytosolic double-stranded DNA, cGAS (cyclic GMP-AMP synthase) produces cGAMP, a second messenger that activates STING [167]. Recently, cGAMP-loaded pulmonary surfactant-mimetic nanoparticles have been developed as inhalable STING agonists. These nanoparticles effectively deliver cGAMP to the pulmonary immune system, activating the STING pathway to enhance both mucosal and systemic immunity against influenza [168]. Studies in animal models have demonstrated that this approach can significantly improve immune responses against influenza, highlighting the potential of STING agonists as nanocarrier-based adjuvants. Although STING agonists have been clinically tested in cancer vaccines, their application in influenza vaccines is still in the early stages of research.

Manganese, an essential micronutrient, plays critical roles in physiological and immune processes. In anti-influenza vaccines, manganese-based nano-adjuvants enhance cGAS sensitivity to DNA [169]. Mn^{2+} significantly strengthens cGAS-DNA binding, enabling efficient detection of even low concentrations of mtDNA or nuclear DNA fragments [170,171]. Studies also indicate that Mn^{2+} acts as a co-activator of STING, promoting its conformational changes and TBK1-IRF3 signaling independent of DNA [172]. This makes manganese-based nanoparticles a promising tool in immune modulation and adjuvant development for influenza vaccines. In addition to enhancing cGAS-STING pathway activation, manganese-based nanoparticles also enhance antigen presentation and immune cell activation. Mn_3O_4 nanoparticles, for instance, promote dendritic cell (DC) antigen uptake, upregulate MHC-II and co-stimulatory molecules (CD80/CD86), and amplify T cell activation [173]. Manganese adjuvants also induce B cell surface markers (CD69, CD86) and the transcription factor Blimp-1, driving antibody secretion [174]. Beyond virology, manganese reverses tumor multidrug resistance (MDR) in cancer microenvironments by inducing reactive oxygen species (ROS) and inhibiting ABC transporters, enhancing chemotherapy efficacy [175]. Similar mechanisms may regulate antiviral immune evasion. Manganese nano-adjuvants (e.g., MnJ) stimulate mucosal immunity in the respiratory or intestinal tract, inducing secretory IgA (sIgA) to form localized barriers against influenza and other mucosal pathogens [176].

5. Towards Universal Influenza Vaccines: Challenges, Innovations, and the Path Forward

5.1. The Urgency and Challenges in the Development of Broad-Spectrum Influenza Vaccines

The high mutability of influenza viruses is the primary factor limiting the long-term efficacy and broad protection of vaccines. Each year, antigenic drift and shift drive the continuous evolution of circulating strains, necessitating frequent updates of seasonal vaccines. Current influenza vaccines mainly target the hemagglutinin (HA) and neuraminidase (NA) antigens and rely heavily on predictions of prevalent strains. However, these vaccines typically confer limited strain-specific protection and often fail to cover rapidly emerging viral lineages. In cases of mismatched predictions or substantial viral mutations, their protective efficacy can be significantly reduced. Moreover, influenza viruses employ various immune evasion mechanisms, such as glycosylation modification, interference with antigen presentation, and the induction of immunosuppression, further undermining the breadth of protection. For example, the glycosylation of HA epitopes can hinder antibody recognition and reduce the efficiency of humoral immune responses. Therefore, in addition to inducing neutralizing antibodies, effective vaccines must also elicit robust memory T cell responses to clear infected cells and compensate for immune escape, thereby achieving broader and more durable protection.

5.2. Key Strategies for Universal Vaccine Design

To counter the challenges of antigenic variation and immune evasion, vaccine strategies targeting conserved epitopes have emerged as a promising approach for achieving broad-spectrum protection. Conserved regions such as the HA stem, M2e, and NP have demonstrated strong cross-reactivity potential; however, their practical application remains challenging. These conserved epitopes are often structurally shielded and inherently poorly immunogenic, making it difficult to induce potent immune responses. Furthermore, their conformational stability and presentation efficiency across different delivery platforms directly influence vaccine performance. Current strategies, including structure-based optimization, multivalent design, and epitope reconfiguration, are widely used to enhance epitope exposure and immune recognition. Nevertheless, balancing structural stability with immunological accessibility remains a key technical bottleneck. Several vaccine candidates based on conserved epitopes have entered clinical trials and show promising potential; yet, their applicability across diverse populations, consistency in immune responses, and durability of protection remain to be validated. Adjuvants play a crucial role in enhancing the immunogenicity of vaccines, particularly in the context of broad-spectrum influenza vaccine development. Pattern recognition receptor (PRR) agonists, including those targeting TLR7/8, STING, and RIG-I pathways, are widely used to potentiate innate immunity, promote Th1-biased responses, and facilitate the formation of memory T cells. These adjuvants have shown significant potential in augmenting the immunogenicity of conserved epitopes. Some are also capable of inducing mucosal immunity, providing an added layer of defense against respiratory infections. However, the clinical translation of these potent immunostimulatory adjuvants still faces considerable hurdles. These include strong dose dependency, the risk of excessive inflammatory responses, limited suitability across different population groups, and variable delivery efficiency depending on the vaccine platform. Achieving controllable immune activation, particularly ensuring safety in high-risk populations, remains a critical challenge that must be addressed to facilitate its broader clinical application.

5.3. Translational Bottlenecks and Practical Challenges

Although various universal influenza vaccine strategies have shown promising results in animal models, transitioning from laboratory research to clinical application remains fraught with challenges. Firstly, there is a lack of comprehensive systems for evaluating cross-protective efficacy. The diversity and rapid evolution of influenza viruses render traditional immune protection metrics insufficient to accurately assess a vaccine's broad-spectrum performance across multiple subtypes. Currently, no unified and systematic criteria exist to evaluate the breadth of protection provided by different vaccine candidates, which limits clinical advancement. Secondly, significant differences exist between animal models and the human immune system. While mouse and ferret models are widely used for preliminary validation, their immune system structure and response patterns do not fully mimic those of humans. Some vaccines elicit robust T cell or antibody responses in animal studies but perform poorly in human trials. Thus, improving the predictive power of preclinical models—such as through humanized models and AI-based tools—has become a critical step toward enhancing clinical translation. Thirdly, population adaptability remains unresolved. Individuals of different ages, health conditions, and immune statuses respond differently to vaccination. In particular, elderly individuals, infants, and immunocompromised patients often exhibit weaker immune responses. A universal influenza vaccine must be designed with these varied immunological characteristics in mind and should be tested across diverse demographic groups during clinical trials.

Moreover, the industrialization of broad-spectrum influenza vaccines presents substantial technical hurdles. Each vaccine type faces distinct challenges in large-scale production, quality control, and supply chain logistics. For subunit protein vaccines, conserved epitopes often exhibit inherently low immunogenicity and expression efficiency. Their complex three-dimensional structures are difficult to replicate accurately in expression systems, complicating antigen purification and leading to variability between production batches, making consistency and stability in scaled production difficult to control. For viral vector vaccines, despite their strong immunogenic potential, issues such as the genetic stability of the vector, maintaining high viral titers, and compatibility with different host systems increase manufacturing demands. As for mRNA vaccines, the encoding of conserved antigens requires precise nucleic acid sequence optimization to enhance translation efficiency and accurate protein folding. These vaccines also rely heavily on lipid nanoparticle (LNP) delivery systems, which pose challenges including batch-to-batch variability, stringent cold-chain requirements, and high manufacturing costs—major barriers to commercialization.

5.4. Future Directions and Technological Frontiers

Vaccine development is typically a long and complex process, encompassing antigen selection, platform design, animal testing, and clinical trials. In the case of influenza vaccines, even those showing promising results in animals often fall short in human trials. One key reason is the discrepancy between animal and human immune responses, leading to divergent outcomes across species. Additionally, varying levels of vaccine responsiveness among different age groups and immune statuses further complicate development. To accelerate vaccine development and improve success rates, computational epitope prediction and structure-based design are emerging as powerful tools. Computational modeling enables researchers to identify conserved surface epitopes early and optimize their immunogenicity. For instance, simulations can quickly pinpoint and assess multiple potential conserved epitopes and guide antigen optimization based on the virus's 3D structure. This approach not only improves immunogenicity but also helps streamline vaccine platform selection, shortening design timelines. Moreover, advances in structural biology and artificial intelligence are driving more precise and efficient vaccine design. Structure-guided antigen design helps overcome issues like low immunogenicity and epitope masking seen in traditional approaches. In virus-like particle (VLP) or nanoparticle delivery systems, structural optimization ensures the effective presentation of multiple conserved epitopes, boosting cross-protective capacity. Beyond computational strategies, humanized mouse models and organoid platforms offer new opportunities in vaccine development. These systems more accurately mimic human immune responses and enable more precise evaluations during early development stages. Humanized mice incorporate human immune cells, making their responses more reflective of human physiology. Similarly, organoids can replicate microenvironments of human tissues, such as the respiratory tract, providing valuable insights into localized immune responses. These tools may significantly reduce the time from laboratory research to clinical application.

The route of vaccine administration also significantly influences immune outcomes. Traditional influenza vaccines are mostly delivered via intramuscular injection, which primarily induces systemic IgG responses in peripheral blood. However, this method weakly stimulates local respiratory immunity, resulting in low antibody concentrations in the airway and insufficient frontline defense. In contrast, intranasal vaccines can directly target the respiratory mucosa, inducing strong local IgA responses and activating T cell immunity. Mucosal immunity not only forms a robust first line of defense against influenza viruses but also enables rapid immune responses upon re-exposure, thanks to memory T cells. This delivery method offers unique potential in preventing respiratory infections and could be particularly beneficial in controlling influenza.

In summary, the development of a universal influenza vaccine is both an urgent need in response to viral evolution and a major public health challenge. With continuous advances in technologies such as computational modeling, structural biology, and innovative delivery strategies, we have reason to believe that future broad-spectrum vaccines will overcome current technical hurdles and offer more effective, long-lasting protection. Cross-disciplinary collaboration and innovation will be key to achieving breakthroughs and ensuring comprehensive protection against diverse influenza strains.

Author Contributions: Conceptualization, M.-Q.Z. and J.-W.B.; Writing—original draft preparation, M.-Q.Z. and J.-W.B.; Data curation, M.-Q.Z. and J.-W.B.; Project administration, Z.-G.W. and S.-L.L.; Funding acquisition, S.-L.L.; Writing—review and editing, M.-Q.Z., Z.-G.W. and S.-L.L. All authors have read and agreed to the published version of the manuscript.

Funding: This work was supported by the National Key Research and Development Program of China (No. 2024YFA1210003), the National Natural Science Foundation of China (Nos. 22293032,

22374138, and 21977054), and the Tianjin Natural Science Foundation (Nos. 24JCZDJC01240 and 23JCYBJC01880).

Institutional Review Board Statement: Not applicable.

Informed Consent Statement: Not applicable.

Data Availability Statement: Not applicable.

Conflicts of Interest: The authors declare no conflict of interest.

References

1. Li, Y.; Wang, L.; Si, H.; Yu, Z.; Tian, S.; Xiang, R.; Deng, X.; Liang, R.; Jiang, S.; Yu, F. Influenza virus glycoprotein-reactive human monoclonal antibodies. *Microb. Infect.* **2020**, *22*, 263–271. [CrossRef] [PubMed]
2. Nair, H.; Brooks, W.A.; Katz, M.; Roca, A.; Berkley, J.A.; Madhi, S.A.; Simmerman, J.M.; Gordon, A.; Sato, M.; Howie, S.; et al. Global burden of respiratory infections due to seasonal influenza in young children: A systematic review and meta-analysis. *Lancet* **2011**, *378*, 1917–1930. [CrossRef] [PubMed]
3. Isolation of Avian Influenza A(H5N1) Viruses from Humans—Hong Kong, May–December 1997. *MMWR Morb. Mortal. Wkly. Rep.* **1997**, *46*, 1204–1207. Available online: https://www.cdc.gov/mmwr/preview/mmwrhtml/00050459.htm (accessed on 12 May 2025).
4. Gao, R.; Cao, B.; Hu, Y.; Feng, Z.; Wang, D.; Hu, W.; Chen, J.; Jie, Z.; Qiu, H.; Xu, K.; et al. Human infection with a novel avian-origin influenza A (H7N9) virus. *N. Engl. J. Med.* **2013**, *368*, 1888–1897. [CrossRef]
5. Wei, C.J.; Crank, M.C.; Shiver, J.; Graham, B.S.; Mascola, J.R.; Nabel, G.J. Next-generation influenza vaccines: Opportunities and challenges. *Nat. Rev. Drug Discov.* **2020**, *19*, 239–252. [CrossRef] [PubMed]
6. Vemula, S.V.; Sayedahmed, E.E.; Sambhara, S.; Mittal, S.K. Vaccine approaches conferring cross-protection against influenza viruses. *Expert Rev. Vaccines* **2017**, *16*, 1141–1154. [CrossRef]
7. Xiong, F.; Zhang, C.; Shang, B.; Zheng, M.; Wang, Q.; Ding, Y.; Luo, J.; Li, X. An mRNA-based broad-spectrum vaccine candidate confers cross-protection against heterosubtypic influenza A viruses. *Emerg. Microbes Infect.* **2023**, *12*, 2256422. [CrossRef]
8. Corti, D.; Voss, J.; Gamblin, S.J.; Codoni, G.; Macagno, A.; Jarrossay, D.; Vachieri, S.G.; Pinna, D.; Minola, A.; Vanzetta, F.; et al. A neutralizing antibody selected from plasma cells that binds to group 1 and group 2 influenza A hemagglutinins. *Science* **2011**, *333*, 850–856. [CrossRef]
9. Ullah, S.; Ross, T.M. Next generation live-attenuated influenza vaccine platforms. *Expert Rev. Vaccines* **2022**, *21*, 1097–1110. [CrossRef]
10. Heiny, A.T.; Miotto, O.; Srinivasan, K.N.; Khan, A.M.; Zhang, G.L.; Brusic, V.; Tan, T.W.; August, J.T. Evolutionarily conserved protein sequences of influenza a viruses, avian and human, as vaccine targets. *PLoS ONE* **2007**, *2*, e1190. [CrossRef]
11. Tan, P.T.; Khan, A.M.; August, J.T. Highly conserved influenza A sequences as T cell epitopes-based vaccine targets to address the viral variability. *Hum. Vaccin.* **2011**, *7*, 402–409. [CrossRef] [PubMed]
12. Gottlieb, T.; Ben-Yedidia, T. Epitope-based approaches to a universal influenza vaccine. *J. Autoimmun.* **2014**, *54*, 15–20. [CrossRef] [PubMed]
13. Jazayeri, S.D.; Poh, C.L. Development of Universal Influenza Vaccines Targeting Conserved Viral Proteins. *Vaccines* **2019**, *7*, 169. [CrossRef] [PubMed]
14. Cheng, Z.; Ma, J.; Zhao, C. Advantages of Broad-Spectrum Influenza mRNA Vaccines and Their Impact on Pulmonary Influenza. *Vaccines* **2024**, *12*, 1382. [CrossRef]
15. Bedi, R.; Bayless, N.L.; Glanville, J. Challenges and Progress in Designing Broad-Spectrum Vaccines Against Rapidly Mutating Viruses. *Annu. Rev. Biomed. Sci.* **2023**, *6*, 419–441. [CrossRef]
16. Lim, C.M.L.; Komarasamy, T.V.; Adnan, N.; Radhakrishnan, A.K.; Balasubramaniam, V. Recent Advances, Approaches and Challenges in the Development of Universal Influenza Vaccines. *Influenza Other Respir. Viruses* **2024**, *18*, e13276. [CrossRef] [PubMed]
17. Vogel, O.A.; Manicassamy, B. Broadly Protective Strategies Against Influenza Viruses: Universal Vaccines and Therapeutics. *Front. Microbiol.* **2020**, *11*, 135. [CrossRef]
18. Viboud, C.; Gostic, K.; Nelson, M.I.; Price, G.E.; Perofsky, A.; Sun, K.; Sequeira Trovão, N.; Cowling, B.J.; Epstein, S.L.; Spiro, D.J. Beyond clinical trials: Evolutionary and epidemiological considerations for development of a universal influenza vaccine. *PLoS Pathog.* **2020**, *16*, e1008583. [CrossRef]
19. Berry, C.M.; Penhale, W.J.; Sangster, M.Y. Passive broad-spectrum influenza immunoprophylaxis. *Influenza Res. Treat.* **2014**, *2014*, 267594. [CrossRef]
20. Kerstetter, L.J.; Buckley, S.; Bliss, C.M.; Coughlan, L. Adenoviral Vectors as Vaccines for Emerging Avian Influenza Viruses. *Front. Immunol.* **2020**, *11*, 607333. [CrossRef]

21. Kozak, M.; Hu, J. The Integrated Consideration of Vaccine Platforms, Adjuvants, and Delivery Routes for Successful Vaccine Development. *Vaccines* **2023**, *11*, 695. [CrossRef] [PubMed]
22. Alqazlan, N.; Astill, J.; Raj, S.; Sharif, S. Strategies for enhancing immunity against avian influenza virus in chickens: A review. *Avian Pathol.* **2022**, *51*, 211–235. [CrossRef] [PubMed]
23. Zhao, T.; Cai, Y.; Jiang, Y.; He, X.; Wei, Y.; Yu, Y.; Tian, X. Vaccine adjuvants: Mechanisms and platforms. *Signal Transduct. Target. Ther.* **2023**, *8*, 283. [CrossRef]
24. Nguyen, H.H.; Tumpey, T.M.; Park, H.J.; Byun, Y.H.; Tran, L.D.; Nguyen, V.D.; Kilgore, P.E.; Czerkinsky, C.; Katz, J.M.; Seong, B.L.; et al. Prophylactic and therapeutic efficacy of avian antibodies against influenza virus H5N1 and H1N1 in mice. *PLoS ONE* **2010**, *5*, e10152. [CrossRef]
25. Tariq, H.; Batool, S.; Asif, S.; Ali, M.; Abbasi, B.H. Virus-Like Particles: Revolutionary Platforms for Developing Vaccines Against Emerging Infectious Diseases. *Front. Microbiol.* **2021**, *12*, 790121. [CrossRef] [PubMed]
26. Hendy, D.A.; Amouzougan, E.A.; Young, I.C.; Bachelder, E.M.; Ainslie, K.M. Nano/microparticle Formulations for Universal Influenza Vaccines. *AAPS J.* **2022**, *24*, 24. [CrossRef]
27. Poria, R.; Kala, D.; Nagraik, R.; Dhir, Y.; Dhir, S.; Singh, B.; Kaushik, N.K.; Noorani, M.S.; Kaushal, A.; Gupta, S. Vaccine development: Current trends and technologies. *Life Sci.* **2024**, *336*, 122331. [CrossRef]
28. Cappellano, G.; Abreu, H.; Casale, C.; Dianzani, U.; Chiocchetti, A. Nano-Microparticle Platforms in Developing Next-Generation Vaccines. *Vaccines* **2021**, *9*, 660. [CrossRef]
29. Chakravarty, M.; Vora, A. Nanotechnology-based antiviral therapeutics. *Drug Deliv. Transl. Res.* **2021**, *11*, 748–787. [CrossRef]
30. Lopez, C.E.; Legge, K.L. Influenza A Virus Vaccination: Immunity, Protection, and Recent Advances Toward A Universal Vaccine. *Vaccines* **2020**, *8*, 434. [CrossRef]
31. Zhu, W.; Dong, C.; Wei, L.; Wang, B.Z. Promising Adjuvants and Platforms for Influenza Vaccine Development. *Pharmaceutics* **2021**, *13*, 68. [CrossRef]
32. Isakova-Sivak, I.; Stepanova, E.; Mezhenskaya, D.; Matyushenko, V.; Prokopenko, P.; Sychev, I.; Wong, P.F.; Rudenko, L. Influenza vaccine: Progress in a vaccine that elicits a broad immune response. *Expert Rev. Vaccines* **2021**, *20*, 1097–1112. [CrossRef] [PubMed]
33. Li, Z.; Zhao, Y.; Li, Y.; Chen, X. Adjuvantation of Influenza Vaccines to Induce Cross-Protective Immunity. *Vaccines* **2021**, *9*, 75. [CrossRef] [PubMed]
34. Ren, H.; Jia, W.; Xie, Y.; Yu, M.; Chen, Y. Adjuvant physiochemistry and advanced nanotechnology for vaccine development. *Chem. Soc. Rev.* **2023**, *52*, 5172–5254. [CrossRef] [PubMed]
35. Goff, P.H.; Hayashi, T.; Martínez-Gil, L.; Corr, M.; Crain, B.; Yao, S.; Cottam, H.B.; Chan, M.; Ramos, I.; Eggink, D.; et al. Synthetic Toll-like receptor 4 (TLR4) and TLR7 ligands as influenza virus vaccine adjuvants induce rapid, sustained, and broadly protective responses. *J. Virol.* **2015**, *89*, 3221–3235. [CrossRef]
36. Cui, Y.; Ho, M.; Hu, Y.; Shi, Y. Vaccine adjuvants: Current status, research and development, licensing, and future opportunities. *J. Mater. Chem. B* **2024**, *12*, 4118–4137. [CrossRef]
37. Zeigler, D.F.; Gage, E.; Clegg, C.H. Epitope-targeting platform for broadly protective influenza vaccines. *PLoS ONE* **2021**, *16*, e0252170. [CrossRef]
38. Nagashima, K.A.; Mousa, J.J. Next-Generation Influenza HA Immunogens and Adjuvants in Pursuit of a Broadly Protective Vaccine. *Viruses* **2021**, *13*, 546. [CrossRef]
39. Sia, Z.R.; Miller, M.S.; Lovell, J.F. Engineered Nanoparticle Applications for Recombinant Influenza Vaccines. *Mol. Pharm.* **2021**, *18*, 576–592. [CrossRef]
40. Boulo, S.; Akarsu, H.; Ruigrok, R.W.H.; Baudin, F. Nuclear traffic of influenza virus proteins and ribonucleoprotein complexes. *Virus Res.* **2007**, *124*, 12–21. [CrossRef]
41. Skehel, J.J.; Wiley, D.C. Receptor binding and membrane fusion in virus entry: The influenza hemagglutinin. *Annu. Rev. Biochem* **2000**, *69*, 531–569. [CrossRef] [PubMed]
42. Garcia, N.K.; Kephart, S.M.; Benhaim, M.A.; Matsui, T.; Mileant, A.; Guttman, M.; Lee, K.K. Structural dynamics reveal subtype-specific activation and inhibition of influenza virus hemagglutinin. *J. Biol. Phys. Chem.* **2023**, *299*, 104765. [CrossRef] [PubMed]
43. Webster, R.G.; Bean, W.J.; Gorman, O.T.; Chambers, T.M.; Kawaoka, Y. Evolution and ecology of influenza A viruses. *Microbiol. Rev.* **1992**, *56*, 152–179. [CrossRef]
44. Rossman, J.S.; Lamb, R.A. Influenza virus assembly and budding. *Virology* **2011**, *411*, 229–236. [CrossRef]
45. Varghese, J.N.; Laver, W.G.; Colman, P.M. Structure of the influenza virus glycoprotein antigen neuraminidase at 2.9 Å resolution. *Nature* **1983**, *303*, 35–40. [CrossRef] [PubMed]
46. Noton, S.L.; Medcalf, E.; Fisher, D.; Mullin, A.E.; Elton, D.; Digard, P. Identification of the domains of the influenza A virus M1 matrix protein required for NP binding, oligomerization and incorporation into virions. *J. Gen. Mol. Virol.* **2007**, *88*, 2280–2290. [CrossRef]

47. Watanabe, K.; Fuse, T.; Asano, I.; Tsukahara, F.; Maru, Y.; Nagata, K.; Kitazato, K.; Kobayashi, N. Identification of Hsc70 as an influenza virus matrix protein (M1) binding factor involved in the virus life cycle. *FEBS Lett.* **2006**, *580*, 5785–5790. [CrossRef]
48. Pinto, L.H.; Lamb, R.A. The M2 proton channels of influenza A and B viruses. *J. Biol. Phys. Chem.* **2006**, *281*, 8997–9000. [CrossRef]
49. Hao, W.; Wang, L.; Li, S. Roles of the Non-Structural Proteins of Influenza A Virus. *Pathogens* **2020**, *9*, 812. [CrossRef]
50. Malik, G.; Zhou, Y. Innate Immune Sensing of Influenza A Virus. *Viruses* **2020**, *12*, 755. [CrossRef]
51. Hale, B.G.; Randall, R.E.; Ortín, J.; Jackson, D. The multifunctional NS1 protein of influenza A viruses. *J. Gen. Virol.* **2008**, *89*, 2359–2376. [CrossRef]
52. Wu, N.C.; Wilson, I.A. Influenza Hemagglutinin Structures and Antibody Recognition. *Cold Spring Harb. Perspect. Med.* **2020**, *10*, a038778. [CrossRef] [PubMed]
53. Wilson, I.A.; Skehel, J.J.; Wiley, D.C. Structure of the haemagglutinin membrane glycoprotein of influenza virus at 3 Å resolution. *Nature* **1981**, *289*, 366–373. [CrossRef]
54. Chen, J.; Skehel, J.J.; Wiley, D.C. N- and C-terminal residues combine in the fusion-pH influenza hemagglutinin HA(2) subunit to form an N cap that terminates the triple-stranded coiled coil. *Proc. Natl. Acad. Sci. USA* **1999**, *96*, 8967–8972. [CrossRef] [PubMed]
55. Guthmiller, J.J.; Han, J.; Utset, H.A.; Li, L.; Lan, L.Y.; Henry, C.; Stamper, C.T.; McMahon, M.; O'Dell, G.; Fernández-Quintero, M.L.; et al. Broadly neutralizing antibodies target a haemagglutinin anchor epitope. *Nature* **2022**, *602*, 314–320. [CrossRef] [PubMed]
56. Khanna, M.; Sharma, S.; Kumar, B.; Rajput, R. Protective immunity based on the conserved hemagglutinin stalk domain and its prospects for universal influenza vaccine development. *Biomed. Res.* **2014**, *2014*, 546274. [CrossRef]
57. Steel, J.; Lowen, A.C.; Wang, T.T.; Yondola, M.; Gao, Q.; Haye, K.; García-Sastre, A.; Palese, P. Influenza virus vaccine based on the conserved hemagglutinin stalk domain. *mBio* **2010**, *1*, 10–1128. [CrossRef]
58. Xu, R.; McBride, R.; Nycholat, C.M.; Paulson, J.C.; Wilson, I.A. Structural characterization of the hemagglutinin receptor specificity from the 2009 H1N1 influenza pandemic. *J. Virol.* **2012**, *86*, 982–990. [CrossRef]
59. Ekiert, D.C.; Bhabha, G.; Elsliger, M.A.; Friesen, R.H.; Jongeneelen, M.; Throsby, M.; Goudsmit, J.; Wilson, I.A. Antibody recognition of a highly conserved influenza virus epitope. *Science* **2009**, *324*, 246–251. [CrossRef]
60. Mair, C.M.; Meyer, T.; Schneider, K.; Huang, Q.; Veit, M.; Herrmann, A. A histidine residue of the influenza virus hemagglutinin controls the pH dependence of the conformational change mediating membrane fusion. *J. Virol.* **2014**, *88*, 13189–13200. [CrossRef]
61. Byrd-Leotis, L.; Galloway, S.E.; Agbogu, E.; Steinhauer, D.A. Influenza hemagglutinin (HA) stem region mutations that stabilize or destabilize the structure of multiple HA subtypes. *J. Virol.* **2015**, *89*, 4504–4516. [CrossRef]
62. Sun, X.; Ma, H.; Wang, X.; Bao, Z.; Tang, S.; Yi, C.; Sun, B. Broadly neutralizing antibodies to combat influenza virus infection. *Antivir. Res.* **2024**, *221*, 105785. [CrossRef] [PubMed]
63. Sui, J.; Hwang, W.C.; Perez, S.; Wei, G.; Aird, D.; Chen, L.M.; Santelli, E.; Stec, B.; Cadwell, G.; Ali, M.; et al. Structural and functional bases for broad-spectrum neutralization of avian and human influenza A viruses. *Nat. Struct. Mol. Biol.* **2009**, *16*, 265–273. [CrossRef] [PubMed]
64. Mallajosyula, V.V.A.; Citron, M.; Ferrara, F.; Lu, X.; Callahan, C.; Heidecker, G.J.; Sarma, S.P.; Flynn, J.A.; Temperton, N.J.; Liang, X.; et al. Influenza hemagglutinin stem-fragment immunogen elicits broadly neutralizing antibodies and confers heterologous protection. *Proc. Natl. Acad. Sci. USA* **2014**, *111*, E2514–E2523. [CrossRef]
65. Yassine, H.M.; Boyington, J.C.; McTamney, P.M.; Wei, C.-J.; Kanekiyo, M.; Kong, W.-P.; Gallagher, J.R.; Wang, L.; Zhang, Y.; Joyce, M.G.; et al. Hemagglutinin-stem nanoparticles generate heterosubtypic influenza protection. *Nat. Med.* **2015**, *21*, 1065–1070. [CrossRef] [PubMed]
66. McAuley, J.L.; Gilbertson, B.P.; Trifkovic, S.; Brown, L.E.; McKimm-Breschkin, J.L. Influenza Virus Neuraminidase Structure and Functions. *Front. Microbiol.* **2019**, *10*, 39. [CrossRef]
67. Jagadesh, A.; Salam, A.A.; Mudgal, P.P.; Arunkumar, G. Influenza virus neuraminidase (NA): A target for antivirals and vaccines. *Arch. Virol* **2016**, *161*, 2087–2094. [CrossRef]
68. Javanian, M.; Barary, M.; Ghebrehewet, S.; Koppolu, V.; Vasigala, V.; Ebrahimpour, S. A brief review of influenza virus infection. *J. Med. Virol.* **2021**, *93*, 4638–4646. [CrossRef]
69. Creytens, S.; Pascha, M.N.; Ballegeer, M.; Saelens, X.; de Haan, C.A.M. Influenza Neuraminidase Characteristics and Potential as a Vaccine Target. *Front. Immunol.* **2021**, *12*, 786617. [CrossRef]
70. Cady, S.D.; Luo, W.; Hu, F.; Hong, M. Structure and function of the influenza A M2 proton channel. *Biochemistry* **2009**, *48*, 7356–7364. [CrossRef]
71. Zharikova, D.; Mozdzanowska, K.; Feng, J.; Zhang, M.; Gerhard, W. Influenza type A virus escape mutants emerge in vivo in the presence of antibodies to the ectodomain of matrix protein 2. *J. Virol.* **2005**, *79*, 6644–6654. [CrossRef] [PubMed]
72. Wang, Y.; Zhou, L.; Shi, H.; Xu, H.; Yao, H.; Xi, X.G.; Toyoda, T.; Wang, X.; Wang, T. Monoclonal antibody recognizing SLLTEVET epitope of M2 protein potently inhibited the replication of influenza A viruses in MDCK cells. *Biochem. Biophys. Res. Commun.* **2009**, *385*, 118–122. [CrossRef]

73. Neirynck, S.; Deroo, T.; Saelens, X.; Vanlandschoot, P.; Jou, W.M.; Fiers, W. A universal influenza A vaccine based on the extracellular domain of the M2 protein. *Nat. Med.* **1999**, *5*, 1157–1163. [CrossRef] [PubMed]
74. Petrovsky, N.; Brusic, V. Computational immunology: The coming of age. *Immunol. Cell Biol.* **2002**, *80*, 248–254. [CrossRef]
75. De Groot, A.S.; Sbai, H.; Aubin, C.S.; McMurry, J.; Martin, W. Immuno-informatics: Mining genomes for vaccine components. *Immunol. Cell Biol.* **2002**, *80*, 255–269. [CrossRef] [PubMed]
76. Moise, L.; Gutierrez, A.; Kibria, F.; Martin, R.; Tassone, R.; Liu, R.; Terry, F.; Martin, B.; De Groot, A.S. iVAX: An integrated toolkit for the selection and optimization of antigens and the design of epitope-driven vaccines. *Hum. Vaccines Immunother.* **2015**, *11*, 2312–2321. [CrossRef] [PubMed]
77. De Groot, A.S.; Moise, L.; Terry, F.; Gutierrez, A.H.; Hindocha, P.; Richard, G.; Hoft, D.F.; Ross, T.M.; Noe, A.R.; Takahashi, Y.; et al. Better Epitope Discovery, Precision Immune Engineering, and Accelerated Vaccine Design Using Immunoinformatics Tools. *Front. Immunol.* **2020**, *11*, 442. [CrossRef]
78. Bounds, C.E.; Terry, F.E.; Moise, L.; Hannaman, D.; Martin, W.D.; De Groot, A.S.; Suschak, J.J.; Dupuy, L.C.; Schmaljohn, C.S. An immunoinformatics-derived DNA vaccine encoding human class II T cell epitopes of Ebola virus, Sudan virus, and Venezuelan equine encephalitis virus is immunogenic in HLA transgenic mice. *Hum. Vaccines Immunother.* **2017**, *13*, 2824–2836. [CrossRef]
79. Eickhoff, C.S.; Terry, F.E.; Peng, L.; Meza, K.A.; Sakala, I.G.; Van Aartsen, D.; Moise, L.; Martin, W.D.; Schriewer, J.; Buller, R.M.; et al. Highly conserved influenza T cell epitopes induce broadly protective immunity. *Vaccine* **2019**, *37*, 5371–5381. [CrossRef]
80. Watson, J.L.; Juergens, D.; Bennett, N.R.; Trippe, B.L.; Yim, J.; Eisenach, H.E.; Ahern, W.; Borst, A.J.; Ragotte, R.J.; Milles, L.F.; et al. De novo design of protein structure and function with RFdiffusion. *Nature* **2023**, *620*, 1089–1100. [CrossRef]
81. Liu, Y.; Pan, J.; Jenni, S.; Raymond, D.D.; Caradonna, T.; Do, K.T.; Schmidt, A.G.; Harrison, S.C.; Grigorieff, N. CryoEM Structure of an Influenza Virus Receptor-Binding Site Antibody-Antigen Interface. *J. Mol. Biol.* **2017**, *429*, 1829–1839. [CrossRef]
82. Taylor, R.M.; Dreguss, M. An Experiment in Immunization Against Influenza with a Formaldehyde-Inactivated Virus. *Am. J. Epidemiol.* **1940**, *31*, 31–35. [CrossRef]
83. Bachmann, M.F.; Bast, C.; Hengartner, H.; Zinkernagel, R.M. Immunogenicity of a viral model vaccine after different inactivation procedures. *Microbiol. Immunol.* **1994**, *183*, 95–104. [CrossRef] [PubMed]
84. Bonnafous, P.; Nicolaï, M.C.; Taveau, J.C.; Chevalier, M.; Barrière, F.; Medina, J.; Le Bihan, O.; Adam, O.; Ronzon, F.; Lambert, O. Treatment of influenza virus with beta-propiolactone alters viral membrane fusion. *Biochim. Biophys. Acta* **2014**, *1838*, 355–363. [CrossRef] [PubMed]
85. Clements, M.L.; Betts, R.F.; Tierney, E.L.; Murphy, B.R. Serum and nasal wash antibodies associated with resistance to experimental challenge with influenza A wild-type virus. *J. Clin. Microbiol.* **1986**, *24*, 157–160. [CrossRef] [PubMed]
86. Belongia, E.A.; Simpson, M.D.; King, J.P.; Sundaram, M.E.; Kelley, N.S.; Osterholm, M.T.; McLean, H.Q. Variable influenza vaccine effectiveness by subtype: A systematic review and meta-analysis of test-negative design studies. *Lancet Infect. Dis.* **2016**, *16*, 942–951. [CrossRef]
87. Kuriakose, T.; Man, S.M.; Malireddi, R.K.; Karki, R.; Kesavardhana, S.; Place, D.E.; Neale, G.; Vogel, P.; Kanneganti, T.D. ZBP1/DAI is an innate sensor of influenza virus triggering the NLRP3 inflammasome and programmed cell death pathways. *Sci. Immunol.* **2016**, *1*, aag2045. [CrossRef]
88. Zhang, N.; Zheng, B.J.; Lu, L.; Zhou, Y.; Jiang, S.; Du, L. Advancements in the development of subunit influenza vaccines. *Microbes Infect.* **2015**, *17*, 123–134. [CrossRef]
89. Zhang, Y.; Gao, J.; Xu, W.; Huo, X.; Wang, J.; Xu, Y.; Ding, W.; Guo, Z.; Liu, R. Advances in protein subunit vaccines against H1N1/09 influenza. *Front. Immunol.* **2024**, *15*, 1499754. [CrossRef]
90. Zhu, X.; Luo, Z.; Leonard, R.A.; Hamele, C.E.; Spreng, R.L.; Heaton, N.S. Administration of antigenically distinct influenza viral particle combinations as an influenza vaccine strategy. *PLoS Pathog.* **2025**, *21*, e1012878. [CrossRef]
91. Puig Barbera, J.; Gonzalez Vidal, D. MF59-adjuvanted subunit influenza vaccine: An improved interpandemic influenza vaccine for vulnerable populations. *Expert Rev. Vaccines* **2007**, *6*, 659–665. [CrossRef] [PubMed]
92. Lee, S.M.; Lee, J.; Kim, D.I.; Avila, J.P.; Nakaya, H.; Kwak, K.; Kim, E.H. Emulsion adjuvant-induced uric acid release modulates optimal immunogenicity by targeting dendritic cells and B cells. *npj Vaccines* **2025**, *10*, 72. [CrossRef]
93. Pan, J.; Wang, Q.; Qi, M.; Chen, J.; Wu, X.; Zhang, X.; Li, W.; Zhang, X.E.; Cui, Z. An Intranasal Multivalent Epitope-Based Nanoparticle Vaccine Confers Broad Protection against Divergent Influenza Viruses. *ACS Nano* **2023**, *17*, 13474–13487. [CrossRef] [PubMed]
94. Sabatino, D. Medicinal Chemistry and Methodological Advances in the Development of Peptide-Based Vaccines. *J. Med. Chem.* **2020**, *63*, 14184–14196. [CrossRef]
95. Hamley, I.W. Peptides for Vaccine Development. *ACS Appl. Bio Mater.* **2022**, *5*, 905–944. [CrossRef]
96. Abbasi, J. FLU-v, a Universal Flu Vaccine Candidate, Advances in Trial. *JAMA* **2020**, *323*, 1336. [CrossRef] [PubMed]
97. Pleguezuelos, O.; James, E.; Fernandez, A.; Lopes, V.; Rosas, L.A.; Cervantes-Medina, A.; Cleath, J.; Edwards, K.; Neitzey, D.; Gu, W.; et al. Efficacy of FLU-v, a broad-spectrum influenza vaccine, in a randomized phase IIb human influenza challenge study. *npj Vaccines* **2020**, *5*, 22. [CrossRef]

98. Pleguezuelos, O.; Dille, J.; de Groen, S.; Oftung, F.; Niesters, H.G.M.; Islam, M.A.; Naess, L.M.; Hungnes, O.; Aldarij, N.; Idema, D.L.; et al. Immunogenicity, Safety, and Efficacy of a Standalone Universal Influenza Vaccine, FLU-v, in Healthy Adults: A Randomized Clinical Trial. *Ann. Intern. Med.* **2020**, *172*, 453–462. [CrossRef]
99. Pleguezuelos, O.; Robinson, S.; Stoloff, G.A.; Caparros-Wanderley, W. Synthetic Influenza vaccine (FLU-v) stimulates cell mediated immunity in a double-blind, randomised, placebo-controlled Phase I trial. *Vaccine* **2012**, *30*, 4655–4660. [CrossRef]
100. Yoon, S.W.; Webby, R.J.; Webster, R.G. Evolution and ecology of influenza A viruses. *Curr. Top. Microbiol. Immunol.* **2014**, *385*, 359–375.
101. Neumann, G.; Noda, T.; Kawaoka, Y. Emergence and pandemic potential of swine-origin H1N1 influenza virus. *Nature* **2009**, *459*, 931–939. [CrossRef]
102. Dou, D.; Revol, R.; Östbye, H.; Wang, H.; Daniels, R. Influenza A Virus Cell Entry, Replication, Virion Assembly and Movement. *Front. Immunol.* **2018**, *9*, 1581. [CrossRef]
103. Wingerath, J.; Ostroumov, D.; Woller, N.; Manns, M.P.; Pinschewer, D.D.; Orlinger, K.; Berka, U.; Kühnel, F.; Wirth, T.C. Recombinant LCMV Vectors Induce Protective Immunity following Homologous and Heterologous Vaccinations. *Mol. Ther.* **2017**, *25*, 2533–2545. [CrossRef]
104. Bonilla, W.V.; Kirchhammer, N.; Marx, A.F.; Kallert, S.M.; Krzyzaniak, M.A.; Lu, M.; Darbre, S.; Schmidt, S.; Raguz, J.; Berka, U.; et al. Heterologous arenavirus vector prime-boost overrules self-tolerance for efficient tumor-specific CD8 T cell attack. *Cell Rep. Med.* **2021**, *2*, 100209. [CrossRef]
105. Liu, L.; Wang, T.; Wang, M.; Tong, Q.; Sun, Y.; Pu, J.; Sun, H.; Liu, J. Recombinant turkey herpesvirus expressing H9 hemagglutinin providing protection against H9N2 avian influenza. *Virology* **2019**, *529*, 7–15. [CrossRef]
106. Yu, B.; Zhou, Y.; Wu, H.; Wang, Z.; Zhan, Y.; Feng, X.; Geng, R.; Wu, Y.; Kong, W.; Yu, X. Seroprevalence of neutralizing antibodies to human adenovirus type 5 in healthy adults in China. *J. Med. Virol.* **2012**, *84*, 1408–1414. [CrossRef]
107. Pandey, A.; Singh, N.; Vemula, S.V.; Couëtil, L.; Katz, J.M.; Donis, R.; Sambhara, S.; Mittal, S.K. Impact of preexisting adenovirus vector immunity on immunogenicity and protection conferred with an adenovirus-based H5N1 influenza vaccine. *PLoS ONE* **2012**, *7*, e33428. [CrossRef]
108. Vannucci, L.; Lai, M.; Chiuppesi, F.; Ceccherini-Nelli, L.; Pistello, M. Viral vectors: A look back and ahead on gene transfer technology. *New Microbiol.* **2013**, *36*, 1–22.
109. Iavarone, C.; O'Hagan, D.T.; Yu, D.; Delahaye, N.F.; Ulmer, J.B. Mechanism of action of mRNA-based vaccines. *Expert Rev. Vaccines* **2017**, *16*, 871–881. [CrossRef]
110. Sandbrink, J.B.; Shattock, R.J. RNA Vaccines: A Suitable Platform for Tackling Emerging Pandemics? *Front. Immunol.* **2020**, *11*, 608460. [CrossRef]
111. Rosa, S.S.; Prazeres, D.M.F.; Azevedo, A.M.; Marques, M.P.C. mRNA vaccines manufacturing: Challenges and bottlenecks. *Vaccine* **2021**, *39*, 2190–2200. [CrossRef]
112. Reina, J. The new generation of messenger RNA (mRNA) vaccines against influenza. *Enferm. Infecc. Microbiol. Clin. (Engl. Ed.)* **2023**, *41*, 301–304. [CrossRef]
113. Rockman, S.; Laurie, K.L.; Parkes, S.; Wheatley, A.; Barr, I.G. New Technologies for Influenza Vaccines. *Microorganisms* **2020**, *8*, 1745. [CrossRef]
114. Karikó, K.; Muramatsu, H.; Ludwig, J.; Weissman, D. Generating the optimal mRNA for therapy: HPLC purification eliminates immune activation and improves translation of nucleoside-modified, protein-encoding mRNA. *Nucleic Acids Res.* **2011**, *39*, e142. [CrossRef]
115. Liang, S.L.; Quirk, D.; Zhou, A. RNase L: Its biological roles and regulation. *IUBMB Life* **2006**, *58*, 508–514. [CrossRef]
116. Tanji, H.; Ohto, U.; Shibata, T.; Taoka, M.; Yamauchi, Y.; Isobe, T.; Miyake, K.; Shimizu, T. Toll-like receptor 8 senses degradation products of single-stranded RNA. *Nat. Struct. Mol. Biol.* **2015**, *22*, 109–115. [CrossRef]
117. Andries, O.; Mc Cafferty, S.; De Smedt, S.C.; Weiss, R.; Sanders, N.N.; Kitada, T. N(1)-methylpseudouridine-incorporated mRNA outperforms pseudouridine-incorporated mRNA by providing enhanced protein expression and reduced immunogenicity in mammalian cell lines and mice. *J. Control. Release* **2015**, *217*, 337–344. [CrossRef]
118. Lee, S.; Ryu, J.-H. Influenza Viruses: Innate Immunity and mRNA Vaccines. *Front. Immunol.* **2021**, *12*, 710647. [CrossRef]
119. Del Giudice, G.; Rappuoli, R.; Didierlaurent, A.M. Correlates of adjuvanticity: A review on adjuvants in licensed vaccines. *Semin. Immunol.* **2018**, *39*, 14–21. [CrossRef]
120. Bilyy, R.; Paryzhak, S.; Turcheniuk, K.; Dumych, T.; Barras, A.; Boukherroub, R.; Wang, F.; Yushin, G.; Szunerits, S. Aluminum oxide nanowires as safe and effective adjuvants for next-generation vaccines. *Mater. Today* **2019**, *22*, 58–66. [CrossRef]
121. Harrison, W.T. Some Observations on the Use of Alum Precipitated Diphtheria Toxoid. *Am. J. Public Health Nations Health* **1935**, *25*, 298–300. [CrossRef]
122. Eisenbarth, S.C.; Colegio, O.R.; O'Connor, W.; Sutterwala, F.S.; Flavell, R.A. Crucial role for the Nalp3 inflammasome in the immunostimulatory properties of aluminium adjuvants. *Nature* **2008**, *453*, 1122–1126. [CrossRef]

123. Kool, M.; Soullié, T.; van Nimwegen, M.; Willart, M.A.; Muskens, F.; Jung, S.; Hoogsteden, H.C.; Hammad, H.; Lambrecht, B.N. Alum adjuvant boosts adaptive immunity by inducing uric acid and activating inflammatory dendritic cells. *J. Exp. Med.* **2008**, *205*, 869–882. [CrossRef]
124. Marichal, T.; Ohata, K.; Bedoret, D.; Mesnil, C.; Sabatel, C.; Kobiyama, K.; Lekeux, P.; Coban, C.; Akira, S.; Ishii, K.J.; et al. DNA released from dying host cells mediates aluminum adjuvant activity. *Nat. Med.* **2011**, *17*, 996–1002. [CrossRef]
125. Cain, D.W.; Snowden, P.B.; Sempowski, G.D.; Kelsoe, G. Inflammation triggers emergency granulopoiesis through a density-dependent feedback mechanism. *PLoS ONE* **2011**, *6*, e19957. [CrossRef]
126. Leroux-Roels, G. Unmet needs in modern vaccinology: Adjuvants to improve the immune response. *Vaccine* **2010**, *28* (Suppl. S3), C25–C36. [CrossRef]
127. Hem, S.L.; Johnston, C.T.; HogenEsch, H. Imject Alum is not aluminum hydroxide adjuvant or aluminum phosphate adjuvant. *Vaccine* **2007**, *25*, 4985–4986. [CrossRef]
128. Marrack, P.; McKee, A.S.; Munks, M.W. Towards an understanding of the adjuvant action of aluminium. *Nat. Rev. Immunol.* **2009**, *9*, 287–293. [CrossRef]
129. O'Hagan, D.T.; Friedland, L.R.; Hanon, E.; Didierlaurent, A.M. Towards an evidence based approach for the development of adjuvanted vaccines. *Curr. Opin. Immunol.* **2017**, *47*, 93–102. [CrossRef]
130. O'Hagan, D.T.; Ott, G.S.; Nest, G.V.; Rappuoli, R.; Giudice, G.D. The history of MF59(®) adjuvant: A phoenix that arose from the ashes. *Expert Rev. Vaccines* **2013**, *12*, 13–30. [CrossRef]
131. Seubert, A.; Monaci, E.; Pizza, M.; O'Hagan, D.T.; Wack, A. The adjuvants aluminum hydroxide and MF59 induce monocyte and granulocyte chemoattractants and enhance monocyte differentiation toward dendritic cells. *J. Immunol.* **2008**, *180*, 5402–5412. [CrossRef] [PubMed]
132. Calabro, S.; Tortoli, M.; Baudner, B.C.; Pacitto, A.; Cortese, M.; O'Hagan, D.T.; De Gregorio, E.; Seubert, A.; Wack, A. Vaccine adjuvants alum and MF59 induce rapid recruitment of neutrophils and monocytes that participate in antigen transport to draining lymph nodes. *Vaccine* **2011**, *29*, 1812–1823. [CrossRef] [PubMed]
133. Cioncada, R.; Maddaluno, M.; Vo, H.T.M.; Woodruff, M.; Tavarini, S.; Sammicheli, C.; Tortoli, M.; Pezzicoli, A.; De Gregorio, E.; Carroll, M.C.; et al. Vaccine adjuvant MF59 promotes the intranodal differentiation of antigen-loaded and activated monocyte-derived dendritic cells. *PLoS ONE* **2017**, *12*, e0185843. [CrossRef]
134. Mastelic Gavillet, B.; Eberhardt, C.S.; Auderset, F.; Castellino, F.; Seubert, A.; Tregoning, J.S.; Lambert, P.H.; de Gregorio, E.; Del Giudice, G.; Siegrist, C.A. MF59 Mediates Its B Cell Adjuvanticity by Promoting T Follicular Helper Cells and Thus Germinal Center Responses in Adult and Early Life. *J. Immunol.* **2015**, *194*, 4836–4845. [CrossRef]
135. Seubert, A.; Calabro, S.; Santini, L.; Galli, B.; Genovese, A.; Valentini, S.; Aprea, S.; Colaprico, A.; D'Oro, U.; Giuliani, M.M.; et al. Adjuvanticity of the oil-in-water emulsion MF59 is independent of Nlrp3 inflammasome but requires the adaptor protein MyD88. *Proc. Natl. Acad. Sci. USA* **2011**, *108*, 11169–11174. [CrossRef] [PubMed]
136. Ko, E.J.; Lee, Y.T.; Kim, K.H.; Jung, Y.J.; Lee, Y.; Denning, T.L.; Kang, S.M. Effects of MF59 Adjuvant on Induction of Isotype-Switched IgG Antibodies and Protection after Immunization with T-Dependent Influenza Virus Vaccine in the Absence of CD4+ T Cells. *J. Virol.* **2016**, *90*, 6976–6988. [CrossRef]
137. Belshe, R.B.; Frey, S.E.; Graham, I.L.; Anderson, E.L.; Jackson, L.A.; Spearman, P.; Edupuganti, S.; Mulligan, M.J.; Rouphael, N.; Winokur, P.; et al. Immunogenicity of avian influenza A/Anhui/01/2005(H5N1) vaccine with MF59 adjuvant: A randomized clinical trial. *J. Am. Med. Assoc.* **2014**, *312*, 1420–1428. [CrossRef]
138. van der Maas, N.; Dijs-Elsinga, J.; Kemmeren, J.; van Lier, A.; Knol, M.; de Melker, H. Safety of vaccination against influenza A (H1N1) during pregnancy in the Netherlands: Results on pregnancy outcomes and infant's health: Cross-sectional linkage study. *BJOG* **2016**, *123*, 709–717. [CrossRef]
139. Bernstein, D.I.; Edwards, K.M.; Dekker, C.L.; Belshe, R.; Talbot, H.K.; Graham, I.L.; Noah, D.L.; He, F.; Hill, H. Effects of adjuvants on the safety and immunogenicity of an avian influenza H5N1 vaccine in adults. *J. Infect. Dis.* **2008**, *197*, 667–675. [CrossRef]
140. Reisinger, K.S.; Holmes, S.J.; Pedotti, P.; Arora, A.K.; Lattanzi, M. A dose-ranging study of MF59(®)-adjuvanted and non-adjuvanted A/H1N1 pandemic influenza vaccine in young to middle-aged and older adult populations to assess safety, immunogenicity, and antibody persistence one year after vaccination. *Hum. Vaccines Immunother.* **2014**, *10*, 2395–2407. [CrossRef]
141. Huang, Z.; Gong, H.; Sun, Q.; Yang, J.; Yan, X.; Xu, F. Research progress on emulsion vaccine adjuvants. *Heliyon* **2024**, *10*, e24662. [CrossRef] [PubMed]
142. Garçon, N.; Di Pasquale, A. From discovery to licensure, the Adjuvant System story. *Hum. Vaccines Immunother.* **2017**, *13*, 19–33. [CrossRef]
143. Garçon, N.; Vaughn, D.W.; Didierlaurent, A.M. Development and evaluation of AS03, an Adjuvant System containing α-tocopherol and squalene in an oil-in-water emulsion. *Expert Rev. Vaccines* **2012**, *11*, 349–366. [CrossRef] [PubMed]
144. Morel, S.; Didierlaurent, A.; Bourguignon, P.; Delhaye, S.; Baras, B.; Jacob, V.; Planty, C.; Elouahabi, A.; Harvengt, P.; Carlsen, H.; et al. Adjuvant System AS03 containing α-tocopherol modulates innate immune response and leads to improved adaptive immunity. *Vaccine* **2011**, *29*, 2461–2473. [CrossRef]

145. McElhaney, J.E.; Beran, J.; Devaster, J.M.; Esen, M.; Launay, O.; Leroux-Roels, G.; Ruiz-Palacios, G.M.; van Essen, G.A.; Caplanusi, A.; Claeys, C.; et al. AS03-adjuvanted versus non-adjuvanted inactivated trivalent influenza vaccine against seasonal influenza in elderly people: A phase 3 randomised trial. *Lancet Infect. Dis.* **2013**, *13*, 485–496. [CrossRef]
146. Madan, A.; Collins, H.; Sheldon, E.; Frenette, L.; Chu, L.; Friel, D.; Drame, M.; Vaughn, D.W.; Innis, B.L.; Schuind, A. Evaluation of a primary course of H9N2 vaccine with or without AS03 adjuvant in adults: A phase I/II randomized trial. *Vaccine* **2017**, *35*, 4621–4628. [CrossRef] [PubMed]
147. Langley, J.M.; Frenette, L.; Chu, L.; McNeil, S.; Halperin, S.; Li, P.; Vaughn, D. A randomized, controlled non-inferiority trial comparing A(H1N1)pmd09 vaccine antigen, with and without AS03 adjuvant system, co-administered or sequentially administered with an inactivated trivalent seasonal influenza vaccine. *BMC Infect. Dis.* **2012**, *12*, 279. [CrossRef]
148. Sarkanen, T.O.; Alakuijala, A.P.E.; Dauvilliers, Y.A.; Partinen, M.M. Incidence of narcolepsy after H1N1 influenza and vaccinations: Systematic review and meta-analysis. *Sleep Med. Rev.* **2018**, *38*, 177–186. [CrossRef]
149. Nazareth, I.; Tavares, F.; Rosillon, D.; Haguinet, F.; Bauchau, V. Safety of AS03-adjuvanted split-virion H1N1 (2009) pandemic influenza vaccine: A prospective cohort study. *BMJ Open* **2013**, *3*, e001912. [CrossRef]
150. Xiang, M.; Fan, J. Pattern recognition receptor-dependent mechanisms of acute lung injury. *Mol. Med.* **2010**, *16*, 69–82. [CrossRef]
151. Murugaiah, V.; Tsolaki, A.G.; Kishore, U. Collectins: Innate Immune Pattern Recognition Molecules. *Adv. Exp. Med. Biol.* **2020**, *1204*, 75–127.
152. Suresh, R.; Mosser, D.M. Pattern recognition receptors in innate immunity, host defense, and immunopathology. *Adv. Physiol. Educ.* **2013**, *37*, 284–291. [CrossRef] [PubMed]
153. Skwarczynski, M.; Toth, I. Recent advances in peptide-based subunit nanovaccines. *Nanomedicine* **2014**, *9*, 2657–2669. [CrossRef]
154. Wicherska-Pawłowska, K.; Wróbel, T.; Rybka, J. Toll-Like Receptors (TLRs), NOD-Like Receptors (NLRs), and RIG-I-Like Receptors (RLRs) in Innate Immunity. TLRs, NLRs, and RLRs Ligands as Immunotherapeutic Agents for Hematopoietic Diseases. *Int. J. Mol. Sci.* **2021**, *22*, 13397. [CrossRef]
155. Coler, R.N.; Baldwin, S.L.; Shaverdian, N.; Bertholet, S.; Reed, S.J.; Raman, V.S.; Lu, X.; DeVos, J.; Hancock, K.; Katz, J.M.; et al. A synthetic adjuvant to enhance and expand immune responses to influenza vaccines. *PLoS ONE* **2010**, *5*, e13677. [CrossRef] [PubMed]
156. Behzad, H.; Huckriede, A.L.; Haynes, L.; Gentleman, B.; Coyle, K.; Wilschut, J.C.; Kollmann, T.R.; Reed, S.G.; McElhaney, J.E. GLA-SE, a synthetic toll-like receptor 4 agonist, enhances T-cell responses to influenza vaccine in older adults. *J. Infect. Dis.* **2012**, *205*, 466–473. [CrossRef] [PubMed]
157. Taylor, D.N.; Treanor, J.J.; Sheldon, E.A.; Johnson, C.; Umlauf, S.; Song, L.; Kavita, U.; Liu, G.; Tussey, L.; Ozer, K.; et al. Development of VAX128, a recombinant hemagglutinin (HA) influenza-flagellin fusion vaccine with improved safety and immune response. *Vaccine* **2012**, *30*, 5761–5769. [CrossRef]
158. Campbell, J.D. Development of the CpG Adjuvant 1018: A Case Study. *Methods Mol. Biol.* **2017**, *1494*, 15–27.
159. Hoxie, I.; Vasilev, K.; Clark, J.J.; Bushfield, K.; Francis, B.; Loganathan, M.; Campbell, J.D.; Yu, D.; Guan, L.; Gu, C.; et al. A recombinant N2 neuraminidase-based CpG 1018(R) adjuvanted vaccine provides protection against challenge with heterologous influenza viruses in mice and hamsters. *Vaccine* **2024**, *42*, 126269. [CrossRef]
160. Ichinohe, T.; Watanabe, I.; Ito, S.; Fujii, H.; Moriyama, M.; Tamura, S.; Takahashi, H.; Sawa, H.; Chiba, J.; Kurata, T.; et al. Synthetic double-stranded RNA poly(I:C) combined with mucosal vaccine protects against influenza virus infection. *J. Virol.* **2005**, *79*, 2910–2919. [CrossRef]
161. Le, C.T.T.; Ahn, S.Y.; Ho, T.L.; Lee, J.; Lee, D.H.; Hwang, H.S.; Kang, S.M.; Ko, E.J. Adjuvant effects of combination monophosphoryl lipid A and poly I:C on antigen-specific immune responses and protective efficacy of influenza vaccines. *Sci. Rep.* **2023**, *13*, 12231. [CrossRef] [PubMed]
162. Martinez-Gil, L.; Goff, P.H.; Hai, R.; Garcia-Sastre, A.; Shaw, M.L.; Palese, P. A Sendai virus-derived RNA agonist of RIG-I as a virus vaccine adjuvant. *J. Virol.* **2013**, *87*, 1290–1300. [CrossRef]
163. Petsch, B.; Schnee, M.; Vogel, A.B.; Lange, E.; Hoffmann, B.; Voss, D.; Schlake, T.; Thess, A.; Kallen, K.J.; Stitz, L.; et al. Protective efficacy of in vitro synthesized, specific mRNA vaccines against influenza A virus infection. *Nat. Biotechnol.* **2012**, *30*, 1210–1216. [CrossRef]
164. Abe, T.; Marutani, Y.; Shoji, I. Cytosolic DNA-sensing immune response and viral infection. *Microbiol. Immunol.* **2019**, *63*, 51–64. [CrossRef] [PubMed]
165. Amurri, L.; Horvat, B.; Iampietro, M. Interplay between RNA viruses and cGAS/STING axis in innate immunity. *Front. Cell. Infect. Microbiol.* **2023**, *13*, 1172739. [CrossRef] [PubMed]
166. Ishikawa, H.; Ma, Z.; Barber, G.N. STING regulates intracellular DNA-mediated, type I interferon-dependent innate immunity. *Nature* **2009**, *461*, 788–792. [CrossRef]
167. Wu, J.; Sun, L.; Chen, X.; Du, F.; Shi, H.; Chen, C.; Chen, Z.J. Cyclic GMP-AMP is an endogenous second messenger in innate immune signaling by cytosolic DNA. *Science* **2013**, *339*, 826–830. [CrossRef]

168. Wang, J.; Li, P.; Yu, Y.; Fu, Y.; Jiang, H.; Lu, M.; Sun, Z.; Jiang, S.; Lu, L.; Wu, M.X. Pulmonary surfactant-biomimetic nanoparticles potentiate heterosubtypic influenza immunity. *Science* **2020**, *367*, eaau0810. [CrossRef]
169. Rozenberg, J.M.; Kamynina, M.; Sorokin, M.; Zolotovskaia, M.; Koroleva, E.; Kremenchutckaya, K.; Gudkov, A.; Buzdin, A.; Borisov, N. The Role of the Metabolism of Zinc and Manganese Ions in Human Cancerogenesis. *Biomedicines* **2022**, *10*, 1072. [CrossRef]
170. Liu, J.; Zhang, X.; Wang, H. The cGAS-STING-mediated NLRP3 inflammasome is involved in the neurotoxicity induced by manganese exposure. *Biomed. Pharmacother.* **2022**, *154*, 113680. [CrossRef]
171. Lv, M.; Chen, M.; Zhang, R.; Zhang, W.; Wang, C.; Zhang, Y.; Wei, X.; Guan, Y.; Liu, J.; Feng, K.; et al. Manganese is critical for antitumor immune responses via cGAS-STING and improves the efficacy of clinical immunotherapy. *Cell Res.* **2020**, *30*, 966–979. [CrossRef] [PubMed]
172. Sun, X.; Zhang, Y.; Li, J.; Park, K.S.; Han, K.; Zhou, X.; Xu, Y.; Nam, J.; Xu, J.; Shi, X.; et al. Amplifying STING activation by cyclic dinucleotide-manganese particles for local and systemic cancer metalloimmunotherapy. *Nat. Nanotechnol.* **2021**, *16*, 1260–1270. [CrossRef] [PubMed]
173. Sheng, Y.; Li, Z.; Lin, X.; Wang, L.; Zhu, H.; Su, Z.; Zhang, S. In situ bio-mineralized Mn nanoadjuvant enhances anti-influenza immunity of recombinant virus-like particle vaccines. *J. Control. Release* **2024**, *368*, 275–289. [CrossRef] [PubMed]
174. Cui, C.; Wang, S.; Lu, W.; Wang, Y.; Li, J.; Qu, K.; Yang, M.; Wang, L.; Yu, Y. The adjuvanticity of manganese for microbial vaccines via activating the IRF5 signaling pathway. *Biochem. Pharmacol.* **2021**, *192*, 114720. [CrossRef]
175. Pei, M.; Liu, K.; Qu, X.; Wang, K.; Chen, Q.; Zhang, Y.; Wang, X.; Wang, Z.; Li, X.; Chen, F.; et al. Enzyme-catalyzed synthesis of selenium-doped manganese phosphate for synergistic therapy of drug-resistant colorectal cancer. *Nanobiotechnology* **2023**, *21*, 72. [CrossRef]
176. Zhang, R.; Wang, C.; Guan, Y.; Wei, X.; Sha, M.; Yi, M.; Jing, M.; Lv, M.; Guo, W.; Xu, J.; et al. Manganese salts function as potent adjuvants. *Cell. Mol. Immunol.* **2021**, *18*, 1222–1234. [CrossRef]

targets

MDPI

Review

Application of Machine Learning in Cell Detection

Xinyue Liu, Xiaoyuan Wang and Ruocan Qian *

School of Chemistry and Molecular Engineering, East China University of Science and Technology, Shanghai 200237, China; y30230278@mail.ecust.edu.cn (X.L.); xiaoyuanwang@mail.ecust.edu.cn (X.W.)
* Correspondence: ruocanqian@ecust.edu.cn

Abstract: In recent years, machine learning algorithms have seen extensive application in chemical science, especially in cell detection technologies. Machine learning, a branch of artificial intelligence, is designed to automatically discover patterns in data. This review provides an overview of cell detection methods such as bright-field microscopy (BL), dark-field microscopy (DL), surface-enhanced Raman scattering (SERS), and fluorescence detection (FL). We highlight key computational models like support vector machines and convolutional neural networks that significantly enhance the precision and efficiency of automated cell detection. Relevant research applications are discussed, along with future prospects for machine learning in cell analysis.

Keywords: machine learning; cell detection; detection methods

1. Introduction

Machine learning is widely applied across various scientific disciplines, often in tandem with big data analytics and artificial intelligence [1]. It offers exceptional advantages in cellular data processing, with increased flexibility, efficiency, and precision in solving practical problems. In this review, we focus on the evolution of cell detection technology, examining how traditional techniques have transitioned into modern approaches using deep learning models.

Prior to 2012, classical machine learning algorithms, such as the viola jones detector (VJ detector) [2], histogram of oriented gradients (HOG) [3], and deformable part models (DPM) [4], dominated the landscape. However, the introduction of deep learning architectures like AlexNet [5], R-CNN [6], and YOLO [7] after 2012 led to significant advancements in the field. These deep learning models have since been integrated into cell detection technologies, enhancing the intelligence and accuracy of automated detection systems. Traditional detection algorithms primarily relied on feature extraction and image segmentation, using methods based on object region or color. With the advent of deep learning and the growth of cell data, two-stage detection models like R-CNN [8] and single-stage models such as YOLO and SSD [9] have emerged as standard tools. More recently, algorithms based on the Transformer architecture have gained traction, representing the next wave of innovation in object detection [10]. Machine learning in cell detection continues to evolve, with diverse algorithmic approaches improving the accuracy and reliability of detection results. This review explores the trajectory of these advancements and discusses their applications in cell detection.

As a software tool for building models, machine learning is trainable and reliable, which can facilitate researchers in cell analysis. With the advancement of machine learning algorithms and the development of cell detection technology, we can automatically analyze large amounts of data through machine learning models, so as to predict and analyze more complex cell behaviors and provide new insights for the treatment of diseases; this is expected to be applied in the fields of chemistry, biology, and medicine and to promote interdisciplinary cooperation.

2. Machine Learning Algorithms for Cell Detection

Early cell detection techniques primarily utilized two types of machine learning algorithms: kernel methods and ensemble methods. Kernel methods address the challenge of mapping non-linear, low-dimensional data into a higher-dimensional space, with support vector machines (SVMs) serving as a classical model. For instance, Svensson et al. achieved an 88% accuracy rate in identifying circulating tumor cells using an SVM coupled with a naive Bayes classifier [11]. This algorithm is reliable and automated and plays a key role in the early diagnosis of diseases. However, the kernel method is often limited to specific types of target models, leading to the development of more generalized ensemble approaches.

Ensemble methods combine multiple machine learning algorithms to transform a series of weak learners into strong predictors. Key examples include the AdaBoost and random forest algorithms [12]. Pereira et al. demonstrated the utility of random forest models in brain tumor segmentation, achieving superior precision and stability compared to earlier segmentation techniques [13]. We believe that the method can be further improved by combining the model with other techniques, resulting in data with higher accuracy and more stable results. Despite these advancements, traditional machine learning algorithms struggle with complex, large-scale cell data, and detection accuracy can be limited by the sample size.

Recent developments in machine learning have introduced deep learning algorithms that simulate the neural networks of the human brain. In particular, a convolutional neural network (R-CNN) is a region-based, two-stage target detection algorithm, which has been shown to be effective in cell detection tasks. Zeune et al. used a CNN combined with visualization techniques to achieve over 96% accuracy in detecting circulating tumor cells, while also identifying novel cell subtypes [14]. This strategy combines a variety of technologies, more in-depth research based on machine learning, and higher accuracy for cell detection. In addition, YOLO is a border-based single-stage target detection algorithm, which has been successfully applied to cervical cancer and breast cancer cell detection [15]. We believe that the single-stage object detection algorithm has lower computational complexity than the two-stage object detection algorithm, so the detection speed is faster. These advancements suggest that deep learning algorithms will continue to play a pivotal role in the future of cell detection technologies.

3. The Application of Machine Learning to Cell Detection Methods

The diversity of machine learning and cell detection methods can help us better understand the similarities and differences in cell data for tasks such as cell data visualization, dimensionality reduction, clustering, and feature selection. In practical applications, it is necessary to select the appropriate cell detection means according to the specific needs, determine the machine learning algorithms and parameters, and evaluate and optimize the results to achieve the best results. Therefore, according to the means of cell detection, we can roughly divide this into four categories: bright-field microscopic detection, dark-field microscopic detection, surface-enhanced Raman scattering, and fluorescence detection.

3.1. Bright-Field Microscopic Detection

Bright-field microscopy, a cost-effective optical method, allows for the analysis of cell shape, size, and morphology [16]. By applying machine learning algorithms to bright-field images, researchers can rapidly and accurately identify cellular characteristics, improving detection and analysis outcomes.

The early diagnosis and evaluation of cancer are crucial to the treatment of patients with the disease. Chemotherapy is an important method used in the treatment of leukemia, but cancer cells of different patients have different resistances to the treatment effect [17]. Uelu et al. used the computer vision algorithm to quantitatively detect the immunomagnetic beads and leukemia cells by using the cell images collected by the high-power objective lens under the bright-field microscope (Figure 1a), and the accuracy reached 91.6% under the 40-fold objective lens [18]. This method provides convenience for the realization of on-site cell analysis.

Circulating tumor cells (CTCs) are an important biomarker of cancer, and their count can predict the survival of patients, but their identification is difficult [19]. Wang et al. used convolutional neural networks to detect CTCs in blood under a bright-field microscope (Figure 1b) and counted the CTCs simply and quickly through cell images with high detection accuracy [20]. This method is smart and could be used to detect rare cells in the future.

Due to the low contrast of images obtained under bright-field microscopy and the possibility of cell overlap, automatic segmentation cannot be performed, and manual acquisition is time-consuming and complicated [21]. Asha et al. developed a remarkable and spherically driven U-shaped network (SBU-net), which can accurately segment cells in bright-field microscopic images (Figure 1c) and obtain cell structure information [22]. This model has strong cell segmentation performance and greatly promotes the development of the automatic segmentation of microscopic images.

Mesenchymal stem cells (MSCs) are a kind of stem cell with diverse differentiation abilities, widely used in the research of immune diseases. The aging of MSCs will cause adverse reactions to the human body, so the count of senescent cells in MSCs is very important [23,24]. Celebi et al. used the model combined with self-supervised learning and mask R-CNNs to automatically segment and count senescent cells in bright-field microscopy images (Figure 1d), with an accuracy of more than 80% [25]. The model can be further applied to the detection of other cell types.

3.2. Dark-Field Microscopic Detection

Because of the low contrast that is characteristic of bright-field images, the performance of the segmentation algorithm is affected to some extent, and there are some limitations to cell detection. With a dark-field microscope, the condenser collects diffracted light by passing light through the aperture of the lens so that the background of the image is black but the cells are bright, and a high-contrast image is obtained [26]. The combination of machine learning and dark-field microscopy has improved the accuracy of cell detection due to the enhanced performance of the algorithm.

In cell detection studies, the growth state of cells is extremely important, among which cell density and activity are the most important [27]. Wei et al. developed a probe based on the support vector machine model (SVM), which was used to analyze yeast cell images under dark-field microscopy (Figure 2a), so as to separate living and dead cells and detect cell density and activity with high precision [28]. This method has high precision, good stability, and good application prospects in distinguishing living cells from dead cells.

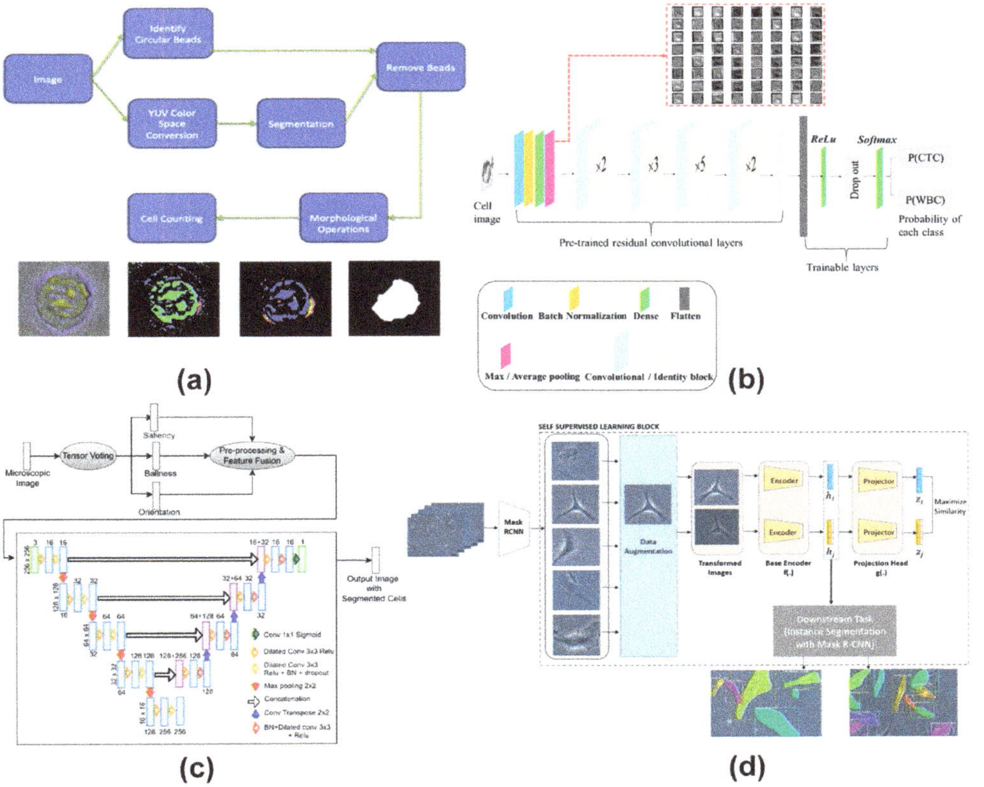

Figure 1. Application of machine learning in bright-field microscopic detection of cells. (**a**) Quantitative detection of immunomagnetic beads and leukemia cells using visual algorithms and bright-field microscopy. Reproduced from Ref. [18] with the permission of Elsevier. (**b**) Using CNN algorithm and bright-field microscope to detect CTCs in blood. Reproduced from Ref. [20] with the permission of Nature. (**c**) Accurate segmentation of cells in bright-field microscopic images using SBU-net. Reproduced from Ref. [22] with the permission of Elsevier. (**d**) Automatic segmentation of senescent cells in bright-field microscopic images using self-supervised learning. Reproduced from Ref. [25] with the permission of Wiley Online Library.

In the process of cell culture, cells are easily affected by the external environment and so are easy to be contaminated, and it is important to develop non-staining and non-harmful cell analysis methods. Based on SVMs in machine learning, Burgemeister et al. proposed a CellViCAM system (Figure 2b), which can estimate the animal cell density and cell differentiation degree without adding any marks when processing dark-field microscopic images [29]. This strategy can better distinguish cell differentiation states such as live cells, necrotic cells, and apoptotic cells.

The change in blood value will lead to the occurrence of many diseases. The count of blood cells includes red blood cells, white blood cells, platelets, and other quantitative indicators [30]. The collection of human blood is conducive to the screening of diseases, in which the morphologic study of red blood cells can know whether anemia is present [31]. Using Mi scattering and machine learning, Chen et al. obtained cell morphological information (Figure 2c) by imaging in the dark field, including the volume, concentration, and distribution range of red blood cells, with high accuracy [32]. This method has a good advantage in detecting anemia.

Gold nanoparticles (AuNPs) have high stability, good biocompatibility, and strong molecular signals, which can provide good imaging [33]. Based on the scattering properties of AuNPs, researchers can construct functional nanosensors [34,35]. However, conventional methods for detecting nanoparticle scattering in living cells are time-consuming and complex. Huang et al. developed U-Net convolutional deep learning neural network technology (Figure 2d), which can identify the scattered light signal of nanoparticles in living cells with high accuracy under complex environments with background interference by using dark-field microscopic imaging technology [36]. This method provides a new idea for the imaging analysis of living cells in the field of chemistry.

In addition to intracellular, we can also detect the degree of AuNPs' aggregation in different saline solutions. Wang et al. used dark-field microscopic imaging to build an AlexNet model of machine learning, which could accurately predict the AuNP aggregation degree in solution (Figure 2e) with an accuracy higher than 96% [37]. This strategy is useful for the predictive analysis of dark-field imaging, and it has potential biological applications.

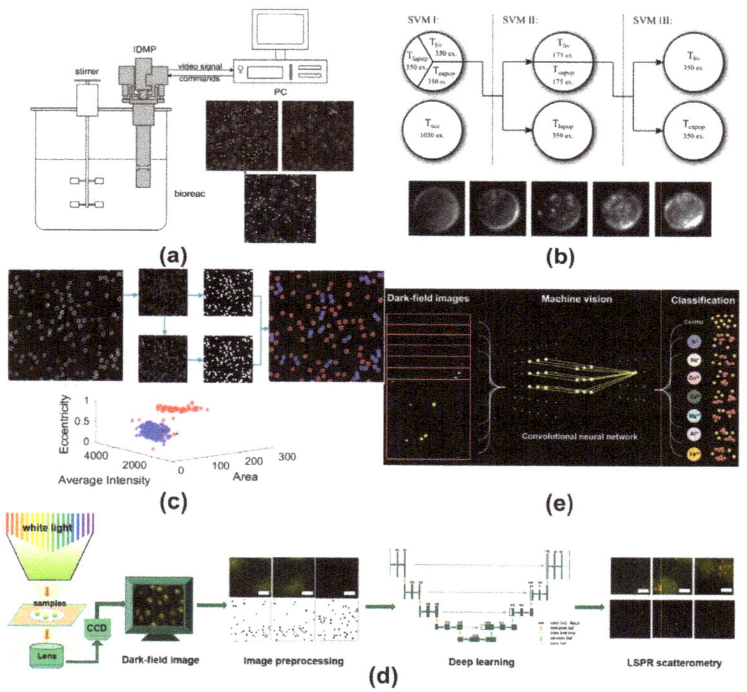

Figure 2. Application of machine learning in dark-field microscopic detection of cells. (**a**) A probe developed based on SVMs for the analysis of yeast cells under dark-field microscopic images. Reproduced from Ref. [28] with the permission of Wiley Analytical Science. (**b**) Cell density in dark-field microscopic images was estimated based on the CellViCAM system. Reproduced from Ref. [29] with the permission of Elsevier. (**c**) The morphologic information of red blood cells was obtained by using machine learning imaging under dark-field microscopy. Red represents single cells and blue represents multiple cells. Reproduced from Ref. [32] with the permission of Optica Publishing Group. (**d**) Detection of nanoparticle scattering in living cells using U-Net convolutional deep learning neural network technology and dark-field microscopic imaging. Reproduced from Ref. [36] with the permission of American Chemical Society. (**e**) Dark-field microscopy and an AlexNet model were used to predict AuNPs aggregation in salt solution. Reproduced from Ref. [37] with the permission of American Chemical Society.

3.3. Surface-Enhanced Raman Scattering

Nowadays, surface-enhanced Raman scattering (SERS) technology is used to study the spectrum of biomolecules in cells. The Raman effect is caused by the vibration of the molecule, and this technique can identify differences in the composition of the sample by the characteristic peak location and intensity. This method has high detection speed, high accuracy, and no sample pretreatment, so it has been more and more widely used in cell detection [38]. Different cells have different feature types, and the collected Raman spectral data are different. Machine learning can be used to quickly analyze the structure and composition of different feature peaks from a large number of Raman spectral data, so as to find the best feature data and obtain effective information faster. Therefore, after the introduction of machine learning technology, the detection effect of cells is better, the accuracy is higher, and the research value is very meaningful.

Ulcerative colitis (UC) is a chronic inflammatory disease, and evaluating the severity of UC is crucial for the treatment of the disease, but the biomolecular information related to UC cannot be found by conventional methods [39]. Kirchberger-Tolstik et al. used Raman spectroscopy to construct a 1D CNN model (Figure 3a) for predicting the Mayo endoscopic score in biopsies of patients with colonic inflammation [40]. The results show that many molecular changes, such as proteins, DNA, and lipids, occur during inflammation, and this work also has important implications for the diagnosis of other types of diseases.

Cyanobacteria are important microorganisms in photosynthesis, and the selection of mutant cells from them is of great significance in biogenetic studies [41]. Gao et al. used surface-enhanced Raman scattering technology to establish a support vector analyzer (SVC) model (Figure 3b) to distinguish wild-type and mutant cyanobacteria cells with an accuracy of 97% [42]. This high-throughput selection approach provides innovative strategies for genetic and cellular detection in biology.

Changes in cell secretions affect cell death, so the classification of cancer cells is beneficial for cancer treatment, but conventional methods are time-consuming and easily destroy cells [43,44]. Plou et al. used marker-free SERS to combine microfluidic and machine learning to quickly identify cell secretions under different conditions (Figure 3c), with an accuracy of 95% [45]. This scheme lays a good foundation for high-throughput cell detection.

Exosomes are extracellular vesicles with abundant cellular body fluids, which contain a lot of molecular information and are an important biomarker for cancer [46,47]. However, it is difficult to find comprehensive molecular information when detecting exosomes [48]. By using machine learning and SERS, Diao et al. used AuNPs as a basis to accurately detect exosomes in samples with high sensitivity (Figure 3d), and the prediction accuracy of cancer reached 91.1% [49]. This method has a wide application prospect for the diagnosis of cancer and even other diseases in the future.

Dopamine (DA), as a neurotransmitter, plays an important role in functional regulation [50]. However, it is difficult to determine the content of DA in exosomes secreted by cells. Lv et al. built an XGBoost model that can accurately identify exosome signals secreted by different cells (Figure 3e) and used nanotubes as an auxiliary method to accurately measure the DA content in exosomes by using continuous pulse current [51]. This strategy uses the continuous pulse current signal and can obtain a single dimension spectrum similar to the Raman spectrum. The spectrum with signal peaks can also be analyzed by it, which provides a new idea for chemical research.

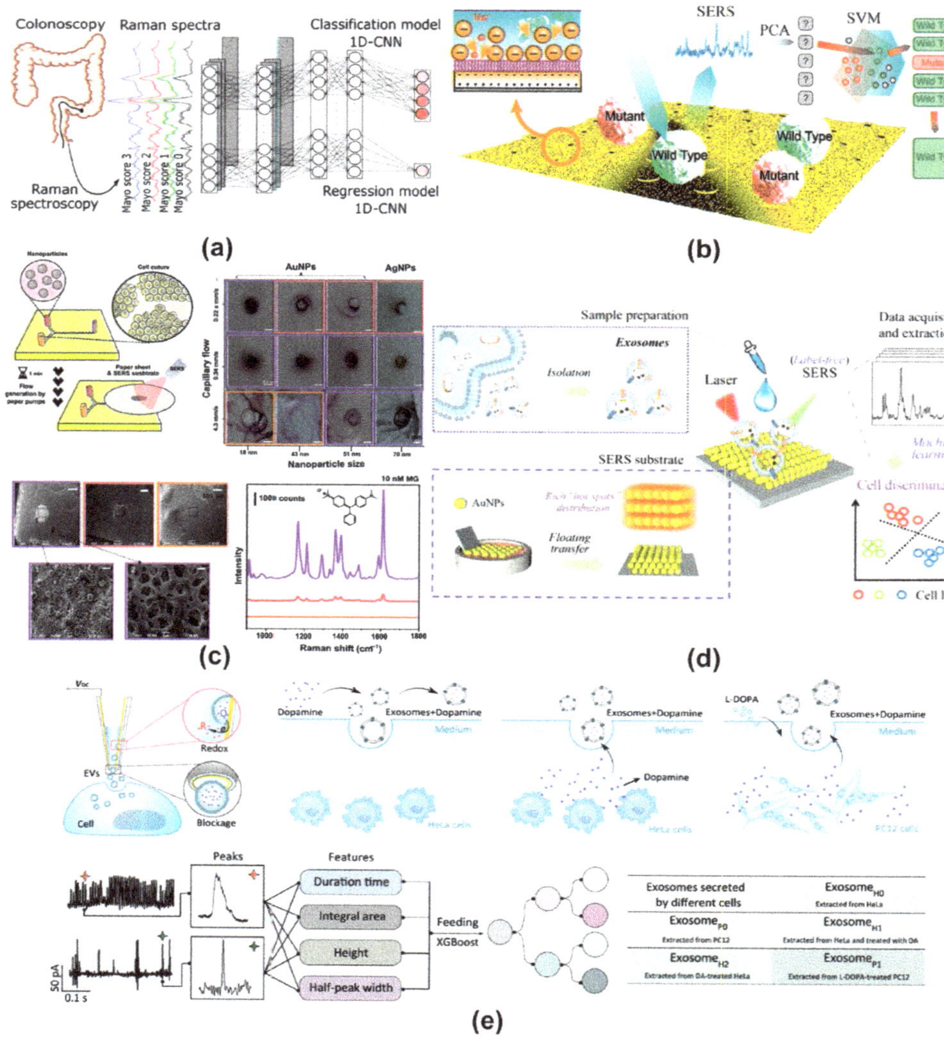

Figure 3. Application of machine learning to cell detection with surface-enhanced Raman scattering technique. (**a**) Raman spectrum and CNN model were used to classify the severity of UC. Reproduced from Ref. [40] with the permission of American Chemical Society. (**b**) SVC model and SERS were used to identify mutant cyanobacteria cells. Reproduced from Ref. [42] with the permission of American Chemical Society. (**c**) Rapid identification of cell secretions using marker-free SERS and machine learning. Reproduced from Ref. [45] with the permission of Wiley Online Library. (**d**) Highly sensitive detection of cellular exosomes using machine learning and SERS. Reproduced from Ref. [49] with the permission of American Chemical Society. (**e**) XGBoost model and continuous pulse current were used to determine the DA content in exosomes. Reproduced from Ref. [51] with the permission of American Chemical Society.

3.4. Fluorescence Detection

For cell detection, the fluorescent probe is also a good detection method, which has good specificity, high sensitivity, and strong tissue penetration [52]. However, it is sometimes limited by the high complexity of the sample, and the establishment of an algorithm model through machine learning can break through this limitation, so as to obtain a fluorescence probe with higher sensitivity and specificity, which makes the cell detection more accurate, allowing it to be widely used in the detection of various biological cells. In addition, fluorescent labeling methods based on optical imaging can make specific cells show fluorescent signals in order to clearly observe the behavior of cells. By using machine learning algorithms, fluorescent images can be automatically analyzed, specific objects can be identified after multiple training, quantitative information can be collected, and reliable data can be obtained quickly, which is conducive to cell analysis.

Non-melanoma skin cancer (NMSC) is a type of skin lesion, and its early diagnosis is crucial for the treatment of the disease [53]. NMSC is divided into various types, and it is difficult to evaluate the growth status of the cells [54]. Chen et al. used the stained cell sections for fluorescence imaging and built a linear support vector machine model (LSVM), which could accurately obtain different characteristics of skin cancer (Figure 4a), so as to detect cancerous cells with high sensitivity [55]. This strategy has great potential for the diagnosis of skin diseases.

Myelin is a lipid on the outside of neurons, and damage to it by heredity and external environments can lead to multiple sclerosis disease [56,57]. Drug treatments that regenerate myelin are urgently needed, so its detection is crucial [58]. Based on machine learning, Yeti et al. proposed a method for detecting myelin in fluorescence microscopic images (Figure 4b), which could extract feature data. In different machine learning technologies, the detection accuracies of the Boosted Trees model and CNN model were up to more than 98% [59]. This method facilitates the rapid screening of drugs for myelin regeneration.

Muscle stem cells (MuSCs) play an important role in the growth of skeletal muscle and contribute to the repair of damaged tissues; that is, they have strong regenerative capacity [60,61]. Therefore, the dynamic regulation of MuSCs contributes to the regeneration of skeletal muscle. Togninalli et al., using machine learning strategies, invented a microscope method for Dual-FLIT imaging (Figure 4c), which can track the dynamic growth of single cells at high resolution, thus regulating the dynamics of MuSCs [62]. This strategy helps us better understand the mechanisms of tissue cell regeneration.

Bladder cancer is a malignant tumor with a high incidence, and the detection of bladder cancer cells is helpful for early diagnosis and treatment [63]. Based on the SVM model in the machine algorithm, Zhang et al. developed a double-fluorescence micro-image flow cytometer (μ-FCM) for the high-throughput detection of bladder cancer cells (Figure 4d) and effectively reduced the overlap of different cells [64]. The system provides a new method for the detection of bladder cancer cells and the diagnosis of the disease.

3.5. Other Methods

The combination of machine learning and cell detection technology is diverse, and in addition to the combination method mentioned above, the research has more directions. Islam et al. developed a multi-head attention-based transformer model for finding plasmodium parasites from blood cell data using the gradient-weighted class activation graph technique with high test accuracy [65]. Karimzadeh et al. developed a multi-task generating artificial intelligence model for analyzing orphan non-coding RNA from patients with non-small cell lung cancer, which can detect early cancer with high sensitivity. Due to the large amount of blood sample data required in the experiment, this study can promote interdisciplinary cooperation in chemistry and medicine [66]. Pastuszak et al. developed a tree-based machine learning model for screening circulating tumor cells by analyzing single-cell RNA sequencing data [67].

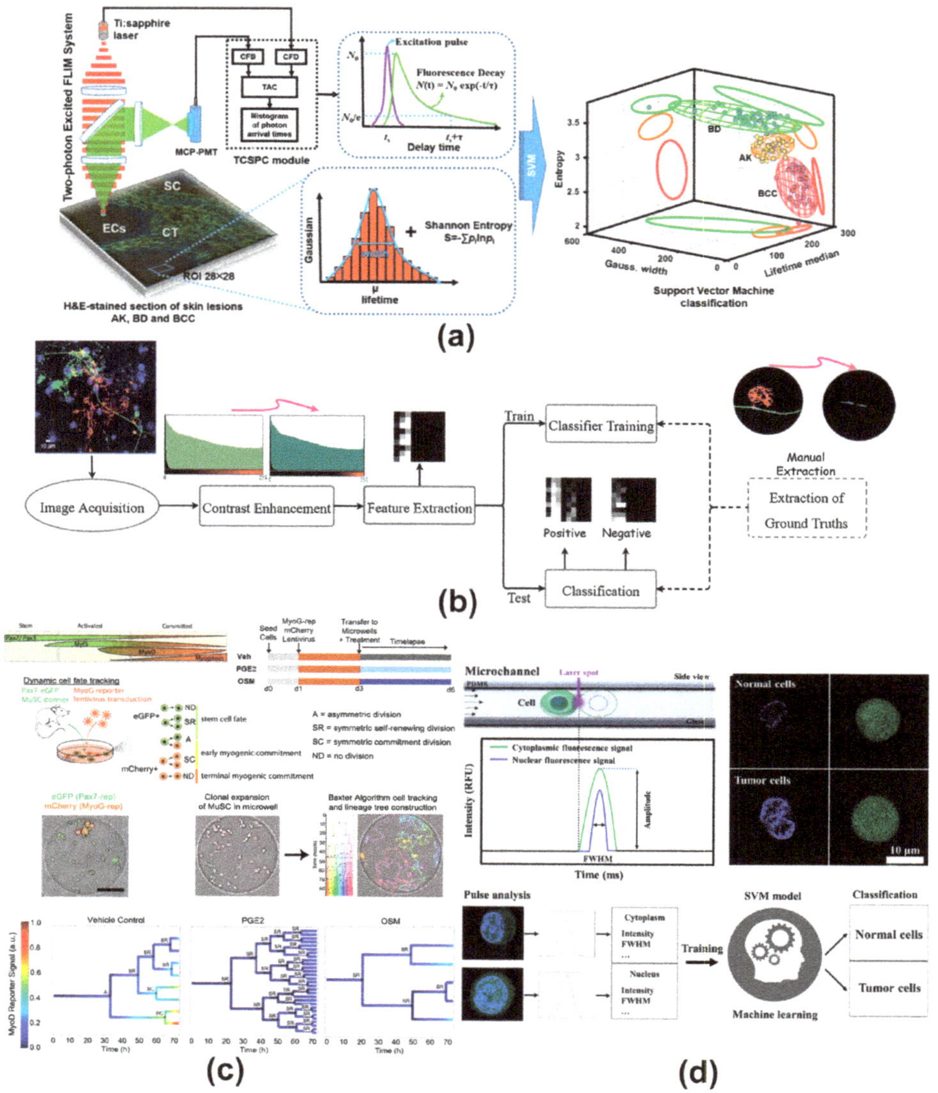

Figure 4. Application of machine learning to fluorescence detection of cells. (**a**) LSVM and fluorescence imaging were used to obtain characteristic information about different skin cancer cells. In the figure on the right, green represents Bowen's disease (BD), yellow represents actinic keratosis (AK) and red represents basal cell carcinoma (BCC). Reproduced from Ref. [55] with the permission of American Chemical Society. (**b**) Detection of myelin using machine learning models and fluorescence. Reproduced from Ref. [59] with the permission of Elsevier. (**c**) Dynamic regulation of MuSC using machine learning and Dual-FLIT. Reproduced from Ref. [62] with the permission of Nature. (**d**) SVM and μ-FCM were used to detect bladder cancer cells. Reproduced from Ref. [64] with the permission of Elsevier.

Combined with the detailed analysis of the latest progress, the machine learning model and detection means are diversified, and the detection means and machine learning model are selected according to the specific detection target. To improve detection accuracy, researchers often improve on the original machine learning model to develop new models that are conducive to cell detection. This type of research uses a variety of analytical chemical assays, cell, and blood sample data, so it can greatly facilitate interdisciplinary collaboration. The applications of machine learning models in chemistry and medicine are promising, but they also pose ethical challenges. The privacy of patient data is the most critical issue, with the risk of data breach or misuse in research. As we analyze data and develop machine learning models, we also need to ensure that patient confidentiality is maintained, either by encrypting or anonymizing the data. In short, the application of machine learning in cell detection has broad prospects and unlimited potential.

4. Discussion

The combination of machine learning and various cell detection methods, including bright-field, dark-field, SERS, and fluorescence techniques, has significantly improved the accuracy and efficiency of cell analysis. While traditional cell analysis methods cannot analyze the heterogeneity between cells, machine learning can analyze the data of individual cells to provide information at the single-cell level. In the single-cell analysis of tumor tissue, machine learning can excavate the function and interaction relationship of different cell groups through cluster analysis, providing new ideas and methods for tumor precision treatment. Machine learning is capable of processing and analyzing complex datasets, thus improving research efficiency and potentially revealing relationships at the cellular and molecular level that cannot be observed with traditional methods. Combined with cell analysis and machine learning, new biological phenomena and laws can be discovered to promote the further development of life science research. However, challenges remain. The performance of machine learning models is highly dependent on the quality and type of training data, and mislabeled or incomplete data can hinder model performance. Additionally, complex machine learning models often lack interpretability, making it difficult to balance accuracy with transparency.

To improve the reliability of machine learning models in cell detection, ensuring high-quality, diverse, and accurately annotated datasets is critical, as it enhances model generalizability and reduces bias. Techniques such as data augmentation and synthetic data generation can further expand dataset variety and prevent overfitting. Employing robust validation strategies like stratified k-fold cross-validation and leveraging ensemble learning can mitigate individual model weaknesses and improve overall robustness. Incorporating transfer learning from pre-trained models accelerates learning while enhancing performance on smaller datasets. For better data interpretation, explainable AI (XAI) techniques can be utilized to visualize and understand model decisions, while feature attribution methods, such as SHAP or Grad-CAM, help link model predictions to meaningful biological insights. Together, these approaches ensure both reliable performance and improved interpretability in cell detection tasks.

The future of machine learning in cell detection holds immense promise in terms of advancements in computational methods, imaging technologies, and biological research converge. Enhanced algorithms, including deep learning techniques like transformers, will improve the accuracy and speed of cell detection, enabling the real-time analysis and more precise identification of cell types and abnormalities. The integration of multimodal data from imaging techniques, genomics, proteomics, and transcriptomics will facilitate holistic cell profiling, linking cellular morphology with molecular states and behaviors. Machine learning will also play a vital role in personalized medicine by analyzing single-cell heterogeneity and predicting cellular responses to therapies, thus accelerating drug discovery and enabling tailored treatments. Automated, high-throughput systems powered by machine learning will handle large-scale cell imaging datasets, while edge computing and IoT devices will decentralize and democratize cell detection across laboratories worldwide. Additionally, machine learning models will improve the detection of rare cell types and complex interactions, providing critical insights into tissue organization, immune responses, and disease progression. Explainable AI (XAI) will enhance the transparency of these models, fostering trust in clinical applications and aiding researchers in uncovering new biological insights. Emerging technologies like quantum computing and neuromorphic chips will further revolutionize machine learning applications by increasing processing speed and energy efficiency. Ethical considerations, such as bias-free models and data privacy, alongside the proliferation of open-source tools and interdisciplinary education, will democratize access to machine learning in cell detection, ensuring global adoption. These advancements will transform not only basic research but also clinical diagnostics, drug development, and personalized medicine, driving novel therapeutic approaches and deeper insights into cellular mechanisms.

5. Conclusions

This review highlights the growing role of machine learning in cell detection technologies, focusing on four primary methods: bright-field, dark-field, SERS, and fluorescence detection. By leveraging advanced algorithms, researchers can now perform rapid, accurate, and high-throughput cell analysis. As artificial intelligence continues to evolve, its integration into cell detection technologies will likely yield new breakthroughs in both scientific research and clinical practice.

Author Contributions: X.L.: literature review, manuscript preparation, and writing—original draft. X.W. and R.Q.: writing—review editing and supervision. All authors have read and agreed to the published version of the manuscript.

Funding: This research was supported by National Natural Science Foundation of China (21977031), Science and Technology Commission of Shanghai Municipality (2018SHZDZX03, 24DX1400200), Shanghai Science and Technology Committee (22ZR1416800, 23ZR1416100), and the Fundamental Research Funds for the Central Universities (222201717003).

Institutional Review Board Statement: Not applicable.

Informed Consent Statement: Not applicable.

Data Availability Statement: No new data were created.

Conflicts of Interest: The authors declare no conflicts of interest.

References

1. Zhou, L.; Pan, S.; Wang, J. Machine learning on big data: Opportunities and challenges. *Neurocomputing* **2017**, *237*, 350–361. [CrossRef]
2. Ren, H.; Li, Z.N. Object detection using boosted local binaries. *Pattern Recogn.* **2016**, *60*, 793–801. [CrossRef]
3. Sangeetha, D.; Deepa, P. A low-cost and high-performance architecture for robust human detection using histogram of edge oriented gradients. *Microprocess. Microsyst.* **2017**, *53*, 106–119. [CrossRef]
4. Cheng, D.; Gong, Y.; Wang, J. Balanced mixture of deformable part models with automatic part configurations. *IEEE Trans. Circuits Syst. Video Technol.* **2016**, *27*, 1962–1973. [CrossRef]
5. Krizhevsky, A.; Sutskever, I.; Hinton, G.E. Imagenet classification with deep convolutional neural networks. *Commun. ACM* **2017**, *60*, 84–90. [CrossRef]
6. Li, Q.; Liang, A.; Liu, H. Hierarchical semantic segmentation of image scene with object labeling. *EURASIP J. Image Vide.* **2018**, *2018*, 15. [CrossRef]
7. Son, J.; Jung, H. Teacher–student model using grounding DINO and you only look once for multi-sensor-based object detection. *Appl. Sci.* **2024**, *14*, 2232. [CrossRef]
8. Xu, X.; Zhao, M.; Shi, P. Crack detection and comparison study based on faster R-CNN and mask R-CNN. *Sensors* **2022**, *22*, 1215. [CrossRef]

9. Ma, W.; Wang, X.; Yu, J. A lightweight feature fusion single shot multibox detector for garbage detection. *IEEE Access* **2020**, *8*, 188577–188586. [CrossRef]
10. Lima, L.R.; Godeiro, L.L. Equity-premium prediction: Attention is all you need. *J. Appl. Economet.* **2023**, *38*, 105–122. [CrossRef]
11. Svensson, C.M.; Krusekopf, S.; Lücke, J. Automated detection of circulating tumor cells with naive Bayesian classifiers. *Cytom. Part A* **2014**, *85*, 501–511. [CrossRef]
12. Freund, Y.; Schapire, R.E. A decision-theoretic generalization of on-line learning andan application to boosting. *J. Comput. Syst. Sci.* **1997**, *55*, 119–139. [CrossRef]
13. Pereira, S.; Pinto, A.; Oliveira, J. Automatic brain tissue segmentation in MR images using random forests and conditional random fields. *J. Neurosci. Meth.* **2016**, *270*, 111–123. [CrossRef] [PubMed]
14. Zeune, L.L.; Boink, Y.E.; Van, D.G. Deep learning of circulating tumor cells. *Nat. Mach. Intell.* **2020**, *2*, 124–133. [CrossRef]
15. Jia, D.; He, Z.; Zhang, C. Detection of cervical cancer cells in complex situation based on improved YOLOv3 network. *Multimed. Tools Appl.* **2022**, *81*, 8939–8961. [CrossRef]
16. Xing, F.; Yang, L. Robust nucleus/cell detection and segmentation in digital pathology and microscopy images: A comprehensive review. *IEEE Rev. Biomed. Eng.* **2016**, *9*, 234–263. [CrossRef] [PubMed]
17. Haldavnekar, R.; Venkatakrishnan, K.; Tan, B. Cancer stem cell derived extracellular vesicles with self-functionalized 3D nanosensor for real-time cancer diagnosis: Eliminating the roadblocks in liquid biopsy. *ACS Nano* **2022**, *16*, 12226–12243. [CrossRef] [PubMed]
18. Uslu, F.; Icoz, K.; Tasdemir, K.; Yilmaz, B. Automated quantification of immunomagnetic beads and leukemia cells from optical microscope images. *Biomed. Signal Proces.* **2019**, *49*, 473–482. [CrossRef]
19. Yamada, T.; Matsuda, A.; Koizumi, M.; Shinji, S.; Takahashi, G.; Iwai, T.; Yoshida, H. Liquid biopsy for the management of patients with colorectal cancer. *Digestion* **2019**, *99*, 39–45. [CrossRef] [PubMed]
20. Wang, S.; Zhou, Y.; Qin, X.; Nair, S.; Huang, X.; Liu, Y. Label-free detection of rare circulating tumor cells by image analysis and machine learning. *Sci. Rep.* **2020**, *10*, 12226. [CrossRef]
21. He, J.; Huisken, J. Image quality guided smart rotation improves coverage in microscopy. *Nat. Commun.* **2020**, *11*, 150. [CrossRef] [PubMed]
22. Asha, S.B.; Gopakumar, G.; Subrahmanyam, G.R.S. Saliency and ballness driven deep learning framework for cell segmentation in bright field microscopic images. *Eng. Appl. Artif. Intel.* **2023**, *118*, 105704. [CrossRef]
23. Zhuang, W.Z.; Lin, Y.H.; Su, L.J. Mesenchymal stem/stromal cell-based therapy: Mechanism, systemic safety and biodistri-bution for precision clinical applications. *J. Biomed. Sci.* **2021**, *28*, 28. [CrossRef]
24. Wiley, C.D.; Campisi, J. The metabolic roots of senescence: Mech-anisms and opportunities for intervention. *Nat. Metab.* **2021**, *3*, 1290–1301. [CrossRef]
25. Çelebi, F.; Boyvat, D.; Ayaz-Guner, S.; Tasdemir, K.; Icoz, K. Improved senescent cell segmentation on bright-field microscopy images exploiting representation level contrastive learning. *Int. J. Imaging Syst. Technol.* **2024**, *34*, e23052. [CrossRef]
26. Gao, P.F.; Lei, G.; Huang, C.Z. Dark-field microscopy: Recent advances in accurate analysis and emerging applications. *Anal. Chem.* **2021**, *93*, 4707–4726. [CrossRef] [PubMed]
27. Höpfner, T.; Bluma, A.; Rudolph, G.; Lindner, P.; Scheper, T. A review of non-invasive optical-based image analysis systems for continuous bioprocess monitoring. *Bioprocess. Biosyst. Eng.* **2010**, *33*, 247–256. [CrossRef] [PubMed]
28. Wei, N.; You, J.; Friehs, K.; Flaschel, E.; Nattkemper, T.W. An in situ probe for on-line monitoring of cell density and viability on the basis of dark field microscopy in conjunction with image processing and supervised machine learning. *Biotechnol. Bioeng.* **2007**, *97*, 1489–1500. [CrossRef]
29. Burgemeister, S.; Nattkemper, T.W.; Noll, T.; Hoffrogge, R.; Flaschel, E. CellViCAM—Cell viability classification for animal cell cultures using dark field micrographs. *J. Biotechnol.* **2010**, *149*, 310–316. [CrossRef]
30. Lafferty, J.D.; Crowther, M.A.; Ali, M.A.; Levine, M. The evaluation of various mathematical RBC indices and their efficacy in discriminating between thalassemic and non-thalassemic microcytosis. *Am. J. Clin. Pathol.* **1996**, *106*, 201–205. [CrossRef] [PubMed]
31. Zhou, Y.; Zhang, J.; Wang, C.; Zhou, L.; Zhou, L.; Ou, D.; Peng, D. Application of HbA2 levels and red cell indices-based new model in the differentiation of thalassemia traits from iron deficiency in hypochromic microcytic anemia Cases. *Int. J. Lab. Hematol.* **2020**, *42*, 526–532. [CrossRef]
32. Chen, X.; Luo, P.; Hu, C.; Yan, S.; Lu, D.; Li, Y.; Smith, Z.J. Nanometer precise red blood cell sizing using a cost-effective quantitative dark field imaging system. *Biomed. Opt. Express* **2020**, *11*, 5950–5966. [CrossRef] [PubMed]
33. Hafez, M.E.; Ma, H.; Ma, W. Unveiling the intrinsic catalytic activities of single-gold-nanoparticle-based enzyme mimetics. *Angew. Chem.* **2019**, *131*, 6393–6398. [CrossRef]
34. Ma, H.; Chen, J.F.; Wang, H.F. Exploring dynamic interactions of single nanoparticles at interfaces for surface-confined electro-chemical behavior and size measurement. *Nat. Commun.* **2020**, *11*, 2307. [CrossRef]

35. Zhang, Y.; Li, Q.; Zhou, K. Identification of specific N6-methyladenosine RNA demethylase FTO inhibitors by single-quantum-dot-based FRET nanosensors. *Anal. Chem.* **2020**, *92*, 13936–13944. [CrossRef] [PubMed]
36. Song, M.K.; Chen, S.X.; Hu, P.P.; Huang, C.Z.; Zhou, J. Automated plasmonic resonance cattering imaging analysis via deep learning. *Anal. Chem.* **2021**, *93*, 2619–2626. [CrossRef] [PubMed]
37. Wang, X.Y.; Hong, Q.; Zhou, Z.R.; Jin, Z.Y.; Li, D.W.; Qian, R.C. Holistic prediction of AuNP aggregation in diverse aqueous suspensions based on machine vision and dark-field scattering imaging. *Anal. Chem.* **2024**, *96*, 1506–1514. [CrossRef]
38. Aubertin, K.; Desroches, J.; Jermyn, M. Combining high wavenumber and fingerprint Raman spectroscopy for the detection of prostate cancer during radical prostatectomy. *Biomed. Opt. Express* **2018**, *9*, 4294–4305. [CrossRef] [PubMed]
39. Glick, L.R.; Cifu, A.S.; Feld, L. Ulcerative colitis in adults. *JAMA* **2020**, *324*, 1205–1206. [CrossRef] [PubMed]
40. Kirchberger-Tolstik, T.; Pradhan, P.; Vieth, M.; Grunert, P.; Popp, J.; Bocklitz, T.W.; Stallmach, A. Towards an interpretable classifier for characterization of endoscopic Mayo scores in ulcerative colitis using Raman spectroscopy. *Anal. Chem.* **2020**, *92*, 13776–13784. [CrossRef]
41. Sánchez-Baracaldo, P.; Cardona, T. On the origin of oxygenic photosynthesis and cyanobacteria. *New Phytol.* **2020**, *225*, 1440–1446. [CrossRef] [PubMed]
42. Gao, K.; Zhu, H.; Charron, B. Combining dense Au nanoparticle layers and 2D surface-enhanced Raman scattering arrays for the identification of mutant cyanobacteria using machine learning. *J. Phys. Chem. C* **2022**, *126*, 9446–9455. [CrossRef]
43. Medina, C.B.; Mehrotra, P.; Arandjelovic, S.; Perry, J.S.; Guo, Y.; Morioka, S.; Ravichandran, K.S. Metabolites released from apoptotic cells act as tissue messengers. *Nature* **2020**, *580*, 130–135. [CrossRef]
44. Han, S.; Shin, H.; Lee, J.K.; Liu, Z.; Rabadan, R.; Lee, J.; Nam, D.H. Secretome analysis of patient-derived GBM tumor spheres identifies midkine as a potent therapeutic target. *Exp. Mol. Med.* **2019**, *51*, 1–11. [CrossRef]
45. Plou, J.; Valera, P.S.; García, I.; Vila-Liarte, D.; Renero-Lecuna, C.; Ruiz-Cabello, J.; Liz-Marzán, L.M. Machine learning-assisted high-throughput SERS classification of cell secretomes. *Small* **2023**, *19*, 2207658. [CrossRef] [PubMed]
46. Deng, J.Q.; Zhao, S.; Li, J.H.; Cheng, Y.C.; Liu, C.; Liu, Z.; Li, L.L.; Tian, F.; Dai, B.; Sun, J.S. One-step thermophoretic and gate operation on extracellular vesicles improves diagnosis of prostate cancer. *Angew. Chem. Int. Ed.* **2022**, *61*, e202207037. [CrossRef] [PubMed]
47. Wang, H.Z.; Zeng, J.H.; Huang, J.; Cheng, H.; Chen, B.; Hu, X.; He, X.X.; Zhou, Y.; Wang, K.M. A self-serviced-track 3D DNA walker for ultrasensitive detection of tumor exosomes by glycoprotein profiling. *Angew. Chem.* **2022**, *134*, e202116932. [CrossRef]
48. Xie, Y.; Su, X.; Wen, Y.; Zheng, C.; Li, M. Artificial intelligent label-free SERS profiling of serum exosomes for breast cancer diagnosis and postoperative assessment. *Nano Lett.* **2022**, *22*, 7910–7918. [CrossRef] [PubMed]
49. Diao, X.; Li, X.; Hou, S.; Li, H.; Qi, G.; Jin, Y. Machine learning-based label-free SERS profiling of exosomes for accurate fuzzy diagnosis of cancer and dynamic monitoring of drug therapeutic processes. *Anal. Chem.* **2023**, *95*, 7552–7559. [CrossRef] [PubMed]
50. Zhang, X.; Liu, Q.; Liao, Q.; Zhao, Y. Potential roles of peripheral dopamine in tumor immunity. *J. Cancer* **2017**, *8*, 2966. [CrossRef] [PubMed]
51. Lv, J.; Wang, X.Y.; Chang, S.; Xi, C.Y.; Wu, X.; Chen, B.B.; Qian, R.C. Amperometric identification of single exosomes and their dopamine contents secreted by living cells. *Anal. Chem.* **2023**, *95*, 11273–11279. [CrossRef] [PubMed]
52. Liu, Y.; Li, Y.; Koo, S.; Sun, Y.; Liu, Y.; Liu, X.; Pan, Y.; Zhang, Z.; Du, M.; Lu, S.; et al. Versatile types of inorganic/Organic NIR-IIa/IIb fluorophores: From strategic design toward molecular imaging and theranostics. *Chem. Rev.* **2022**, *122*, 209–268. [CrossRef]
53. Finnane, A.; Dallest, K.; Janda, M.; Soyer, H.P. Teledermatology for the diagnosis and management of skin cancer: A systematic review. *JAMA Dermatol.* **2017**, *153*, 319–327. [CrossRef] [PubMed]
54. Srivastava, J.; Rho, O.; DiGiovanni, J. Twist1 regulates UVB-induced epidermal cell proliferation in non-melanoma skin cancer. *Cancer Res.* **2016**, *76*, 2022. [CrossRef]
55. Chen, B.; Lu, Y.; Pan, W.; Xiong, J.; Yang, Z.; Yan, W.; Qu, J. Support vector machine classification of nonmelanoma skin lesions based on fluorescence lifetime imaging microscopy. *Anal. Chem.* **2019**, *91*, 10640–10647. [CrossRef]
56. Reich, D.S.; Lucchinetti, C.F.; Calabresi, P.A. Multiple sclerosis. *N. Engl. J. Med.* **2018**, *378*, 169–180. [CrossRef] [PubMed]
57. Thompson, A.J.; Baranzini, S.E.; Geurts, J.; Hemmer, B.; Ciccarelli, O. Multiple sclerosis. *Lancet* **2018**, *391*, 1622–1636. [CrossRef] [PubMed]
58. Cole, K.L.; Early, J.J.; Lyons, D.A. Drug discovery for remyelination and treatment of MS. *Glia* **2017**, *65*, 1565–1589. [CrossRef]
59. Yetiş, S.Ç.; Çapar, A.; Ekinci, D.A.; Ayten, U.E.; Kerman, B.E.; Töreyin, B.U. Myelin detection in fluorescence microscopy images using machine learning. *J. Neurosci. Meth.* **2020**, *346*, 108946. [CrossRef]
60. Feige, P.; Brun, C.E.; Ritso, M.; Rudnicki, M.A. Orienting muscle stem cells for regeneration in homeostasis, aging, and disease. *Cell Stem Cell* **2018**, *23*, 653–664. [CrossRef] [PubMed]
61. Fukada, S.-I. The roles of muscle stem cells in muscle injury, atrophy and hypertrophy. *J. Biochem.* **2018**, *163*, 353–358. [CrossRef] [PubMed]

62. Togninalli, M.; Ho, A.T.; Madl, C.M.; Holbrook, C.A.; Wang, Y.X.; Magnusson, K.E.; Blau, H.M. Machine learning-based classification of dual fluorescence signals reveals muscle stem cell fate transitions in response to regenerative niche factors. *NPJ Regen. Med.* **2023**, *8*, 4. [CrossRef] [PubMed]
63. Bray, F.; Laversanne, M.; Sung, H.; Ferlay, J. Global cancer statistics 2022: GLOBOCAN estimates of incidence and mortality worldwide for 36 cancers in 185 countries. *A Cancer J. Clin.* **2024**, *74*, 229–263. [CrossRef]
64. Zhang, S.; Han, Z.; Qi, H.; Zhang, Z.; Zheng, Z.; Duan, X. Machine learning assisted microfluidics dual fluorescence flow cytometry for detecting bladder tumor cells based on morphological characteristic parameters. *Anal. Chem. Acta* **2024**, *1317*, 342899. [CrossRef]
65. Islam, M.R.; Nahiduzzaman, M.; Goni, M.O.F.; Sayeed, A.; Anower, M.S.; Ahsan, M.; Haider, J. Explainable transformer-based deep learning model for the detection of malaria parasites from blood cell images. *Sensors* **2022**, *22*, 4358. [CrossRef] [PubMed]
66. Karimzadeh, M.; Momen-Roknabadi, A.; Cavazos, T.B.; Fang, Y.; Chen, N.C.; Multhaup, M.; Goodarzi, H. Deep generative AI models analyzing circulating orphan non-coding RNAs enable detection of early-stage lung cancer. *Nat. Commun.* **2024**, *15*, 10090. [CrossRef]
67. Pastuszak, K.; Sieczczyński, M.; Dzięgielewska, M.; Wolniak, R.; Drewnowska, A.; Korpal, M.; Żaczek, A.J. Detection of circulating tumor cells by means of machine learning using Smart-Seq2 sequencing. *Sci. Rep.* **2024**, *14*, 11057. [CrossRef]

targets

MDPI

Review

Recent Development of Organic Afterglow Probes for Diagnosis and Treatment of Cancer

Meiqin Li, Le Tu, Huiling Wang, Junrong Li * and Yao Sun *

National Key Laboratory of Green Pesticide, College of Chemistry, Central China Normal University, Wuhan 430079, China; 18235519526@163.com (M.L.); tule@mails.ccnu.edu.cn (L.T.); wanghl@ccnu.edu.cn (H.W.)
* Correspondence: junrong.li@ccnu.edu.cn (J.L.); sunyaogbasp@ccnu.edu.cn (Y.S.)

Abstract: Afterglow imaging plays a crucial role in the cancer treatment field. In contrast to inorganic afterglow imaging agents, organic afterglow imaging agents possess easily modifiable structures and exhibit excellent biocompatibility, thereby presenting significant prospects for application in tumor diagnosis and management. In this review, we summarize the design principles and applications of afterglow probes in tumor imaging and therapy. Finally, we discuss the future challenges and prospects of organic afterglow probes in cancer diagnosis and therapy.

Keywords: afterglow probes; cancer diagnosis and therapy; activatable material; cancer disease

1. Introduction

Cancer is one of the leading causes of death worldwide, with approximately 19.3 million newly diagnosed cases and approximately 10 million deaths each year. Early diagnosis and treatment of cancer are beneficial to improve prognosis outcomes. At present, many cancer treatment and detection probes have been developed, such as optical probes, and magnetic probes which can analyze multiple genes in cancer cells and thus identify those mutations that can lead to the disease and so on [1,2]. Among them, optical imaging, which depends on optical probes, such as fluorescence imaging, chemi/bioluminescence imaging, and afterglow imaging, leverages photon detection to reveal molecular and biological processes in a real-time and noninvasive manner. It is very important in monitoring cellular-level physiological and pathological processes [3]. Fluorescence imaging has attracted increasing attention and has achieved real-time visualization in tumor treatment. [4–6]. However, although the development of the second near-infrared region (NIR II, 1000–1700 nm) fluorescence probes can significantly enhance imaging quality, the fluorescence imaging needs real-time light excitation, which limits the detection in depth and treatment efficacy and compromises the sensitivity and signal-to-background ratio (SBR) for in vivo imaging [7–10]. Chemi/bioluminescence imaging, which is also widely used in tumor imaging, often relies on chemical/enzyme-mediated reactions, and the signals can be influenced by the microenvironment within the biological system [11–13]. In contrast, afterglow luminescence, which can trap excitation energy in defects and slowly release photons after cessation of excitation, holds tremendous potential to detect and monitor diseases sensitively due to the elimination of auto-fluorescence [14–16]. With the rapid development of afterglow luminescent probes in recent years, photo-afterglow probes, sono-afterglow probes, and radio-afterglow probes have been applied in tumor diagnosis and treatment [17–19].

Organic afterglow probes usually include singlet oxygen ($^1O_2$) initiators and afterglow substrate. Under the irradiation of energy (including laser, ultrasound and X-ray), the ROS initiator generates ROS, especially singlet oxygen ($^1O_2$), to react with the high-energy chemical bond precursor structure of afterglow substrate to form high-energy intermediates. Finally, the intermediate breaks and generates fragments, then emits persistent luminescence. Recently, inorganic and organic afterglow imaging probes have been used for tumor imaging. However, inorganic afterglow imaging probes often have low targeting ability and contain metal ions that may leak out, raising concerns about systemic toxicity [20,21]. As compared to inorganic agents, organic afterglow luminescent probes demonstrate superior applicability in biomedicine, due to their advantages of excellent biocompatibility and easy functionalization [22–29]. In this review, we summarize the application of afterglow luminescence, which involves different energy sources such as light, ultrasound, and X-ray excitation, in tumor diagnosis and treatment, and describe their working mechanisms (Scheme 1). Finally, we summarize some features of these organic afterglow probes (Table 1) and offer some prospects and insights into these organic afterglow imaging agents.

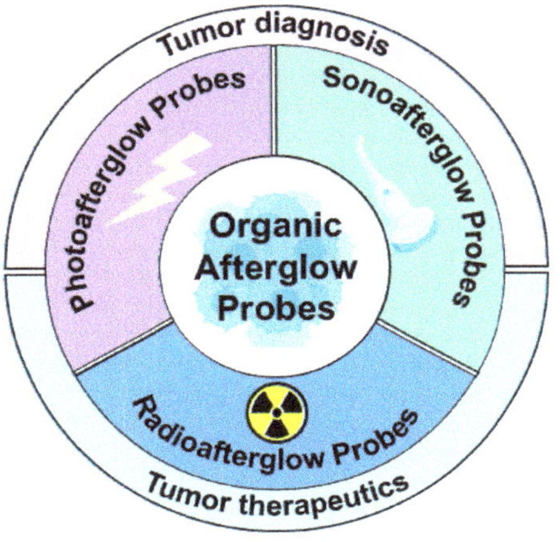

Scheme 1. Overview of organic afterglow probes for cancer diagnosis and treatments.

Table 1. Key characteristics of organic afterglow probes used for cancer diagnosis and treatment.

Excitation Source	Name of Afterglow Probe	Afterglow Emission	Half-Life
Photo-afterglow Probe	NPs-Ce4	680 nm	1.5 h
	PFVA-N-DO	780 nm	-
	AGL-AIE	550–850 nm	48 min
	TPT-DCM	630 nm	-
	TPP-DO	670 nm	-
	$F1^{2+}$-ANP	780 nm	6.6 min
	DBP_{OL}	720 nm	-
	B-AGL-HCPT	400–650 nm	118.5 min
Sono-afterglow Probe	PNCL	710 nm	-
	NPs-Ce6	680 nm	-
	NCBS/DPAs SNAP	780 nm	110 s
	TD	635–650 nm	180 s
Radio-afterglow Probe	IDPAs	624–792 nm	9.2–196.8 min

2. Photo-Afterglow Probes

Surgical resection of tumor lesions demonstrates effective local control and therapeutic outcomes. However, accurately identifying the margins of tumor lesions during surgery remains challenging. Image-guided surgical resection can significantly address this issue [30–33]. Afterglow imaging agents possess the capability to emit photons gradually even after the cessation of light excitation, offering a distinct advantage in the realm of ultra-sensitive detection [34,35]. Recently, a variety of inorganic and organic photo-afterglow imaging probes have been employed for the monitoring and treatment of tumors. Particularly, organic afterglow probes with molecule-level specificity and minimal side effects are attracting increasing attention. Currently, to improve the signal-to-background ratio and sensitivity of imaging, many organic photo-afterglow probes have been developed. In 2022, Miao et al. [36] first reported a series of chlorin-based afterglow agents (Ppa, Cp6, Ce6, Ce4) and then they used an amphiphilic triblock copolymer PEG-b-PPG-b-PEG as a surface-capping agent to enhance the water solubility of the compound (Figure 1A,B). Among them, the NPs-Ce4 exhibited the strongest afterglow luminescence and an ultralong half-life (1.5 h), which was attributed to its low highest occupied molecular orbital (HOMO) levels (−5.08 ev) (Figure 1C). A mechanism study showed that NPs-Ce4 could produce $^1O_2$ under light excitation to oxidize the vinylene bond (C=C) of NPs-Ce4 to form an active intermediate species followed by afterglow luminescence (Figure 1D). An in vivo imaging study indicated that the afterglow intensity was 26 times higher than that of the fluorescence signal after subcutaneous injection of NPs-Ce4. Furthermore, they injected NPs-Ce4 through the tail vein to detect the 4T1 tumors of different sizes. The result shows that the probe can identify tumors as small as 3 mm^3, facilitating imaging-guided surgical resection of the tumor foci. In 2019, Pu et al. [37] selected RB (rose bengal octyl ester), TPP (meso-tetraphenylporphyrin), NCBS (silicon 2,3-naphthalocyanine bis(trihexylsilyloxide) as the afterglow initiators. DO (*N*, *N* dimethyl-4-(3-phenyl-5,6-dihydro-1,4-dioxin-2-yl) aniline), SO (*N*, *N*-dimethyl-4-(2-phenyl-5,6-dihydro-1,4-oxathiin-3-yl) aniline), and HBA (3-((1r,3r,5R,7S)-adamantan-2-ylidene (methoxy) methyl) phenol) were employed as the afterglow substrates. Semiconducting polymers (SPs), small molecule dyes (SMDs), and inorganic fluorophores (IFs) served as relay units for the afterglow process (Figure 1E,F). A series of afterglow luminescent nanoparticles (ALNPs) were synthesized via co-nanoprecipitation utilizing an amphiphilic copolymer, PEG-b-PPG-b-PEG. Among them, PFVA-N-DO was selected to detect tumors in mice. In vivo tumor imaging shows that at one hour post-injection, the signal-to-noise ratio of the afterglow signal at the tumor region (2922 ± 121) was 3 orders of magnitude higher than that of the fluorescence signal (~1) (Figure 1G). Moreover, the ALNPs were mainly cleared by the liver and had good biosafety.

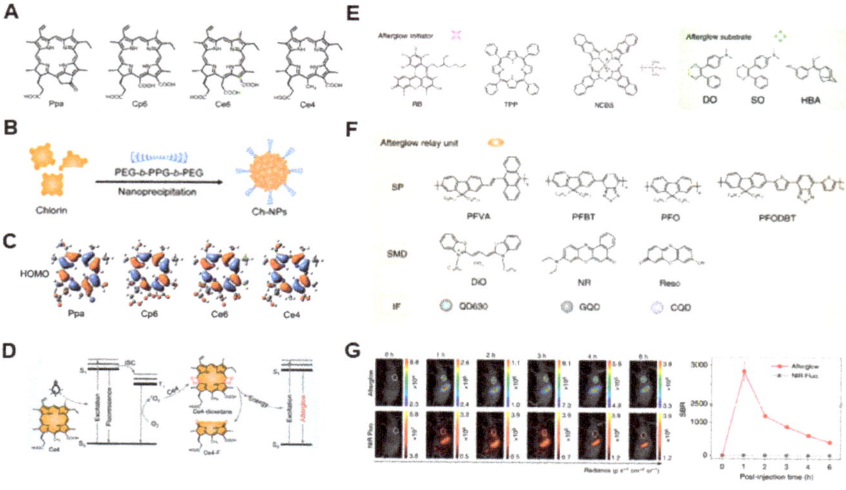

Figure 1. (**A**) Chemical structures of chlorins. (**B**) Synthesis route of Ch-NPs. (**C**) The HOMO levels of chlorins. (**D**) Luminescence mechanism of NPs-Ce4 [36]. Reprinted (adapted) with permission from Ref. [36]. Copyright 2022 American Chemical Society. Chemical structures of (**E**) afterglow initiators, afterglow substrates, and (**F**) afterglow relay unit. (**G**) In vivo tumor afterglow imaging and fluorescence imaging [37]. Reprinted (adapted) with permission from Ref. [37]. Copyright 2019 Springer Nature.

Moreover, aggregation-induced emission (AIE) molecules have been widely used in bioimaging due to their bright luminescence and high photostability [38]. Benefiting from the excellent optical properties of the AIEgens, Ding et al. [39] synthesized an AIE probe with near-infrared afterglow luminescence properties (AGL AIE dots) (Figure 2A). The TPE-Ph-DCM moiety could produce $^1O_2$ under light excitation to oxidize an enol ether precursor, forming an unstable intermediate with high energy and then emitting persistent luminescence. The luminescence could continue more than 10 days after the cessation of light and penetrate up to 7 mm of tissue. For in vivo imaging, the tumor-to-liver ratio of the afterglow signal was nearly 100-fold larger than that of the fluorescence signal, indicating excellent performance in image-guided cancer surgery with the AIE dots. Another example is from the work of Zhang et al. [40] who encapsulated photosensitizer 2-((5-(4(diphenylamino)phenyl)thiophen-2-yl)methylene)malononitrile (TTMN) and afterglow metricspoly[(9,9-di(2-ethylhexyl)-9H-fluorene-2,7-vinylene)-co-(1-methoxy-4-(2-ethylhexyloxy)-2,5phenylenevinylene)] (PFPV) with F127 to construct AIE afterglow imaging probe P-TNPs. They found that resonance energy transfer existed between TTMN and PFPV. After irradiation by a white LED lamp, the afterglow of PFPV-dioxetane intermediates could irradiate TTMN again to produce $^1O_2$ (Figure 2B). This type of closed-loop probe can achieve effective imaging of subcutaneous tumors after injection through the tail vein. In order to improve the luminescence properties of AIE molecules, Liu et al. [41] balanced the twist and conjugation of molecules and reported a luminescent core (TPT), which possesses a higher molar extinction coefficient (ε) and fluorescent brightness than the conventional triphenylamine (TPA) and tetraphenylethene (TPE). Moreover, they introduced a strong electron acceptor into the TPT core to construct a D-A near-infrared AIE molecule (TPT-DCM) with a triazole luminescent core (Figure 2C), featuring a high molar extinction coefficient, high brightness, and efficient reactive oxygen species generation efficiency, which is attributed to the decreased energy gap (ΔE_{ST}). Furthermore, they encapsulated TPT-DCM and Schaap's dioxetane precursor with DSPE-PEG 2000. The TPT-DCM can be light-irradiated to generate $^1O_2$ to convert Schaap's dioxetane precursor to Schaap's dioxetane and then generate an afterglow with long time (up to 20 days), tumor-to-liver signal ratio (up to 187) and deep tissue penetration (1.6 cm). To enhance the effectiveness at the in vivo level, they synthesized a nanoparticle by encapsulating the TPT-DCM and AGL (enol ether precursor of Schapp's dioxetane) with DSPE-PEG 2000; the nanoparticle was termed as TPT-DCM/AGL NPs. Notably, the tumor with a tiny size of <1 mm can be successfully detected by injecting nanoprobes into 4T1 tumor-bearing Balb/c mice through the tail vein. In 2024, Miao et al. [42] synthesized a single molecule (TPP-DO) by the covalent bonding of the afterglow substrate (DO) with the afterglow initiator (mesotetraphenyl porphyrin). To increase the hydrophilicity of TPP-DO, they co-assembled the TPP-DO with an amphiphilic triblock copolymer PEG-b-PPG-b-PEG to form TPP-DO NPs (Figure 2D). The covalent bond between the afterglow substrate and the energy transfer unit enhances intramolecular energy transfer efficiency, resulting in increased afterglow brightness. The mechanism study indicated that the TPP-DO NPs can be light-irradiated for TPP to generate $^1O_2$ to convert the DO to its 1,2-dioxetane intermediate TPP-DO-dioxetane, which decomposed into TPP-DE, which can absorb the released energy to form excited TPP-DE*. The energy transfer excites the TPP moiety, producing bright NIR luminescence. An in vitro study showed that the TPP-DO NPs can penetrate a thickness of 6 cm with a SBR of 35. The in vivo imaging can be conducted even at TPP-DO NP concentrations as low as 0.1 nM. Furthermore, they prepared a GSH-activatable afterglow probe (Q-TPP-DO NPs) by co-precipitating the DSPE-PEG2000-ssBHQ3 and the TPP-DO (Figure 2E) for the detection of subcutaneous tumors as small as 0.048 mm^3.

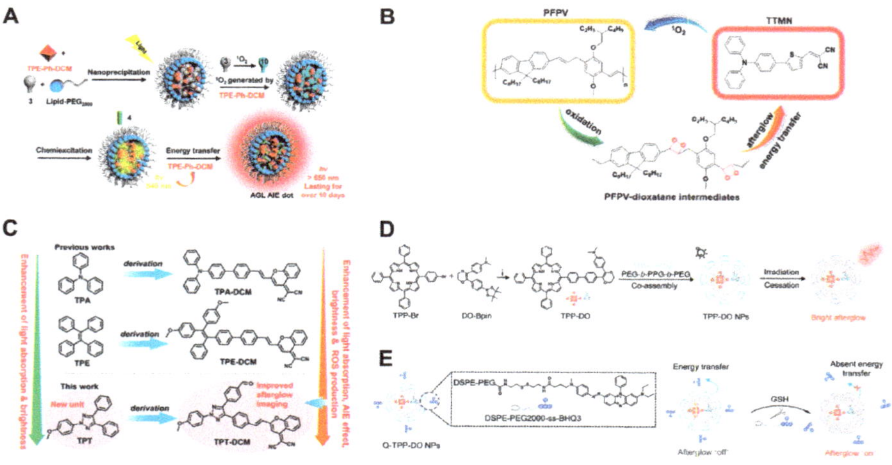

Figure 2. (**A**) Synthetic route and luminescence mechanism of AGL AIE dot [39]. Reprinted (adapted) with permission from Ref. [39]. Copyright 2019 American Chemical Society. (**B**) Closed-loop luminescence mechanism of P-TNPs [40]. Reprinted (adapted) with permission from Ref. [40]. Copyright 2020 Royal Society of Chemistry. (**C**) The chemical structures of TPA, TPE, and TPT [41]. Reprinted (adapted) with permission from Ref. [41]. Copyright 2023 Wiley-VCH. The synthetic route of (**D**) TPP-DO NPs and (**E**) Q-TPP-DO NPs [42]. Reprinted (adapted) with permission from Ref. [42]. Copyright 2024 Wiley-VCH.

Aiming at achieving precise tumor imaging with high SBR of tumors, it is essential to enhance the in vivo targeting ability of imaging probes. Ye et al. [43] developed an H_2S-activated afterglow probe ($F1^{2+}$-ANP) by integrating organic electrochromic material (EM $F1^{2+}$) into a NIR photosensitizer (silicon 2,3-naphthalocyanine bis(trihexylsilyloxide) and (poly[2-methoxy-5-(2-ethylhexyloxy)1,4-phenylenevinylene]) containing nanoparticles (Figure 3A). In the presence of H_2S, the FRET (fluorescent resonance energy transfer) process was eliminated, leading to the recovery of $^1O_2$ production in F2-ANP under 808 nm light irradiation. In addition, the $F1^{2+}$-ANP possessed positive reduction potential E^{red} = +0.40 V, resulting in a faster reaction to H_2S with a second-order reaction rate $k_2 = 3600 \pm 130\ M^{-1}s^{-1}$. Under the 808 nm irradiation, the NIR775 produced $^1O_2$ to oxidize the vinylene bond of MEH-PPV, accompanied by the emission of photons at 580 nm. This process was followed by energy transfer to NIR775, which ultimately activates the NIR afterglow luminescence at 780 nm. To enable precise detection of H_2S in HCC HepG2 cells, the β-Gal was introduced to the surface of $F1^{2+}$-ANP to prepare $F1^{2+}$-ANP-Gal (Figure 3B). They further used the $F1^{2+}$-ANP-Gal probe for the imaging of subcutaneous and orthotopic HepG2 tumors (<3 mm in diameter) in mice. It is worth noting that $F1^{2+}$-ANP-Gal precisely delineated tumor margins in excised hepatic cancer specimens, which may be used for intraoperative guidance in liver cancer surgery. To break the limitation of energy transfer distance between the donor and the acceptor, Miao et al. [44] developed an organic PH-activatable upconversion afterglow luminescence cocktail of nanoparticles (ALCNs) consisting of an "energy" donor (AIN) and an "energy" acceptor (ASN) (Figure 3C). In the acidic microenvironment, the amide bond of ASN was broken and the surface charge of the nanoparticles changed from negative to positive, which reduced the transfer distance of singlet oxygen ($^1O_2$) in the aqueous solution, so that the afterglow substrate was oxidized to intermediate, and then produced afterglow. An in vitro afterglow imaging study proved that the intensity of the afterglow signal was proportional to the number of 4T1 cells and HepG2 cells. Furthermore, the ALCNs successfully achieved afterglow imaging of 4T1 subcutaneous tumors and orthotopic liver tumors. Notably, bio-ALCNs exhibited superior tumor-targeting capability and enhanced tumor-imaging performance compared to ALCNs.

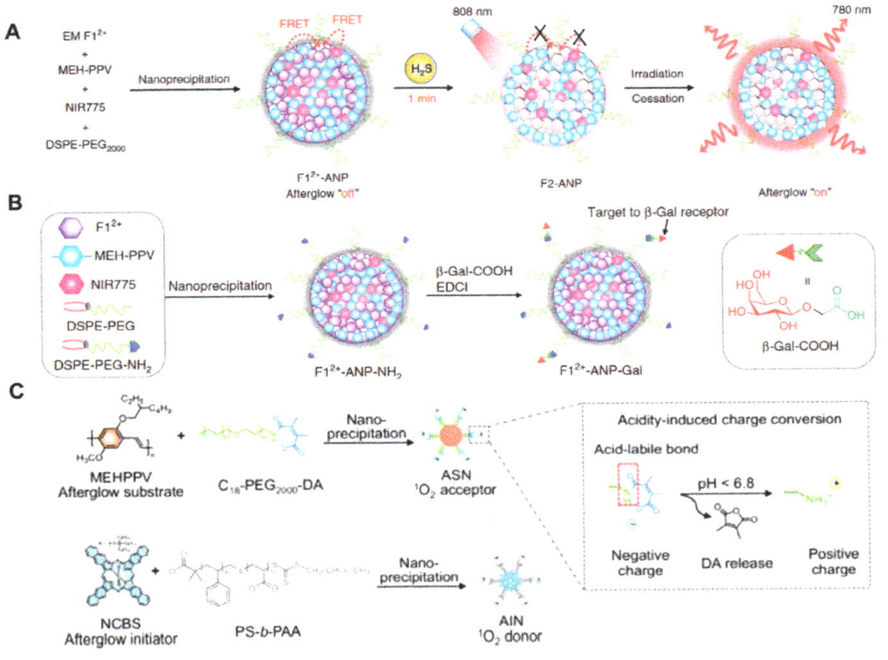

Figure 3. Synthesis route and response mechanism of (**A**) F1^{2+}-ANP and (**B**) F1^{2+}-ANP-Gal to hydrogen sulfide [43]. Reprinted (adapted) with permission from Ref. [43]. Copyright 2020 Springer Nature. (**C**) Synthetic route and PH response mechanism of ALCNs [44]. Reprinted (adapted) with permission from Ref. [44]. Copyright 2024 Springer Nature.

Due to the generation of ROS, afterglow luminescent probes can also be used for photodynamic therapy of tumors. As an exploration, Zeng successfully integrated the acceptor moiety TBQ, which possessed $^1O_2$ generation capabilities, with the donor moiety DCL featuring an $ONOO^-$-responsive group, into a nanoparticle [45] (Figure 4A). This innovative construct enabled in vitro detection of ONOO- with a limit of detection (LOD) of 46.1 nM and achieved a penetration depth of 2 cm, thereby facilitating effective photodynamic therapy in vivo. Tang et al. [46] reported two novel photochemical afterglow luminescent nano-photosensitizers (ABEI-TPA and Iso-TPA) (Figure 4B). Two PSs with extended afterglow were created using TMAH and isoluminol derivatives through conjugate and non-conjugate links. While ABEI-TPA possessed higher photoluminescence quantum efficiency, Iso-TPA exhibited better effects of luminescent bioimaging such as high SNR and deep tissue penetration (32 mm), which were attributed to the higher ROS generation efficiency. Consequently, in vivo experiments indicated that two PS-based NPs possessed effective PDT therapeutic efficacy. Furthermore, activatable phototheranostic probes, characterized by specific "turn-on" signals, minimized tissue auto-fluorescence and side effects, have been used for phototheranostics of the tumor. In 2023, Pu et al. [47] synthesized a chemiluminophore (DBP_O) by replacing the malononitrile group of dicyanomethylene-4H-benzopyran-phenoxyl-dioxetane (DBP) with A1,3-dimethyl barbituric acid (an electron-withdrawing group (EWGs) (Figure 4C,D). They further introduced the lysyl oxidase (LOX)-responsive group propylamine into DBP_O to construct the LOX-activatable luminescent probe (DBP_{OL}) with NIR afterglow and $^1O_2$ production after laser irradiation. In the presence of lysyl oxidase (LOX), the CL signal and photodynamic activity could be turned on, enabling imaging-guided PDT therapy in vivo. As an exploration, Ding et al. [48] developed an afterglow $ONOO^-$-activatable theranostic nanoprobe AIE/B -AGL-HCPT NPs for inducing Immunogenic Cell Death (ICD), monitoring the process of ICD and the cold-to-hot tumor transformation. The AIE/B -AGL-HCPT NPs consisted of afterglow hydroxycamptothecin prodrug (B-AGL-HCPT) with the $ONOO^-$-activatable group and AIE photosensitizer ((TPE-DPA)$_2$-Py). Under the irradiation of light, the AIE photosensitizer generated $^1O_2$ to evoke ICD and oxidize the B-AGL-HCPT, forming a 1,2-dioxetane derivative, which had neither afterglow nor therapeutic effect except to retain the fluorescent signal (Figure 4E,F). With the development of ICD, the cold tumor can be converted to a hot tumor with increased $ONOO^-$ levels. The $ONOO^-$ can cleave the B-AGL-HCPT, thereby uncaging the afterglow unit, which can transfer the energy to (TPE-DPA)$_2$-Py to emit NIR afterglow (>650 nm), while also releasing the HCPT to enhance the PDT-mediated ICD process. Furthermore, in vivo experiments demonstrated that AIE/B -AGL-HCPT NPs can effectively eliminate tumors and prevent tumor recurrence in the 4T1 tumor-bearing mice.

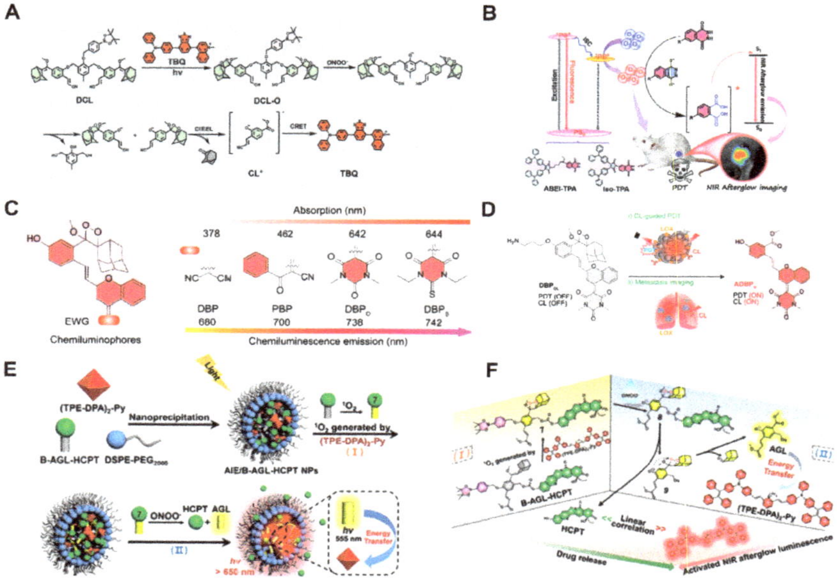

Figure 4. (**A**) Illustration of afterglow luminescence of DCL/TBQ activated by $ONOO^-$ [45]. Reprinted (adapted) with permission from Ref. [45]. Copyright 2024 Wiley-VCH. (**B**) Schematic of in vivo luminescence imaging and PDT therapy of ABEI-TPA and ISO-TPA [46]. Reprinted (adapted) with permission from Ref. [46]. Copyright 2024 Wiley-VCH. (**C**) Molecular design of different chemiluminophores. (**D**) Illustration of LOX-activated afterglow imaging and PDT treatment [47]. Reprinted (adapted) with permission from Ref. [47]. Copyright 2023 Wiley-VCH. (**E**) Illustration of afterglow luminescence of AIE/B-AGL-HCPT NPs activated by $ONOO^-$. (**F**) The molecular structure conversion for the afterglow luminescence processes [48]. Reprinted (adapted) with permission from Ref. [48]. Copyright 2022 Wiley-VCH.

3. Sono-Afterglow Probes

Despite the elimination of auto-fluorescence of tissue, limited penetration (1~2 cm), light-scattering, and reabsorption in tissue still exist in photo-afterglow imaging. Conversely, ultrasound-triggered luminescence, which is characterized by light emission resulting from the energy-releasing process of piezoelectric inorganic materials, has great potential for application in biological imaging owing to its noninvasiveness, high spatiotemporal, deep penetrability (up to 10 cm), precise localization of lesions, and low cost [49–52].

To achieve ultrasound-activated luminescence imaging of tumor sites, Song et al. [53] reported an ultrasound-activated NIR chemiluminescence PNCL, which was prepared with sonosensitizer protoporphyrin IX (PpIX) and an enol ether precursor of Schaap's dioxetane containing adicyanomethyl chromone (DCMC) acceptor scaffold (NCL) through click reaction. The PNCL could generate singlet oxygen under ultrasound irradiation (1 MHz). Oxidation of the enol ether of the NCL moiety by $^1O_2$ resulted in the forming of active intermediates and then the producing of luminescence. The strongest afterglow luminescence peaked at 710 nm (Figure 5A), enabling US-activated afterglow imaging in deep tissue (~2 cm). Furthermore, bright chemiluminescence was detected in vivo after tail vein injection of PNCL (Figure 5B). This strategy achieves ultrasound-induced luminescence imaging of the tumor. Aiming at mitigating the hypoxic microenvironment present in tumors and enhancing therapeutic outcomes for tumor treatment, Chen et al. [54] constructed a probe by combining Synechococcus elongatus PCC 7942 (PCC) with sono-afterglow Ce6 nanoparticles (NPs-Ce6). Under ultrasound stimulation, the afterglow emitted by NPs-Ce6 facilitates photodynamic conversion to generate oxygen, thereby alleviating the hypoxic microenvironment within the tumor. Concurrently, the singlet oxygen produced through ultrasound exposure can induce ferroptosis in tumor cells (Figure 5C).

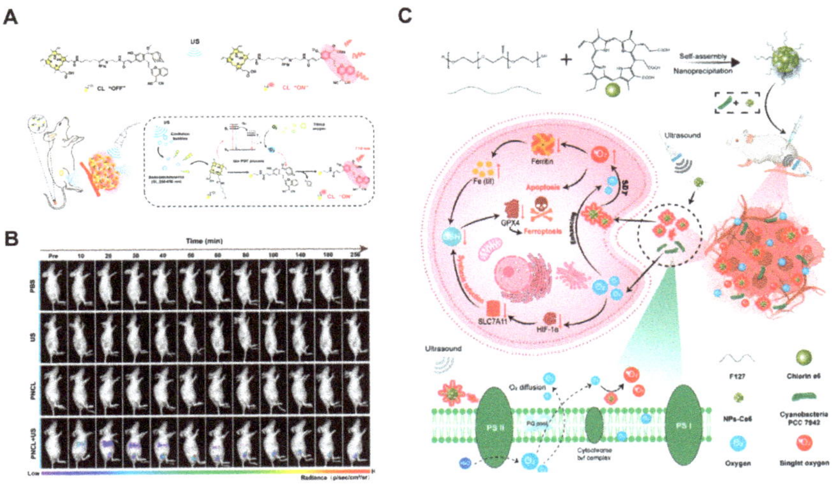

Figure 5. (**A**) Mechanism illustration of sono-afterglow of PNCL for in vivo imaging. (**B**) Sono-afterglow imaging of 4T1 tumor at different times [53]. Reprinted (adapted) with permission from Ref. [53]. Copyright 2023 American Chemical Society. (**C**) Schematic illustration of preparation of NPs-Ce6 and antitumor mechanisms [54]. Reprinted (adapted) with permission from Ref. [54]. Copyright 2024 Wiley-VCH.

The immune system can maintain homeostasis in tumor initiation and development, and many activatable molecular sono-afterglow probes have been reported to evaluate immunotherapy and early diagnosis of tumors [55,56]. In 2022, Pu et al. [57] reported a NIR-I molecular sono-afterglow probe SNAP (peaked at 780 nm), which was designed by silicon 2,3-naphthalocyanine bis(trihexylsilyloxide) (NCBS) as a sonosensitizer moiety, and dicyanomethylene-4H-benzothiopyran-phenoxyl-adamantylidene (DPAs) as a sono-afterglow substrate (Figure 6A). When the ultrasound condition was 2.0 W cm^{-2} for 30 s, the mass ratio between NCBS and DPAs was 1:5, and the sono-afterglow signal of SNAP could be strong enough to penetrate tissue to a depth of 4 cm. To further evaluate the performance of biomarker-activated sono-afterglow imaging, they synthesized an activatable sono-afterglow probe SNAP-M, comprising a singlet oxygen initiator (NCBS) and a sono-afterglow substrate (DPAs), which is caged with $ONOO^-$ responsive moiety (Pro-DPAs) (Figure 6B). The afterglow substrate could be activated by overproduced $ONOO^-$ in the pro-inflammatory tumor microenvironment and subsequently reacted with $^1O_2$, generated from NCBS under ultrasound irradiation, to produce long-lasting and deep tissue luminescence, which could be used to specifically and sensitively (LOD down to 0.3 μM) detect $ONOO^-$ in deep tissues (4 cm). To further accelerate the in vivo use of sono-afterglow nanoparticles, SCAN, consisting of Pro-MB, Pro-R837, and DPAs, has been reported to achieve tumor immunotherapy. The MB could be activated in the $ONOO^-$ overproduced tumor microenvironment and produced $^1O_2$ under ultrasound condition (5 min, 2.0 W cm^{-2}). Due to the $^1O_2$ effect, R837 and DPAs were liberated from SCAN in situ for immunotherapy and accurate sono-afterglow imaging of subtle $ONOO^-$ molecular changes in deep tumor tissue, respectively (Figure 6C). Moreover, Tan et al. [58] synthesized luminescence nanoparticle TD NPs, which enable early detection of tumors and monitoring cancer treatment. Due to the piezocatalytic effect of ultrasound, TD NPs could generate a large amount of ROS ($^1O_2$ and HO•) under ultrasonic excitation and then react with the TD to produce luminescence with a peak at 625~650 nm, which exhibited excellent tissue penetration (up to 2.2 cm) (Figure 6D). Whether in delayed or real-time imaging mode, the TD had excellent sono-afterglow performance. In the delayed imaging mode, its luminescence intensity was 2389 times higher than that of H_2O, and in the real-time imaging mode, it was 1428 times higher. To investigate the luminescence ability in vivo, orthotopic GBM-bearing mice and pancreatic tumor-bearing mice were intraperitoneally injected with TD NPs, and then dynamically enhanced signals were observed. Moreover, the nanoparticle could produce strong sono-afterglow in metastatic tumors and lymph nodes. These results demonstrated the ability of in vivo imaging. Furthermore, based on the critical role of granzyme B in cell death through cytotoxic lymphocytes, the team further constructed a granzyme B-activated TD-Grz-BHQ probe for the precise immune response differentiation of different tumor types (Figure 6E).

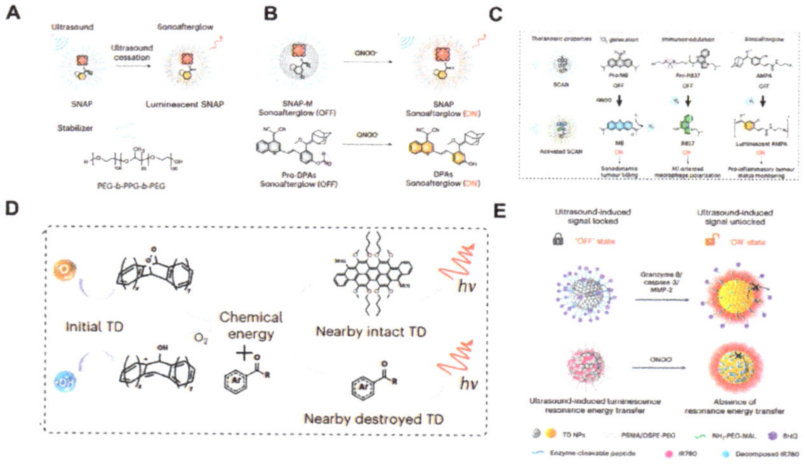

Figure 6. Mechanism illustration of sono-afterglow imaging for (**A**) SNAP, (**B**) SNAP-M, (**C**) SCAN [57]. Reprinted (adapted) with permission from Ref. [57]. Copyright 2022 Springer Nature. Molecule mechanism of (**D**) TD NPs (**E**) TD-Grz-BHQ for sono-afterglow imaging [58]. Reprinted (adapted) with permission from Ref. [58]. Copyright 2024 Springer Nature.

4. Radio-Afterglow Probes

As invisible high-energy rays, X-rays have no penetration limitation, making X-ray-activated afterglow imaging useful for diagnostic imaging. Moreover, similar to photodynamics, X-ray can also enable radiodynamic therapy (RDT) of radio-afterglow probes by initiating the photodynamic process of optical agents [59–66]. However, it is still limited to a few inorganic nanophosphors [67]. Recently, Pu et al. [68] reported a series of organic radio afterglow probes with tunable afterglow emission (624 nm~792 nm) and $^1O_2$ generation for cancer radiodynamic theranostics (Figure 7A,B). Mechanistic studies showed that the generation of $^1O_2$ by IDPA$_{Su}$ was the highest, which enabled stronger radio afterglow. The IDPA$_{Su}$ exhibited deep tissue penetration (15 cm) in vitro and in vivo. To allow precise cancer diagnosis and therapy, the authors synthesized an organic radio afterglow dynamic probe (MRAP), comprising a tumor-targeting moiety (cyclic arginine-glycine-aspartate (cRGD)), a theranostic moiety and a CatB-cleavable peptide moiety (Cit-Val). In the CatB-upregulated microenvironment, the peptide moiety was specifically cleaved, and then the radio afterglow signal and radiodynamic of MRAP could be activated. It should be noted that the probe enables the detection of a diminutive tumor. This work provided a new avenue for precision theranostics.

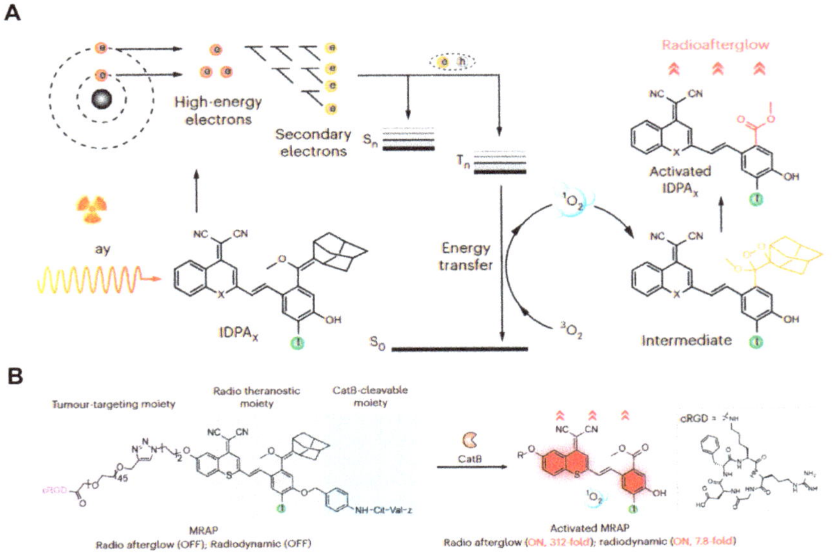

Figure 7. (**A**) Molecule mechanism of IDPA$_{Su}$ for radio-afterglow imaging. (**B**) Molecule mechanism of MRAP for radio-afterglow imaging and RDT [68]. Reprinted (adapted) with permission from Ref. [68]. Copyright 2023 Springer Nature.

5. Conclusions and Perspective

Afterglow imaging plays a crucial role in the field of biomedicine. At present, compared with inorganic afterglow probes, organic afterglow probes have advantages such as good biocompatibility, low toxicity, and easy adjustment of structure, which provide broad prospects for development in tumor imaging and treatment. The imaging effect of photo-afterglow probes is often affected by the unavoidable absorption of light by tissues. In contrast, due to the superior deep penetration capabilities of ultrasound and X-ray, both ultrasound-activated afterglow imaging and X-ray-activated afterglow imaging can effectively address this issue. Despite significant advancements, the design of promising organic afterglow imaging probes still faces many challenges.

Firstly, sono-afterglow probes and radio-afterglow probes can effectively enable deep tissue imaging. However, the development and reporting of organic sono-afterglow probes and radio-afterglow probes remain relatively underexplored, which highlights a significant opportunity for further investigation and innovation in this field [69,70].

Secondly, the reported probes are often composed of multiple components (stimuli-responsive units, singlet oxygen-generating units, luminescent units) encapsulated together. The construction process of these organic probes is relatively cumbersome, and the component ratio is not easily controlled precisely. Developing an activatable organic afterglow probe that integrates multiple units in one molecule is very meaningful.

Thirdly, the half-life of singlet oxygen which plays a crucial role in the process of afterglow imaging is very short (4 ms). This issue affects the efficiency of singlet oxygen transfer from the singlet oxygen generator to the afterglow substrate. The current solution is to prepare the imaging probe into a small nanoparticle. Exploring more methods to improve the efficiency of singlet oxygen transfer is urgent.

Fourthly, there are still very few organic luminescent probes with near infrared luminescence, especially in the second region of the near infrared. To improve the imaging resolution and depth, it is very necessary to develop the second region of NIR II afterglow probes.

Finally, the development of organic afterglow probes with near-infrared emission, long luminescence duration, bright afterglow intensity (more than 10^5 ps^{-1} cm^{-2} sr^{-1} measured by an IVIS imaging system) and multi-target detection is still in its initial stage. There is an urgent need to explore new organic afterglow substrates.

In summary, there are still many issues with probes used for precise imaging and treatment of tumors. We hope that this review will serve as a valuable resource, offering insightful and practical information about this specialized field. We aim to encourage and facilitate the development and eventual clinical application of afterglow imaging agents.

Author Contributions: Conceptualization, M.L. and L.T.; methodology, M.L.; software, M.L. and H.W.; validation, M.L. and H.W.; formal analysis, M.L. and L.T.; investigation, M.L. and L.T.; resources, M.L. and H.W; data curation, M.L. and L.T.; writing—original draft preparation, M.L. and L.T.; writing—review and editing, M.L. and L.T.; visualization, M.L.; supervision, Y.S. and J.L.; project administration, Y.S. and J.L.; funding acquisition, Y.S. and J.L. All authors have read and agreed to the published version of the manuscript.

Funding: This work was funded by the National Natural Science Foundation of China (22374055, 22022404, 22074050, 82172055), the National Natural Science Foundation of Hubei Province (22022CFA033), and the Fundamental Research Funds for the Central Universities (CCNU24JCPT001, CCNU24JCPT020). The authors have reported that they have no relationships relevant to the contents of this paper to disclose.

Institutional Review Board Statement: Not applicable.

Informed Consent Statement: Not applicable.

Data Availability Statement: No new data were created or analyzed in this study.

Conflicts of Interest: The authors declare no conflicts of interest.

References

1. Winkler, R.; Ciria, M.; Ahmad, M.; Plank, H.; Marcuello, C. A Review of the Current State of Magnetic Force Microscopy to Unravel the Magnetic Properties of Nanomaterials Applied in Biological Systems and Future Directions for Quantum Technologies. *Nanomaterials* **2023**, *13*, 2585. [CrossRef] [PubMed]
2. Luthra, R.; Chen, H.; Roy-Chowdhuri, S.; Singh, R.R. Next-Generation Sequencing in Clinical Molecular Diagnostics of Cancer: Advantages and Challenges. *Cancers* **2015**, *7*, 2023–2036. [CrossRef] [PubMed]
3. Razgulin, A.; Ma, N.; Rao, J. Strategies for in vivo imaging of enzyme activity: An overview and recent advances. *Chem. Soc. Rev.* **2011**, *40*, 4186–4216. [CrossRef]
4. Zhang, J.; Chai, X.; He, X.P.; Kim, H.J.; Yoon, J.; Tian, H. Fluorogenic Probes for Disease-Relevant Enzymes. *Chem. Soc. Rev.* **2019**, *48*, 683–722. [CrossRef]

5. Peng, Y.; Wang, Y.; Zhou, C.; Mei, W.; Zeng, C. PI3K/Akt/mTOR Pathway and Its Role in Cancer Therapeutics: Are We Making Headway? *Front. Oncol.* **2022**, *12*, 819128. [CrossRef]
6. Huang, J.; Li, J.; Lyu, Y.; Miao, Q.; Pu, K. Molecular Optical Imaging Probes for Early Diagnosis of Drug-Induced Acute Kidney Injury. *Nat. Mater.* **2019**, *18*, 1133–1143. [CrossRef]
7. Wang, S.; Fan, Y.; Li, D.; Sun, C.; Lei, Z.; Lu, L.; Wang, T.; Zhang, F. Anti-Quenching NIR-II Molecular Fluorophores for in Vivo High-Contrast Imaging and pH Sensing. *Nat. Commun.* **2019**, *10*, 1058–1068. [CrossRef]
8. Moore, C.; Chen, F.; Wang, J.; Jokerst, J.V. Listening for the Therapeutic Window: Advances in Drug Delivery Utilizing Photoacoustic Imaging. *Adv. Drug Deliv. Rev.* **2019**, *144*, 78–89. [CrossRef]
9. Huang, Z.; Yu, P.; Tang, J. Characterization of Triple-Negative Breast Cancer MDA-MB-231 Cell Spheroid Model. *OncoTargets Ther.* **2020**, *13*, 5395–5405. [CrossRef]
10. Shuhendler, A.J.; Pu, K.; Cui, L.; Uetrecht, J.P.; Rao, J. Real-time imaging of oxidative and nitrosative stress in the liver of live animals for drug-toxicity testing. *Nat. Biotechnol.* **2024**, *32*, 373–380. [CrossRef]
11. Badr, C.E.; Tannous, B.A. Bioluminescence imaging: Progress and applications. *Trends Biotechnol.* **2011**, *29*, 624–633. [CrossRef] [PubMed]
12. Suzuki, K.; Nagai, T. Recent progress in expanding the chemiluminescent toolbox for bioimaging. *Curr. Opin. Biotechnol.* **2017**, *48*, 135–141. [CrossRef] [PubMed]
13. Jiang, Y.; Pu, K. Multimodal Biophotonics of Semiconducting Polymer Nanoparticles. *Acc. Chem. Res.* **2018**, *51*, 1840–1849. [CrossRef]
14. Sun, S.K.; Wang, H.F.; Yan, X.P. Engineering Persistent Luminescence Nanoparticles for Biological Applications: From Biosensing/Bioimaging to Theranostics. *Acc. Chem. Res.* **2018**, *51*, 1131–1143. [CrossRef]
15. Xu, Y.; Li, C.; An, J.; Ma, X.; Yang, J.; Luo, L.; Deng, Y.; Kim, J.S.; Sun, Y. Construction of a 980 nm laser-activated Pt(II) metallacycle nanosystem for efficient and safe photo-induced bacteria sterilization. *Sci. China Chem.* **2023**, *66*, 155–163. [CrossRef]
16. Duan, Q.; Zhao, Z.; Zhang, Y.; Fu, L.; Yuan, Y.; Du, J.; Wang, J. Activatable fluorescent probes for real-time imaging-guided tumor therapy. *Adv. Drug Deliv. Rev.* **2023**, *196*, 114793. [CrossRef]
17. Zeng, W.; Wu, L.; Ishigaki, Y.; Harimoto, T.; Hu, Y.; Sun, Y.; Wang, Y.; Suzuki, T.; Chen, H.; Ye, D. An Activatable Afterglow/MRI Bimodal Nanoprobe with Fast Response to H_2S for In Vivo Imaging of Acute Hepatitis. *Angew. Chem. Int. Ed.* **2022**, *134*, e202111759. [CrossRef]
18. Yang, J.; Yin, W.; Van, R.; Yin, K.; Wang, P.; Zheng, C.; Zhu, B.; Ran, K.; Zhang, C.; Kumar, M.; et al. Turn-on Chemiluminescence Probes and Dual-Amplification of Signal for Detection of Amyloid Beta Species in Vivo. *Nat. Commun.* **2020**, *11*, 4052. [CrossRef]
19. Zhen, X.; Xie, C.; Pu, K. Temperature-Correlated Afterglow of a Semiconducting Polymer Nanococktail for Imaging-Guided Photothermal Therapy. *Angew. Chem. Int. Ed.* **2018**, *57*, 3938–3942. [CrossRef]
20. Zhang, X.; Zhang, T.; Tuo, W.; Liu, Y.; Umar, T.; Chen, Y.; Wu, Z.; Zhou, Q.; Li, X.; Deng, G.; et al. Rising interest in the accurate and controllable anticancer strategy: Based on photon-evoked pyroptosis engineering perspective. *Coord. Chem. Rev.* **2024**, *501*, 215588. [CrossRef]
21. Xu, Y.; Pang, Y.; Luo, L.; Sharma, A.; Yang, J.; Li, C.; Liu, S.; Zhan, J.; Sun, Y. De Novo Designed Ru(II) Metallacycle as a Microenvironment-Adaptive Sonosensitizer and Sonocatalyst for Multidrug-Resistant Biofilms Eradication. *Angew. Chem. Int. Ed.* **2024**, *63*, E202319966. [CrossRef] [PubMed]
22. Zhang, Z.; Ye, H.; Cai, F.; Sun, Y. Recent advances on the construction of long-wavelength emissive supramolecular coordination complexes for photo-diagnosis and therapy. *Dalton Trans.* **2023**, *52*, 15193–15202. [CrossRef] [PubMed]
23. Su, Y.; Phua, S.Z.F.; Li, Y.; Zhou, X.; Jana, D.; Liu, G.; Lim, W.Q.; Ong, W.K.; Yang, C.; Zhao, Y. Ultralong Room Temperature Phosphorescence from Amorphous Organic Materials toward Confidential Information Encryption and Decryption. *Sci. Adv.* **2018**, *4*, eaas9732. [CrossRef] [PubMed]
24. Xie, C.; Zhen, X.; Miao, Q.; Lyu, Y.; Pu, K. Self-Assembled Semiconducting Polymer Nanoparticles for Ultrasensitive Near-Infrared Afterglow Imaging of Metastatic Tumors. *Adv. Mater.* **2018**, *30*, 1801331. [CrossRef]
25. Li, C.; Tu, L.; Xu, Y.; Li, M.; Du, J.; Stang, P.J.; Sun, Y.; Sun, Y. A NIR-Light-Activated and Lysosomal-Targeted Pt(II) Metallacycle for Highly Potent Evoking of Immunogenic Cell Death that Potentiates Cancer Immunotherapy of Deep-Seated Tumors. *Angew. Chem. Int. Ed.* **2024**, *63*, e202406392. [CrossRef]
26. Li, Y.; Gecevicius, M.; Qiu, J. Long Persistent Phosphors-from Fundamentals to Applications. *Chem. Soc. Rev.* **2016**, *45*, 2090–2136. [CrossRef]
27. Chuang, Y.J.; Zhen, Z.; Zhang, F.; Liu, F.; Mishra, J.P.; Tang, W.; Chen, H.; Huang, X.; Wang, L.; Chen, X.; et al. Photostimulable Near-Infrared Persistent Luminescent Nanoprobes for Ultrasensitive and Longitudinal Deep-Tissue Bio-Imaging. *Theranostics* **2014**, *4*, 1112–1122. [CrossRef]
28. Wu, S.; Yang, C.; Yan, X. A dual-functional persistently luminescent nanocomposite enables engineering of mesenchymal stem cells for homing and gene therapy of glioblastoma. *Adv. Funct. Mater.* **2017**, *27*, 1604992. [CrossRef]
29. Zhou, H.; Zeng, X.; Li, A.; Zhou, W.; Tang, L.; Hu, W.; Fan, Q.; Meng, X.; Deng, H.; Duan, L.; et al. Upconversion NIR-II Fluorophores for Mitochondria-Targeted Cancer Imaging and Photothermal Therapy. *Nat. Commun.* **2020**, *11*, 6183. [CrossRef]
30. Ren, J.; He, J.; Zhang, H.; Xia, Y.; Hu, Z.; Loughran, P.; Billiar, T.; Huang, H.; Tsung, A. Platelet TLR4-ERK5 Axis Facilitates NET-Mediated Capturing of Circulating Tumor Cells and Distant Metastasis after Surgical Stress. *Cancer Res.* **2021**, *81*, 2373–2385. [CrossRef]

31. Ma, L.; Fei, B. Comprehensive review of surgical microscopes: Technology development and medical applications. *J. Biomed. Opt.* **2021**, *26*, 010901. [CrossRef] [PubMed]
32. Bortot, B.; Mangogna, A.; Lorenzo, G.; Stabile, G.; Ricci, G.; Biffi, S. Image-guided cancer surgery: A narrative review on imaging modalities and emerging nanotechnology strategies. *J. Nanobiotechnol.* **2023**, *21*, 155. [CrossRef] [PubMed]
33. Miao, Q.; Xie, C.; Zhen, X.; Lyu, Y.; Duan, H.; Liu, X.; Jokerst, J.V.; Pu, K. Molecular afterglow imaging with bright, biodegradable polymer nanoparticles. *Nat. Biotechnol.* **2017**, *35*, 1102–1110. [CrossRef] [PubMed]
34. Pei, P.; Chen, Y.; Sun, C.; Fan, Y.; Yang, Y.; Liu, X.; Lu, L.; Zhao, M.; Zhang, H.; Zhao, D.; et al. X-ray-activated persistent luminescence nanomaterials for NIR-II imaging. *Nat. Nanotechnol.* **2021**, *16*, 1011–1018. [CrossRef]
35. Hananya, N.; Shabat, D. A Glowing Trajectory between Bio- and Chemiluminescence: From Luciferin-Based Probes to Triggerable Dioxetanes. *Angew. Chem. Int. Ed.* **2017**, *56*, 16454–16463. [CrossRef]
36. Wan, C.; Zhang, Y.; Li, Q.; Jiang, Y.; Zhou, H.; Liu, Y.; Miao, Q.; Gao, M. Near-Infrared Afterglow Luminescence of Chlorin Nanoparticles for Ultrasensitive In Vivo Imaging. *J. Am. Chem. Soc.* **2022**, *144*, 6719–6726.
37. Jiang, Y.; Huang, J.; Zhen, X.; Li, J.; Xie, C.; Miao, Q.; Chen, J.; Chen, P.; Pu, K. A generic approach towards afterglow luminescent nanoparticles for ultrasensitive in vivo imaging. *Nat. Commun.* **2019**, *10*, 2064. [CrossRef]
38. Niu, G.; Zhang, R.; Shi, X.; Park, H.; Xie, S.; Kwok, R.T.K.; Lam, J.W.Y.; Tang, B.Z. AIE Luminogens as Fluorescent Bioprobes. TrAC. *Trends Anal. Chem.* **2020**, *123*, 115769. [CrossRef]
39. Ni, X.; Zhang, X.; Duan, X.; Zheng, H.L.; Xue, X.S.; Ding, D. Near-Infrared Afterglow Luminescent Aggregation-Induced Emission Dots with Ultrahigh Tumor-to-Liver Signal Ratio for Promoted Image-Guided Cancer Surgery. *Nano Lett.* **2019**, *19*, 318–330. [CrossRef]
40. Xu, Y.; Yang, W.; Yao, D.; Bian, K.; Zeng, W.; Liu, K.; Wang, D.; Zhang, B. An Aggregation-Induced Emission Dye-Powered Afterglow Luminogen for Tumor Imaging. *Chem. Sci.* **2019**, *11*, 419–428. [CrossRef]
41. Li, J.; Zhang, G.; Zhang, Y.; Tang, Y.; Ding, D.; Li, W.; Liu, Q. Building Highly Light-Harvesting Near-Infrared AIEgens Using Triazole-Based Luminescent Core for Improved Intravital Afterglow Imaging. *Adv. Funct. Mater.* **2023**, *33*, 2212380. [CrossRef]
42. Yang, L.; Zhao, M.; Chen, W.; Zhu, J.; Xu, W.; Li, Q.; Pu, K.; Miao, Q. A Highly Bright Near-Infrared Afterglow Luminophore for Activatable Ultrasensitive In Vivo Imaging. *Angew. Chem. Int. Ed.* **2024**, *63*, e202313117. [CrossRef] [PubMed]
43. Wu, L.; Ishigaki, Y.; Hu, Y.; Sugimoto, K.; Zeng, W.; Harimoto, T.; Sun, Y.; He, J.; Suzuki, T.; Jiang, X.; et al. H_2S-Activatable near-Infrared Afterglow Luminescent Probes for Sensitive Molecular Imaging in Vivo. *Nat. Commun.* **2020**, *11*, 446. [CrossRef]
44. Jiang, Y.; Zhao, M.; Miao, J.; Chen, W.; Zhang, Y.; Miao, M.; Yang, L.; Li, Q.; Miao, Q. Acidity-activatable upconversion afterglow luminescence cocktail nanoparticles for ultrasensitive in vivo imaging. *Nat. Commun.* **2024**, *15*, 2124. [CrossRef]
45. Huang, S.; Bai, S.; Luo, T.; Zheng, F.; Fan, D.; Zhou, Y.; Dong, J.; Chen, F.; Zeng, W. Chemiluminescent Afterglow Material for Enhanced Tumor Diagnosis and Photodynamic Therapy. *Adv. Funct. Mater.* **2024**. [CrossRef]
46. Li, Y.; He, D.; Zheng, Q.; Tang, R.; Wan, Q.; Tang, B.; Wang, Z. Single-Component Photochemical Afterglow Near-Infrared Luminescent Nano-Photosensitizers: Bioimaging and Photodynamic Therapy. *Adv. Healthc. Mater.* **2024**, *13*, 2304392. [CrossRef]
47. Huang, J.; Zhang, C.; Wang, X.; Wei, X.; Pu, K. Near-Infrared Photodynamic Chemiluminescent Probes for Cancer Therapy and Metastasis Detection. *Angew. Chem. Int. Ed.* **2023**, *62*, e202303982. [CrossRef]
48. Gao, Z.; Jia, S.; Ou, H.; Hong, Y.; Shan, K.; Kong, X.; Wang, Z.; Feng, G.; Ding, D. An Activatable Near-Infrared Afterglow Theranostic Prodrug with Self-Sustainable Magnification Effect of Immunogenic Cell Death. *Angew. Chem. Int. Ed.* **2022**, *61*, e202209793. [CrossRef]
49. Tong, Y.; Li, M.; Huang, H.; Long, S.; Sun, W.; Du, J.; Fan, J.; Wang, L.; Liu, B.; Peng, X. Urea-Bond Scission Induced by Therapeutic Ultrasound for Biofunctional Molecule Release. *J. Am. Chem. Soc.* **2022**, *144*, 16799–16807. [CrossRef]
50. Dou, Y.; Wang, Y.; Tian, S.; Song, Q.; Deng, Y.; Zhang, Z.; Chen, P.; Sun, Y. Metal–organic framework (MOF)-based materials for pyroptosis-mediated cancer therapy. *Chem. Commun.* **2024**, *60*, 6476–6487. [CrossRef]
51. Liu, T.; Zhang, N.; Wang, Z.; Wu, M.; Chen, Y.; Ma, M.; Chen, H.; Shi, J. Endogenous Catalytic Generation of O_2 Bubbles for In Situ Ultrasound-Guided High Intensity Focused Ultrasound Ablation. *ACS Nano* **2017**, *11*, 9093–9102. [CrossRef] [PubMed]
52. Zhang, Y.; Zhang, X.; Yang, H.; Yu, L.; Xu, Y.; Sharma, A.; Yin, P.; Li, X.; Kim, J.S.; Sun, Y. Advanced Biotechnology-Assisted Precise Sonodynamic Therapy. *Chem. Soc. Rev.* **2021**, *50*, 11227–11248. [CrossRef] [PubMed]
53. Wu, R.; Yao, Z.; Chen, Z.; Ge, X.; Su, L.; Wang, S.; Wu, Y.; Song, J. Ultrasound-Activated NIR Chemiluminescence for Deep Tissue and Tumor Foci Imaging. *Anal. Chem.* **2023**, *95*, 11219–11226. [CrossRef]
54. Yang, Z.; Shen, X.; Jin, J.; Jiang, X.; Pan, W.; Wu, C.; Yu, D.; Li, P.; Feng, W.; Chen, Y. Sonosynthetic Cyanobacteria Oxygenation for Self-Enhanced Tumor-Specific Treatment. *Adv. Sci.* **2024**, *11*, 2400251. [CrossRef]
55. De Visser, K.E.; Eichten, A.; Coussens, L.M. Paradoxical Roles of the Immune System during Cancer Development. *Nat. Rev. Cancer* **2006**, *6*, 24–37. [CrossRef]
56. Schmidt, R.E.; Grimbacher, B.; Witte, T. Autoimmunity and Primary Immunodeficiency: Two Sides of the Same Coin? *Nat. Rev. Rheumatol.* **2018**, *14*, 7–18. [CrossRef]
57. Xu, C.; Huang, J.; Jiang, Y.; He, S.; Zhang, C.; Pu, K. Nanoparticles with Ultrasound-Induced Afterglow Luminescence for Tumour-Specific Theranostics. *Nat. Biomed. Eng.* **2022**, *7*, 298–312. [CrossRef]
58. Wang, Y.; Yi, Z.; Guo, J.; Liao, S.; Li, Z.; Xu, S.; Yin, B.; Liu, Y.; Feng, Y.; Rong, Q.; et al. In Vivo Ultrasound-Induced Luminescence Molecular Imaging. *Nat. Photonics* **2024**, *18*, 334–343. [CrossRef]

59. Xu, J.; Chen, Q.; Tian, K.; Liang, R.; Chen, T.; Gong, A.; Mathy, N.; Yu, T.; Chen, X. m6A methyltransferase METTL3 maintains colon cancer tumorigenicity by suppressing SOCS2 to promote cell proliferation. *Oncol. Rep.* **2020**, *44*, 973–986. [CrossRef]
60. Zhang, Q.; Wang, N.; Ma, M.; Luo, Y.; Chen, H. Transferrin Receptor-Mediated Sequential Intercellular Nanoparticles Relay for Tumor Deep Penetration and Sonodynamic Therapy. *Adv. Ther.* **2019**, *2*, 1800152. [CrossRef]
61. Chen, W.; Zhang, J. Using Nanoparticles to Enable Simultaneous Radiation and Photodynamic Therapies for Cancer Treatment. *J. Nanosci. Nanotechnol.* **2006**, *6*, 1159–1166. [CrossRef] [PubMed]
62. Gao, W.; Liu, Y.; Zhang, H.; Wang, Z. Electrochemiluminescence Biosensor for Nucleolin Imaging in a Single Tumor Cell Combined with Synergetic Therapy of Tumor. *ACS Sens.* **2020**, *5*, 1216–1222. [CrossRef] [PubMed]
63. Wang, X.; Sun, W.; Shi, H.; Ma, H.; Niu, G.; Li, Y.; Zhi, J.; Yao, X.; Song, Z.; Chen, L.; et al. Organic Phosphorescent Nanoscintillator for Low-Dose X-Ray-Induced Photodynamic Therapy. *Nat. Commun.* **2022**, *13*, 5091. [CrossRef] [PubMed]
64. Chen, X.; Song, J.; Chen, X.; Yang, H. X-Ray-Activated Nanosystems for Theranostic Applications. *Chem. Soc. Rev.* **2019**, *48*, 3073–3101. [CrossRef]
65. Chen, Z.; Wang, L.; Manoharan, D.; Lee, C.; Wu, L.; Huang, W.; Huang, E.; Su, C.; Sheu, H.; Yeh, C. Low Dose of X-Ray-Excited Long-Lasting Luminescent Concave Nanocubes in Highly Passive Targeting Deep-Seated Hepatic Tumors. *Adv. Mater.* **2019**, *31*, 1905087. [CrossRef]
66. Shi, T.; Sun, W.; Qin, R.; Li, D.; Feng, Y.; Chen, L.; Liu, G.; Chen, X.; Chen, H. X-Ray-Induced Persistent Luminescence Promotes Ultrasensitive Imaging and Effective Inhibition of Orthotopic Hepatic Tumors. *Adv. Funct. Mater.* **2020**, *30*, 2001166. [CrossRef]
67. Huang, K.; Le, N.; Wang, J.S.; Huang, L.; Zeng, L.; Xu, W.; Li, Z.; Li, Y.; Han, G. Designing Next Generation of Persistent Luminescence: Recent Advances in Uniform Persistent Luminescence Nanoparticles. *Adv. Mater.* **2022**, *34*, 2107962. [CrossRef]
68. Huang, J.; Su, L.; Xu, C.; Ge, X.; Zhang, R.; Song, J.; Pu, K. Molecular Radio Afterglow Probes for Cancer Radiodynamic Theranostics. *Nat. Mater.* **2023**, *22*, 1421–1429. [CrossRef]
69. Wilkinson, F.; Helman, W.P.; Ross, A.B. Rate Constants for the Decay and Reactions of the Lowest Electronically Excited Singlet State of Molecular Oxygen in Solution: An Expanded and Revised Compilation. *J. Phys. Chem. Ref. Data* **1995**, *24*, 663–677. [CrossRef]
70. Lu, C.; Zhang, C.; Wang, P.; Zhao, Y.; Yang, Y.; Wang, Y.; Yuan, H.; Qu, S.; Zhang, X.; Song, G.; et al. Light-free Generation of Singlet Oxygen through Manganese-Thiophene Nanosystems for pH-Responsive Chemiluminescence Imaging and Tumor Therapy. *Chem* **2020**, *6*, 2314–2334. [CrossRef]

targets

Review

Design of Cell-Specific Targeting Peptides for Cancer Therapy

Xiaoyong Chen, Di Wang, Yun-Bao Jiang and Tao Jiang *

The MOE Key Laboratory of Spectrochemical Analysis and Instrumentation, Department of Chemistry, College of Chemistry and Chemical Engineering, Xiamen University, Xiamen 361005, China; xiaoyongchen@stu.xmu.edu.cn (X.C.); 20520230156666@stu.xmu.edu.cn (D.W.); ybjiang@xmu.edu.cn (Y.-B.J.)
* Correspondence: tjiang6@xmu.edu.cn

Abstract: The landscape of cancer therapy has gained major impetus through the development of materials capable of selectively targeting cancer cells while sparing normal cells. Synthetic peptides are appealing as scaffolds for the creation of such materials. They are small in size, amenable to chemical synthesis and functionalization, and possess diverse chemical and structural space for modulating targeting properties. Here, we review some fundamental insights into the design, discovery, and evolution of peptide-based targeting agents, with a particular focus on two types of cancer cell targets: unique/overexpressed surface receptors and abnormal physiological properties. We highlight the cutting-edge strategies from the literature of the last two decades that demonstrate innovative approaches to constructing receptor-specific cyclic binders and stimulus-responsive targeting materials. Additionally, we discuss potential future directions for advancing this field, with the aim of pushing the frontiers of targeted cancer therapy forward.

Keywords: peptide-based binder; responsive material; targeting; cancer therapy

1. Introduction

Cancer remains one of the leading causes of death worldwide, with 20 million new cases expected annually by 2025. This underscores the need for highly effective cancer therapies [1]. Surgery, radiation therapy, chemotherapy, and combinations of these therapies dominate clinical practice. However, they often result in severe side effects and unintended toxicity due to their non-selective action against normal cells and tissues [2–4]. Immunotherapy is a transformative approach in this field by deploying our body's own immune system as a personalized medicine to fight against cancer [5,6]. This can be achieved by targeting inhibitory pathways with immune checkpoint inhibitors, or activating pathways with chimeric antigen receptor immune cells or cell engagers. Compared to chemotherapy or radiation, immunotherapy typically has better action specificity with fewer side effects. In some cases, immunotherapy can induce durable responses, contrasting with conventional therapies where cancer may recur more frequently. Clearly, in almost all cancer therapies, it is critical to ensure therapeutic activity against cancer cells while sparing normal cells from harm. Paul Ehrlich's "magic bullet" concept in the 1890s laid the foundation for targeted therapies by emphasizing the importance of selective-targeting capabilities [7]. Developing molecules or materials that specifically target cancer cells has thus become a highly rewarding endeavor. Cancer cells have a large number of unique or overexpressed surface receptors compared to normal cells. These proteins serve as prime molecular targets for the design or discovery of receptor-specific binding molecules. These binders can facilitate targeted cargo delivery, enable immune checkpoint inhibition, and recruit immune and effector cells, thereby advancing both cancer therapy and diagnostics. In addition, cancer cells exhibit aberrant physiological properties such as overexpressed enzymes, elevated redox potentials, and acidic pH conditions. Effectively targeting these features, as opposed to the receptor-based approaches, is also critical for distinguishing cancer cells from normal cells in therapeutic interventions.

Monoclonal antibodies (mAbs) are widely used because of their exceptional binding specificity and high affinity for cell-surface receptors, making them valuable for generating cell-targeting binders. However, their biological production is costly and can lead to variability in potency between batches. The large size of mAbs may also pose challenges in vivo, such as inadequate pharmacokinetics and limited tissue penetration [8–11]. In contrast, nucleic acid aptamers—a short, single strand of DNA or RNA—function like chemical antibodies. Like antibodies, aptamers can specifically recognize molecular targets based on defined nucleotide sequences and conformations. The SELEX random library technique has been used to generate numerous high-affinity aptamers that selectively bind to various cellular targets [12–14]. Aptamer-based antagonists and aptamer–drug/toxin conjugates have been developed for cancer treatment, as reviewed in detail elsewhere [15–17].

Here, we highlight peptides as promising candidates for cancer cell-targeting agents. Compared to mAbs, peptides offer distinct advantages: they are small in size, amenable to chemical synthesis and functionalization, exhibit minimal batch-to-batch variability, possess low immunogenicity, and have an extended shelf life [18–20]. Peptides also offer greater chemical diversity for modulating targeting properties compared to nucleic acid aptamers, thanks to the wide range of natural and unnatural amino acids available. Their diverse secondary and higher order structures further expand the scope for tailoring chemical distribution and folding or assembly behaviors. As shown in recent reviews, peptides have been increasingly explored for cancer treatments and diagnostics [20–24]. Previous reviews have systematically discussed their therapeutic effects in different cell types and ways to improve their in vivo efficacy. However, limited attention has been paid to the approaches for molecular design and the search for cancer cell-targeting peptides. Practical methods to correlate peptide sequence-dependent folding and assembly behaviors with cancer-related physiological properties remain elusive [25–27]. This mini-review illustrates selected studies to show cutting-edge strategies in the discovery and design of peptide-based binders and dynamic peptide materials, and their receptors and physiological signals, respectively. By leveraging the rich molecular codes encoded in peptides, this review promises to reliably and predictably design targeting agents, advancing targeted cancer therapy.

2. Receptor-Specific Binders

2.1. Linear Binders

Cancer cells possess oncogenic aberrations that promote abnormal proliferation, migration, and the evasion of immune surveillance [28,29]. In particular, unique/overexpressed membrane receptors have become prime targets for distinguishing cancer cells from their normal counterparts [20,30–32]. Early efforts in developing receptor-specific binding molecules centered on structure-guided design by exploring natural proteins that engage with target receptors. For example, integrins play a central role in cell adhesion to the extracellular matrix (ECM), facilitating cellular motility and invasion [33]. The well-known RGD peptide was originally derived from the sequence of fibronectin, an abundant ECM protein [34,35]. RGD has been identified as a key interacting motif with integrin heterodimers, such as $\alpha_5\beta_1$, $\alpha_V\beta_3$, $\alpha_V\beta_6$, and $\alpha_{II}\beta_3$. Consequently, a variety of synthetic RGD peptides and their derivatives have been constructed as potent binders for integrin-overexpressed cancer cells in malignancies, such as melanoma, glioblastoma, and breast, prostate, and ovarian malignancies [36–38]. This approach is time-consuming and highly dependent on the availability of high-resolution structural information on the protein-receptor complex. In contrast, techniques such as phage display [39], and mRNA display [40] enable the screening of large numbers of random peptides against nearly any given molecular target. For example, the overexpression of human epidermal growth factor receptor 2 (HER2) on cancer cells triggers receptor homodimerization and clustering. This activates downstream MAPK and PI3K pathways to drive cell proliferation, growth, and anti-apoptosis [41]. Quinn and coworkers used a random 6-amino-acid peptide bacteriophage display library to find the HER2-specific peptide (KCCYSL) with a dissociation constant (K_D) in the range of hundreds of micromolar [42]. In a separate study, Sioud and colleagues used a phage display biopanning technique to identify a HER2 binder (LTVSPWY) with a remarkable K_D value in the nanomolar range [43].

2.2. Cyclic Binders

Cyclic peptides dominate in cell-related molecular binders based on the literature of the last two decades [44,45]. Studies suggest that cyclization of linear RGD peptides can significantly increase their binding affinity to $\alpha_V\beta_3$ receptors from several micromolar to nanomolar levels [46,47]. The increase is attributed to reduced chain flexibility, which facilitates peptide–target engagements by minimizing entropic loss during binding. In addition, cyclization of many linear peptide-based binders has resulted in improved proteolytic resistance during in vivo cancer cell targeting. This is due to constrained backbone conformations that impede access to protease catalytic sites. Clearly, cyclization has emerged as a facile route to enhancing the targeting performance of peptide-based binders.

Numerous bioactive cyclic peptides found in nature maintain their three-dimensional structures through intramolecular disulfide bonds. Wu and coworkers were inspired to develop a series of cysteine-rich motifs, such as CXC and CPPC, to direct peptide cyclization. These motifs enable the programmable formation of intramolecular disulfide bonds, leading to single-, bi-, and multi-cyclic architectures with predictable topologies (Figure 1A) [48,49]. In addition to the disulfide-based approach, various chemical conjugation and crosslinking strategies have been developed, allowing the creation of cyclic scaffolds with exceptional efficiency [50]. The screening library (such as phage [39] and mRNA display [40]) has been extensively used to search for unnatural cyclic binders that target given molecular entities. For example, Heinis, Winter and coworkers presented a strategy using phage display to screen and isolate cyclic binders for various molecular targets [45,51]. Cysteine-containing peptides presented on phage tips were cyclized using thiol-reactive chemical linkers, resulting in redox-stable conformations as opposed to disulfide-bridged cyclic structures (Figure 1B) [52]. Suga and coworkers established the random nonstandard peptide integrated discovery (RaPID) platform to generate the library of thioether-closed macrocyclic peptides containing non-proteinogenic amino acids (Figure 1C) [53,54]. This platform integrates genetic code reprogramming using a flexible in vitro translation (FIT) system with mRNA display, greatly expanding the repertoire of cyclic binders [55]. An increasing number of cyclic peptides with nanomolar and even picomolar binding affinities have been generated for a wide range of targets, including cancer cell-surface receptors.

Peptide-based binders facilitate cancer therapy by guiding the delivery of chemotherapeutics, radioisotopes, and other cytotoxic agents. They also serve to inhibit and antagonize cell-surface receptors as well as intracellular proteins [56–58]. Among them, many binder-based drugs and peptide–drug/toxin conjugates have been approved or are under clinical evaluation (Table 1). For instance, romidepsin [59], a bicyclic peptide isolated from natural fermentation products, has received FDA approval for the treatment of cutaneous T-cell lymphoma. This prodrug features an intrapeptide disulfide bond that is reduced within the cell. The active peptide form is then released to specifically target and inhibit histone deacetylase enzymes. Another notable advancement is ^{177}Lu DOA-TATE [60], the first FDA-approved radiopharmaceutical for the treatment of gastroenteropancreatic neuroendocrine tumors. The molecule conjugates a radionuclide to octreotide, and a cyclic peptide and somatostatin analogue. The octreotide selectively delivers ionizing radiation to cancer cells bearing overexpressed somatostatin receptors. Similarly, ^{177}Lu-AB-3PRGD is in a Phase I trial to determine its effectiveness in various advanced solid tumors (NCT06375564) [61]. This conjugate employs a dimeric RGD peptide with 3 PEG_4 linkers for high-avidity multivalent targeting of $\alpha_V\beta_3$ on cancer cells. In addition to these targeting peptides derived from natural protein sources, screened synthetic binders are also being explored in preclinical or clinical trials. A notable example is the bicycle toxin conjugate (BTC) BT8009, which combines a nectin-4 targeting bicyclic peptide with the cytotoxin monomethyl auristatin E [62]. This conjugate has shown promising anticancer activity in patients with advanced or metastatic malignancies, including urothelial cancer (NCT04561362, NCT06225596). Comprehensive lists of binder-based therapeutics in clinical trials and approved for marketing have been presented elsewhere [63,64]. Concurrent with the rapid development of cancer immunotherapy, peptide-based binders have also been harnessed as cell engagers to orchestrate cancer elimination using immune effector cells. For example, Wang and coauthors designed a bispecific triblock peptide AKMGEGGWGANDY-GNNQQNY-RGD to facilitate interactions between T cells and cancer cells [65]. The first and third blocks of the peptide selectively targeted CD3 on T cells and integrins on MCF-7 cells, respectively. Upon RGD-integrin interaction, the fibril-forming sequence (the second block) drove the clustering of the cell-surface peptides, activating T cells and culminating in the cytolysis of cancer cells.

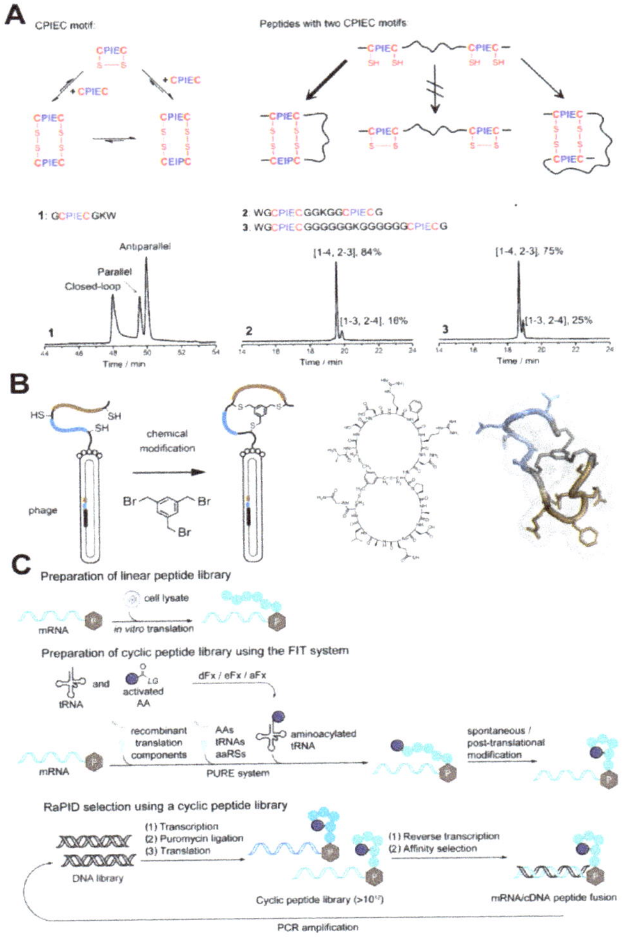

Figure 1. (**A**) Schematics of disulfide pairing of CPIEC motifs and the oxidation of peptides containing two CPIEC motifs [49]. Reprinted (adapted) with permission from 49. Copyright 2023 American Chemical Society. (**B**) Phage selection of bicyclic peptides using a thiol-reactive chemical linker [52]. Reprinted (adapted) with permission from 52. Copyright 2017 American Chemical Society. (**C**) mRNA selection of cyclic peptides on the RaPID platform [54]. Reprinted (adapted) with permission from 54. Copyright 2019 American Chemical Society.

Table 1. Representative cell-targeting peptides that have been approved or are in clinical trials.

Name	Targeting Feature	Mechanism of Action	Development Stage	Clinical Trials Registry	Application	Reference
Romidepsin	Histone deacetylase enzymes	Inhibiting histone deacetylase	Approved	NCT03742921	Cutaneous T-cell lymphoma	[59]
^{177}Lu DOA-TATE	Somatostatin receptors	^{177}Lu irradiation	Approved	NCT01578239	Gastroenteropancreatic neuroendocrine tumors	[60]
^{177}Lu-AB-3PRGD	$\alpha_V\beta_3$	^{177}Lu irradiation	Phase I	NCT06375564	Solid tumors	[61]
BT8009	Nectin-4	Disrupting mitotic via monomethyl auristatin E	Phase I/II	NCT04561362	Advanced or metastatic malignancies	[62,64]
BT8009	Nectin-4	Pembrolizumab	Phase II	NCT06225596	Metastatic urothelial cancer	[62]
CBX-12-101	pH-induced membrane insertion	DNA damage via exatecan	Phase I/II	NCT04902872	Ovarian cancer, solid tumors	[66]
CBX-12	pH-induced membrane insertion	DNA damage via exatecan	Phase I	NCT05691517	Solid tumors	[66]
pHLIP ICG	pH-induced membrane insertion	ICG based fluorescence imaging	Phase I/II	NCT05130801	Breast cancer	[67]
^{18}F-Var3	pH-induced membrane insertion	^{18}F based imaging	Phase I	NCT04054986	Breast cancer	[68]

3. Physiological Stimulus–Responsive Peptide Assemblies

In addition to targeting cell-surface receptors, the aberrant physiological characteristics inherent to cancer cells can be an alternative type of cancer cell target [69,70]. Tumor microenvironments, both extracellular and intracellular, exhibit overexpressed enzymes, elevated redox potentials, and acidic pH conditions [71,72]. Aiming at these physiological attributes, targeting cancer cells require approaches distinct from the binder-receptor interaction mode. A straightforward strategy to create the targeting materials is to use these physiological traits to alter the physicochemical properties of self-assembling peptides (i.e., hydrophobicity, hydrophilicity, size, and charge state). As demonstrated by the following examples, this provides a facile route to correlate the changes in folding and or assembly behaviors of the peptides with cancer cell-specific cues (Table 2).

3.1. *Enzyme-Responsive Materials*

Elevated levels of enzymes in both the extra- and intracellular environments of cancer have been documented, including alkaline phosphatases, matrix metalloproteinases (MMPs), cathepsin B, carbonic anhydrases, and many others [73–75]. The catalytic abilities of these enzymes to cleave chemical bonds have been leveraged to link peptide assembly behavior to the presence of enriched enzymes from cancer cells [73,76,77]. Xu and colleagues have pioneered the construction of enzyme-instructed peptide self-assembly systems [78,79]. They first showed the use of alkaline phosphatase (ALP) to induce molecular self-assembly [80]. By dephosphorylating an FMOC-tyrosine phosphate, the balance of molecular hydrophobicity and hydrophilicity was altered, resulting in the transformation of the precursor into a hydrogenator. This innovative strategy was subsequently used to direct peptide fibrillization in the pericellular space of cancer cells [81]. Surface and secretory ALPs from HeLa and MES-SA cancer cells catalytically dephosphorylated a naphthalene-capped peptide, $^{D}F^{D}Fp^{D}Y$. With enhanced hydrophobicity and decreased charge repulsion, the resultant tripeptide $^{D}F^{D}F^{D}Y$ self-assembled into hydrogel-like structures selectively surrounding cancer cells. This action effectively inhibited cancer cell growth and metastasis by impeding cellular mass exchange and inducing apoptosis (Figure 2A).

Recently, Yang, Gao, and their colleagues extended the concept of ALP-responsive peptide assembly structures to develop a bis-specific cell engager (Supra-BiCE) for cancer immunotherapy [82]. The Supra-BiCE platform comprised two self-assembly peptides: Ada-$G^{D}F^{D}Fp^{D}YG$ conjugated to a ^{D}PPA-1 peptide $^{D}N^{D}Y^{D}S^{D}K^{D}P^{D}T^{D}D^{D}R^{D}Q^{D}Y^{D}H^{D}F$ to target PD-L1 on cancer cells and a ^{D}TBP-3 peptide, $GG^{D}Y^{D}T^{D}F^{D}H^{D}W^{D}H^{D}R^{D}L^{D}N^{D}P$. The two peptide segments selectively targeted the Ig and ITIM domains on T and NK cells, respectively. Upon ALP dephosphorylation, the nanoribbons formed by the co-assembly of these two peptides underwent a morphology transformation into long nanofibers. This structural alternation enhanced the binding affinity of the assemblies to both immune and cancer cells, consequently activating T and NK cells via checkpoint blockade. In vivo studies demonstrated that this in situ peptide assembly effectively suppressed colon carcinoma models in mice by facilitating the targeting, enrichment, and retention of T and NK cells at cancer cell sites.

Enzymatic cleavage peptide sequences by enzymes represents another prevalent approach in the design of enzyme-responsive peptide assemblies. For example, Lou and coworkers designed a peptide-conjugated probe (DMFA), capable of undergoing matrix metalloproteinase-2 (MMP-2)-induced morphological changes for cancer therapy [83]. As shown in Figure 2B, DMFA comprised three consecutive segments: a charged, amphiphilic α-helical peptide with the amino-acid sequence of GRFKRFRKKFKKLFKKLSPVIPLLHL, an MMP-2 cleavable peptide PLGLAG, and an amyloid-forming sequence KLVFF. DMFA formed nanoparticles with the α-helical peptide exposed to the aqueous phase. Following cleavage by overexpressed MMP-2 on cancer cells, the resultant α-helical portion disrupted the phospholipid bilayers of the cells, while the KLVFF fragment self-assembled into extracellular nanofibers. This morphological transformation of peptide assemblies has demonstrated efficacy in suppressing tumor growth and metastasis by facilitating Ca^{2+} influx and disrupting Na^{+}/K^{+}-ATPase, consequently inhibiting the PI3K-Akt signaling pathway.

In addition to extracellular enzymes within the pericellular space, intracellular enzymes also represent promising targets for peptide assembly-based therapies. For example, Liang and coworkers reported an apoptosis-amplified assembly of a peptide analogue for cancer photodynamic therapy (PDT) (Figure 2C) [84]. They synthesized a peptide-porphyrin conjugate, Ac-DEVDD-TPP, by incorporating a caspase-3 cleavable sequence DEVD to a TPP-based PDT photosensitizer. Upon cleavage by endogenous caspase-3, the conjugate yielded D-TPP, which self-assembled into nanofibrils around the mitochondrion. This located assembly enabled the in situ generation of singlet oxygen $^{1}O_2$ upon laser irradiation, inducing mitochondrion damage and triggering cell apoptosis. Concurrently, more generated caspase-3 continued to convert the conjugate into nanofibrils, thereby amplifying cancer cell apoptosis. Similarly, Wang and colleagues demonstrated the conversion of drug-appended peptide nanoparticles into nanofibers via overexpressed lysosomal protease, cathepsin B, in cancer cells [85]. As depicted in Figure 2D, the peptide-based prodrug was synthesized by conjugating an integrin-targeting peptide RGD to the C-terminus of a chemodrug camptothecin (CPT)-appended β-sheet forming peptide LVFF, linked by an enzyme-cleavable sequence GFLG and a hydrophilic PEG chain. The resultant amphiphilic peptide conjugates self-assembled into nanoparticles. The small size of the particles (~30 nm) enabled the internalization of the assemblies into HeLa cell lysosomes via integrin-assisted cellular endocytosis. Subsequent shedding of the hydrophilic PEG-RGD segment by cathepsin B increased the hydrophobicity of the molecules, leading to the formation of CPT-LVFF nanofibers in cancer cells. These preformed fibrous drugs then acted as seeds to promote drug accumulation in cancer cells, effectively inhibiting cancer progression.

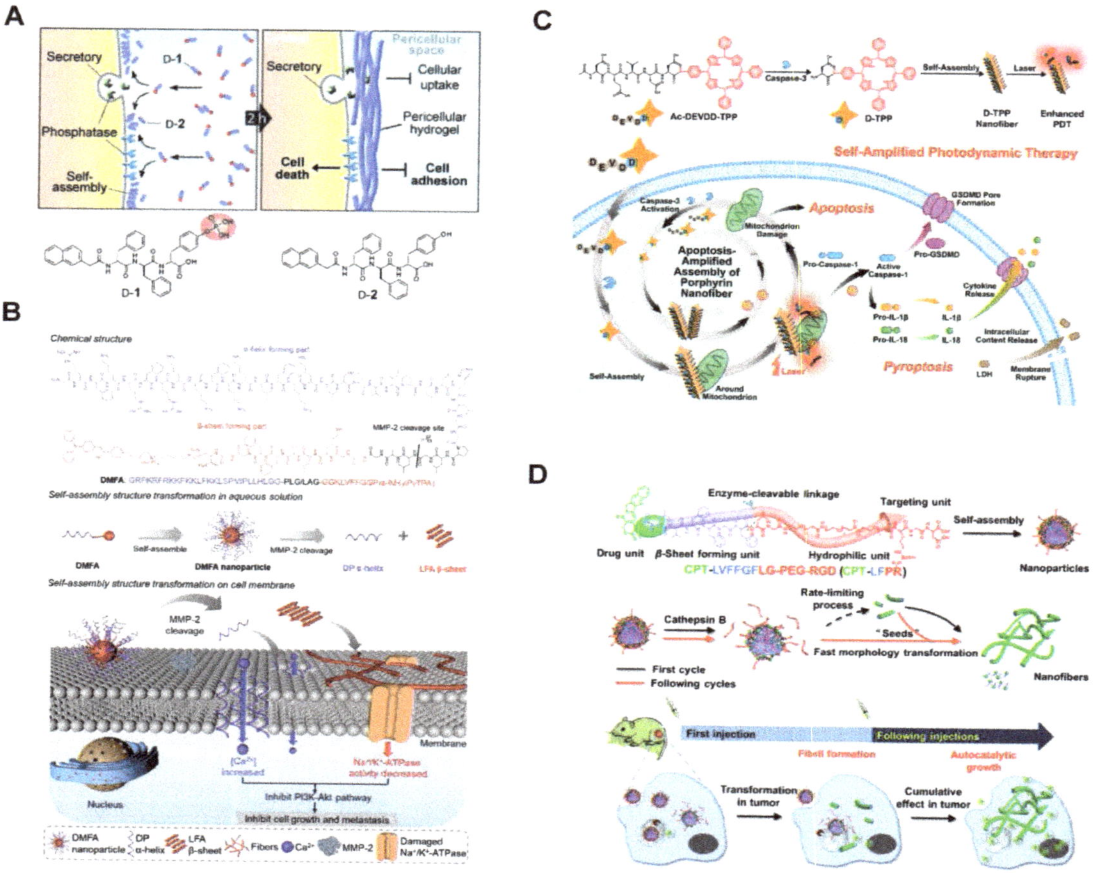

Figure 2. (**A**) Enzyme-catalyzed formation of pericellular hydrogels/nanoparticles to induce cell death [81]. Reproduced with permission from 81, © 2014 WILEY-VCH Verlag GmbH & Co. KGaA, Weinheim. (**B**) Scheme of the peptide-coupled probe DMFA with division-induced changes on cell membrane morphology [83]. Reproduced with permission from 83, © 2023 WILEY-VCH Verlag GmbH & Co. KGaA, Weinheim. (**C**) Self-amplification of Ac-DEVDD-TPP by caspase-3 activation enhances PDT mechanism and induction of apoptosis to enhance pyroptosis generated by porphyrin nanofibers via apoptosis-amplified assembly [84]. Reprinted (adapted) with permission from 84. Copyright 2023 American Chemical Society. (**D**) Schematic representation of enzyme-triggered morphological transformation and autocatalytic growth of nanofibers [85]. Reprinted (adapted) with permission from 85. Copyright 2019 American Chemical Society.

3.2. Redox-Responsive Materials

The aberrant redox environment is another characteristic feature of cancer cells. These cells produce higher basal levels of reactive oxygen species (ROS) compared to normal cells, through the mitochondrial respiratory chain and nicotinamide adenine dinucleotide phosphate oxidase [86–89]. Wang, Qiao, and coworkers constructed assemblies of a polymer–peptide conjugate (PPC) to target the excessive ROS generated in cancer cell mitochondria [90]. Poly (vinyl alcohol) (PVA) was coupled with β-sheet forming peptide KLVFF, tethered with PEG via ROS-cleavable thioketal bond, and the mitochondria-targeting peptide $(KLAKLAK)_2$, respectively (Figure 3A). The resultant amphiphilic PPCs self-assembled into nanoparticles capable of cellular entry through endocytosis. Upon ROS-induced thioketal cleavage, which increased the molecular hydrophobicity, the conjugates transformed into long fibers inside the cells. This increased exposure of the KLAK peptide along the fibers facilitated enhanced multivalent interactions between the assemblies and mitochondria, inducing organelle dysfunction-based cytotoxicity against cancer cells. The ROS-responsive materials exhibited tumor suppression efficacy in mice.

Cancer cells possess antioxidants such as glutathione (GSH) to regulate ROS levels. As a result, a highly reductive environment can be exploited to design GSH-responsive therapeutic assemblies [91–94]. For example, Yu and colleagues demonstrated a GSH-triggered self-sorting assembly of two peptides in cancer cells [95]. As shown in Figure 3B, a fiber-forming peptide amphiphile E3C16 was linked to a hydrophilic peptide sequence EEEEEE via a disulfide bond. The resultant E3C16E6 formed irregular nanostructures. Additionally, seleno-methionine residues were incorporated into a fiber-forming bola-amphiphile EVM to prevent 1D self-assembly of EVM^{SeO} by oxidizing the selenide to selenoxide groups. Upon cellular internalization of two peptides, intracellular GSH cleaved the disulfide bond of E3C16E6 and reduced the selenoxide group of EVM^{SeO}. As a result, E3C16 and EVM^{Se} self-assembled into self-sorted nanofibers. The twisted E3C16 fibers located around the Golgi apparatus due to the reduced thiol from the peptides, targeting the cysteine-rich proteins on the organelle. Meanwhile, the EVM^{Se} fibers grew near the endoplasmic reticulum (ER) due to a p-toluene sulfonamide (Ts) moiety that conferred ER-targeting capability. These mechanisms culminated in combinatorial organelle dysfunction and subsequent cancer cell death.

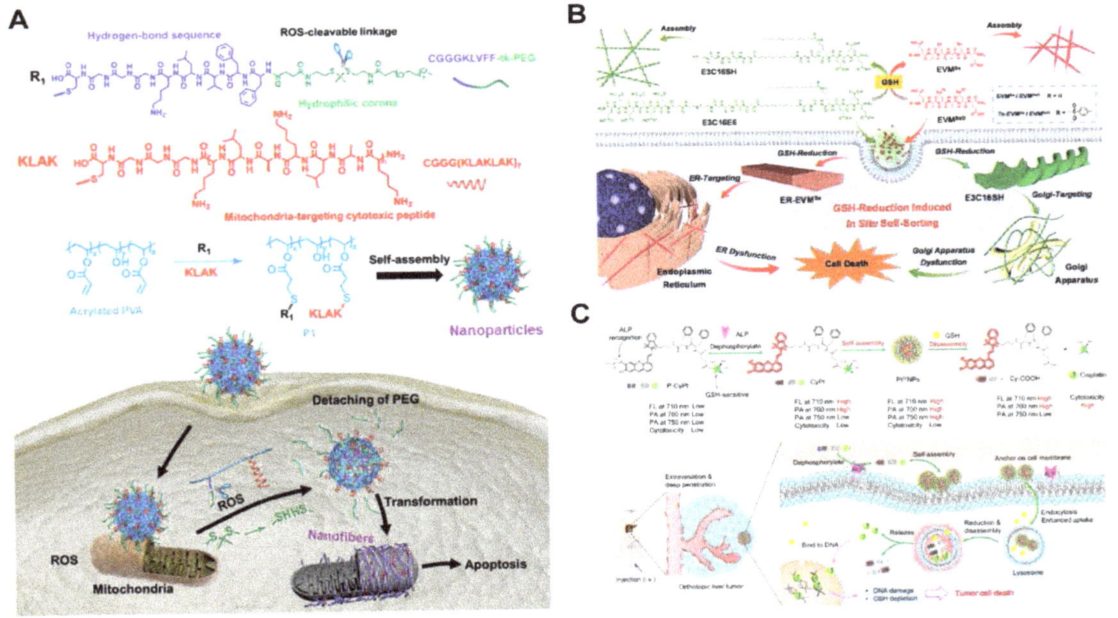

Figure 3. (**A**) Synthesis pathway of ROS-sensitive PPCs and morphological transformation of mitochondrial positions [90]. Reprinted (adapted) with permission from Ref. [90]. Copyright 2019 American Chemical Society. (**B**) Schematic representation of in situ self-sorting peptide assembly within living cells [95]. Reprinted (adapted) with permission from Ref. [95]. Copyright 2022 American Chemical Society. (**C**) Overall design and proposed mechanism of P-CyPt in cancer therapy [96]. Reproduced with permission from Ref. [96], Copyright © 2023.

In addition to morphological changes, intracellular GSH is a key player in inducing the disassembly of therapeutic peptide nanostructures. Ye, Guo, Liu, and coworkers recently demonstrated the controlled self-assembly and disassembly of a peptide-based cisplatin prodrug for cancer therapy [96]. Prodrug P-CyPt consisted of a hydrophobic DFDF dipeptide, a near-infrared (NIR) merocyanine fluorophore capped with an ALP-sensitive phosphate group, and a GSH-reducible cisplatin prodrug (Pt (IV)) (Figure 3C). Pericellular ALP enzymes in the cancer cells catalyzed the conversion of P-CyPt to CyPt, turning on the NIR fluorescence signal and facilitating the assembly of Pt (IV) nanoparticles due to the increased molecular hydrophobicity. Upon internalization of locally formed Pt^{IV}NPs, abundant endogenous GSH reduced Pt^{IV} into Pt^{II}. This promoted disassembly of the resultant Cy-COOH nanoparticles and release of the cytotoxic cisplatin. The ROS-responsive disassembly process showed promising anti-cancer efficacy by increasing the intracellular cisplatin concentration while decreasing the GSH level.

3.3. *pH-Responsive Materials*

Cancer cells typically exhibit a slightly acidic extracellular environment due to the deregulated metabolism and the accumulation of lactic acid [72,97,98]. Additionally, following endocytosis, the pH value within the endosomes and lysosomes falls within the range of 4–6. These factors have led researchers to extensively explore pH as a physiological stimulus of cancer for the development of responsive therapeutic assemblies [99–101]. For example, Weil, Ng, and their coworkers demonstrated a depsipeptide undergoing multistage transformations upon exposure to cancer cells [102]. As shown in Figure 4A, the peptide comprised a self-assembling motif ISA and a cell-penetrating TAT peptide RRRQRRRKKRGY, featuring a pH-dependent boronic acid-salicylhydroxamate crosslinker. Upon endocytosis facilitated by TAT, the acidic environment cleaved the linker, releasing the pro-assembling peptide. Subsequently, elevated or endogenous H_2O_2 within the cancer cells removed the boronic acid cage, inducing an O→N acyl shift. This shift further generated the ISA motif, prompting self-assembly into intracellular fibrillar structures that triggered cell apoptosis.

In addition to incorporating acid–labile bonds, the protonation of residues in self-assembling peptides is another powerful strategy [100,101]. For example, Wang and coworkers designed a pH-responsive laminin mimetic peptide (LMMP) that specifically formed an occlusion in tumor blood vessels [103]. LMMP was composed of a bispyrene-capped fibril-forming peptide sequence KLVFF, a thrombus-targeting sequence $(PEG)_8$-CREKA, and an His6 sequence designed to respond to the acidic environment of cancer cells (Figure 4B). After intravenous administration, LMMP nanoparticles adhered to the microthrombi in tumor blood vessels and underwent transformation into nanofibers through His protonation, altering the molecule's hydrophilic–hydrophobic balance. These laminin-like fibers effectively captured red blood cells, leading to rapid occlusion within the tumor blood vessels. The action provided a pH-responsive, tumor blood vessel-specific therapeutic approach for cancer treatment. Similarly, Wu and colleagues exploited residue protonation to make acid-specific targeting materials [104]. They prepared nanospheres by mixing a cytolytic melittin peptide GIGAVLKVLTTGLPALISWIKRKRQQ, an NIR-absorbing photothermal molecule cypate, and a tumor-targeting hyaluronic acid (HA) (Figure 4C). Subsequent exposure to the acidic environment caused the complexes to transform into nanofibers. The pH-induced morphology change facilitated the prolonged retention of the hemolytic melittin peptide in tumor tissues and inhibited the mobility and metastasis of the cancer cells. This transformation was proposed to be associated with the protonation of the peptide amine groups and carboxyl groups in the cypate, which increased peptide hydrophilicity and cypate hydrophobicity. Subsequent laser irradiation further resulted in the formation of smaller nanoparticles, presumably due to the disturbed internal structures of the fibers after the photodegradation of cypate.

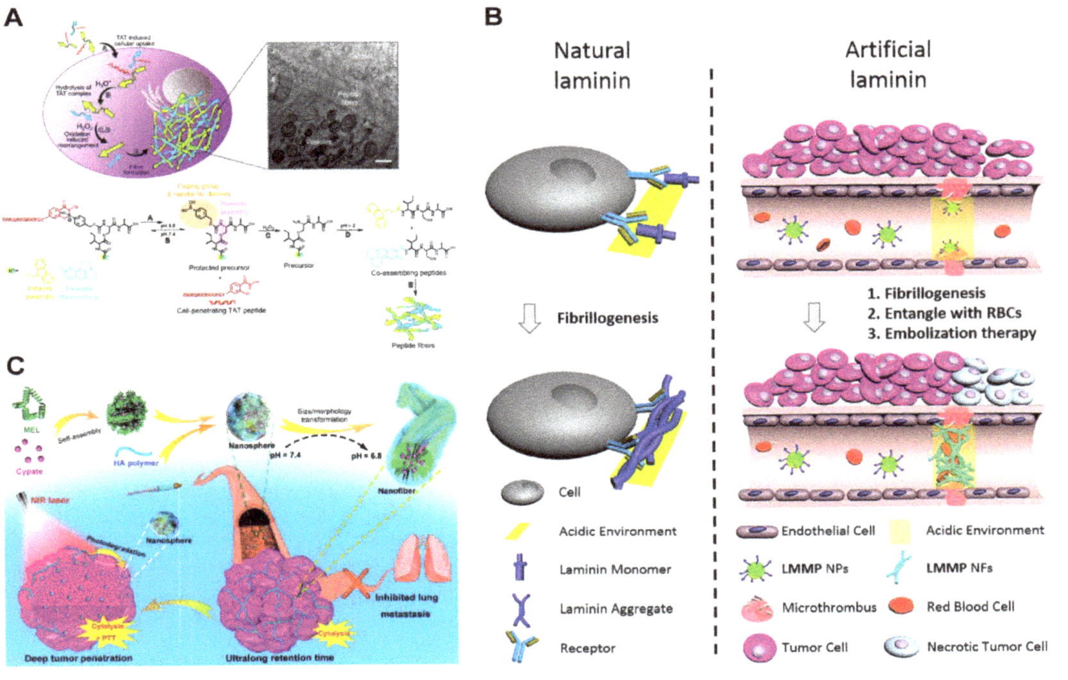

Figure 4. (**A**) Functionalized ISA peptide assembly process in cells [102]. Reprinted (adapted) with permission from Ref. [102]. Copyright 2020 American Chemical Society. (**B**) Schematic diagram of natural laminin and artificial laminin fiber formation [103]. Reprinted (adapted) with permission from Ref. [103]. Copyright 2020 American Chemical Society. (**C**) Preparation of MEL/Cypate@HA complexes and schematic representation of their successive size/morphology transitions under weakly acidic TME and near-infrared laser irradiation [104]. Reprinted (adapted) with permission from Ref. [104]. Copyright 2019 American Chemical Society.

Table 2. Representative stimulus-responsive targeting materials.

Name	Targeting Feature	Mechanism of Action	Application	Reference
$^{D}F^{D}Fp^{D}Y$	Alkaline phosphatase	Inhibiting cellular mass exchange	Cervical cancer, uterine sarcoma	[81]
Supra-BiCE	Alkaline phosphatase	Checkpoint blockade; enriching immune cells at cancer cell sites.	Colon carcinoma, breast cancer	[82]
DMFA	MMP-2	Facilitating Ca^{2+} influx and disrupting Na^{+}/K^{+}-ATPase	Cervical cancer	[83]
Ac-DEVDD-TPP	Caspase-3	Photodynamic therapy	Oral cancer	[84]
CPT-LFPR	Cathepsin B	Sustained release of camptothecin	Cervical cancer	[85]
PPCs	ROS	Inducing organelle dysfunction	Cervical cancer	[90]
E3C16E6 and EVM^{SeO}	GSH	Inducing organelle dysfunction	Cervical cancer	[95]
P-CyPt	GSH and alkaline phosphatase	Cisplatin	Liver cancer, cervical cancer	[96]
depsipeptide	pH and H_2O_2	Fiber-induced cell apoptosis	Lung cancer	[102]
LMMP	pH	Promoting occlusion in tumor blood vessels	Breast cancer	[103]
MEL/Cypate@HA	pH	Retention of melittin	Lung cancer	[104]

Protonatable residues underlie the mechanism of another widely used family of cell-targeting peptides, the pH low-insertion peptide (pHLIP). Originally identified by Engelman and coworkers, the prototype pHLIP sequence was derived from the C-helices of bacteriorhodopsin [105]. This peptide uniquely responds to acidosis at cancer cell surfaces by using protonatable residues to transit to α-helical structure, facilitating its insertion across the plasma membrane. This feature is reflected in the sequence AEQNPIY-WARA**D**WLFTTPLLL**D**LALLV**D**A**D**-**E**GT. It has a polar residue-rich segment in the N-terminus, a central transmembrane segment containing protonatable residues (shown in bold), and a C-terminal flanking region. Under physiological pH conditions, the negatively charged, deprotonated residues hinder membrane insertion and promote less-structured or disordered peptide conformations. Conversely, at lower pH, protonation of these residues increases the peptide's overall hydrophobicity. This restores the helical conformation of the central segment and allows the peptide to partition across cell membranes. Upon insertion, the C-terminus of the peptide is positioned in the cytoplasm while the N-domain remains in the extracellular space. Sequence modifications have led to various pHLIP derivatives, such as Variant 3 (Var3: ADDQN-PWRAYL**D**LLFPT**D**TLLL**D**LLW), which exhibits significantly enhanced cancer-cell insertion efficiency [106]. The pH-responsive folding and membrane insertion properties of pHLIP offer a robust approach to cancer-cell targeting and payload delivery. A recent review summarized the applications of pHLIP technology in cancer treatment [107]. Several pHLIP-related therapeutic agents are advancing in clinical trials (Table 1). For instance, pHLIP-exatecan (CBX-12, NCT04902872, NCT05691517), which carries the topoisomerase inhibitor, is in clinical trials for treating human ovarian cancer and advanced solid tumors [66]. In addition, a pHLIP conjugated to a near-infrared fluorescent dye (indocyanine green, ICG) has been evaluated in a Phase I/II first-in-human clinical study (NCT05130801) [67]. This is expected to facilitate fluorescence-guided surgery in breast cancer patients undergoing breast-conserving surgery. Furthermore, an ^{18}F-labeled Var3 construct is in a Phase I clinical trial as a diagnostic/imaging agent in breast cancer patients (NCT04054986) [68].

4. Conclusions and Perspective

In the field of cancer therapy, both peptide-based molecular binders and responsive targeting materials play crucial roles in facilitating cargo delivery and release [108–110], immune checkpoint inhibition [111], immune-effector cell recruitment [112], and organelle destruction [113]. Significant progress has been made in the search for high-affinity, protein-specific cyclic binders, many of which have advanced to clinical trials as cancer therapeutics. Unlike single-target binders, multicycle structures offer a broader scope for manipulating binding specificity, promising multi-targeting capabilities essential for advanced cancer treatment [45,114]. Moreover, recent advances in computing power and modelling frameworks have revolutionized the exploration of the chemical space of peptide-based binders, utilizing diverse molecular backbones beyond natural peptide chains. This has the potential to expand the binder library beyond the limitations of current biological display techniques [115,116]. Compared to binder-based therapeutics, stimulus-responsive targeting materials are still in their infancy and have a long way to go before reaching clinical trials (Table 2). Challenges persist in multiple aspects. For instance, many endogenous stimuli overlap with the natural physiological processes of cells. They exhibit heterogeneity in normal and pathological contexts, resulting in on-target, off-cancer effects. To address this, the next frontier is to develop next-generation materials with built-in responsiveness to multiple physiological properties or combinations of pathological cues and exogenous stimuli (i.e., light, magnetic, and acoustic energy). Unlike size-defined binders, peptide assemblies such as intracellular fibers exhibit polydispersity in diameter and length, potentially compromising targeting efficacy and reproducibility. There is an urgent need to incorporate molecular codes (i.e., frustration elements) that restrict self-assembly propensity into the design of future targeting assemblies [117,118]. Recent breakthroughs in artificial intelligence, coupled with the rapid accumulation of data on supramolecular structures (i.e., cryo-EM structural reconstruction), greatly improve the computational design of self-assembling peptides [119,120]. These advances hold the promise of creating targeting materials that integrate full molecular code (i.e., composition, sequence, and chemical moiety) within peptides in unprecedented ways. We anticipate that the combination of these areas of progress will enable continuous evolution of peptide-based cyclic binders and dynamic targeting materials to address current challenges in targeted cancer therapy.

Author Contributions: X.C.: literature review, manuscript preparation, and writing—original draft. D.W.: literature review, manuscript preparation, and writing—original draft. Y.-B.J.: review and editing. T.J.: writing—review and editing. All authors have read and agreed to the published version of the manuscript.

Funding: This research was funded by National Natural Science Foundation of China (grants 22074128, 22241503, 92356308) and Fundamental Research Funds for the Central Universities (grants 20720210013, 20720220005).

Conflicts of Interest: The authors declare no conflicts of interest.

References

1. Zugazagoitia, J.; Guedes, C.; Ponce, S.; Ferrer, I.; Molina-Pinelo, S.; Paz-Ares, L. Current challenges in cancer treatment. *Clin. Ther.* **2016**, *38*, 1551–1566. [CrossRef] [PubMed]
2. Norsworthy, K.J.; Ko, C.-W.; Lee, J.E.; Liu, J.; John, C.S.; Przepiorka, D.; Farrell, A.T.; Pazdur, R. FDA approval summary: Mylotarg for treatment of patients with relapsed or refractory CD33-positive acute myeloid leukemia. *Oncologist* **2018**, *23*, 1103–1108. [CrossRef] [PubMed]
3. Shinde-Jadhav, S.; Mansure, J.J.; Rayes, R.F.; Marcq, G.; Ayoub, M.; Skowronski, R.; Kool, R.; Bourdeau, F.; Brimo, F.; Spicer, J.; et al. Role of neutrophil extracellular traps in radiation resistance of invasive bladder cancer. *Nat. Commun.* **2021**, *12*, 2776. [CrossRef]
4. Thundimadathil, J. Cancer treatment using peptides: Current therapies and future prospects. *J. Amino Acids* **2012**, *2012*, 1–13. [CrossRef] [PubMed]
5. Demaria, O.; Cornen, S.; Daëron, M.; Morel, Y.; Medzhitov, R.; Vivier, E. Harnessing innate immunity in cancer therapy. *Nature* **2019**, *574*, 45–56. [CrossRef] [PubMed]
6. Paterson, C.; Denlinger, N.; Yang, Y.P. Recent advances and challenges in cancer immunotherapy. *Cancers* **2022**, *14*, 3972. [CrossRef] [PubMed]
7. Strebhardt, K.; Ullrich, A. Paul ehrlich's magic bullet concept: 100 years of progress. *Nat. Rev. Cancer* **2008**, *8*, 473–480. [CrossRef]
8. Hansel, T.T.; Kropshofer, H.; Singer, T.; Mitchell, J.A.; George, A.J.T. The safety and side effects of monoclonal antibodies. *Nat. Rev. Drug Discov.* **2010**, *9*, 325–338. [CrossRef]
9. Yang, W.; Mixich, L.; Boonstra, E.; Cabral, H. Polymer-based mRNA delivery strategies for advanced therapies. *Adv. Healthc. Mater.* **2023**, *12*, e2202688. [CrossRef]
10. Allen, T.M. Ligand-targeted therapeutics in anticancer therapy. *Nat. Rev. Cancer* **2002**, *2*, 750–763. [CrossRef]
11. Qin, M.; Xia, H.; Xu, W.; Chen, B.; Wang, Y. The spatiotemporal journey of nanomedicines in solid tumors on their therapeutic efficacy. *Adv. Drug Deliv. Rev.* **2023**, *203*, 115137. [CrossRef] [PubMed]
12. Tuerk, C.; Gold, L. Systematic evolution of ligands by exponential enrichment: RNA ligands to bacteriophage T4 DNA polymerase. *Science* **1990**, *249*, 505–510. [CrossRef] [PubMed]
13. Robertson, D.L.; Joyce, G.F. Selection in vitro of an RNA enzyme that specifically cleaves single-stranded DNA. *Nature* **1990**, *344*, 467–468. [CrossRef] [PubMed]
14. Ellington, A.D.; Szostak, J.W. In vitro selection of RNA molecules that bind specific ligands. *Nature* **1990**, *346*, 818–822. [CrossRef] [PubMed]
15. Zhou, J.H.; Rossi, J. Aptamers as targeted therapeutics: Current potential and challenges. *Nat. Rev. Drug Discov.* **2017**, *16*, 181–202. [CrossRef] [PubMed]
16. Kulabhusan, P.K.; Hussain, B.; Yüce, M. Current perspectives on aptamers as diagnostic tools and therapeutic agents. *Pharmaceutics* **2020**, *12*, 646. [CrossRef] [PubMed]
17. Xie, S.T.; Sun, W.D.; Fu, T.; Liu, X.S.; Chen, P.; Qiu, L.P.; Qu, F.L.; Tan, W.H. Aptamer-based targeted delivery of functional nucleic acids. *J. Am. Chem. Soc.* **2023**, *145*, 7677–7691. [CrossRef] [PubMed]
18. Diao, L.; Meibohm, B. Pharmacokinetics and pharmacokinetic-pharmacodynamic correlations of therapeutic peptides. *Clin. Pharmacokinet.* **2013**, *52*, 855–868. [CrossRef]
19. Luo, X.; Wu, Y.; Zhang, X.; Tang, M.; Ju, F.; Qin, Z.; Duns, G.J.; Zhang, W.D.; Qin, J.J.; Luan, X. Peptide-based strategies for overcoming multidrug-resistance in cancer therapy. *Chin. Chem. Lett.* **2024**, 109724, *in press*. [CrossRef]
20. Samec, T.; Boulos, J.; Gilmore, S.; Hazelton, A.; Alexander-Bryant, A. Peptide-based delivery of therapeutics in cancer treatment. *Mater. Today Bio.* **2022**, *14*, 100248. [CrossRef]

21. Li, C.M.; Haratipour, P.; Lingeman, R.G.; Perry, J.J.P.; Gu, L.; Hickey, R.J.; Malkas, L.H. Novel peptide therapeutic approaches for cancer treatment. *Cells* **2021**, *10*, 2908. [CrossRef] [PubMed]
22. Ayo, A.; Laakkonen, P. Peptide-based strategies for targeted tumor treatment and imaging. *Pharmaceutics* **2021**, *13*, 481. [CrossRef] [PubMed]
23. Nhàn, N.T.T.; Yamada, T.; Yamada, K.H. Peptide-based agents for cancer treatment: Current applications and future directions. *Int. J. Mol. Sci.* **2023**, *24*, 12931. [CrossRef] [PubMed]
24. Jalil, A.T.; Abdulhadi, M.A.; Al-Ameer, L.R.; Taher, W.M.; Abdulameer, S.J.; Abosaooda, M.; Fadhil, A.A. Peptide-based therapeutics in cancer therapy. *Mol. Biotechnol.* **2023**. [CrossRef] [PubMed]
25. Chagri, S.; Ng, D.Y.W.; Weil, T. Designing bioresponsive nanomaterials for intracellular self-assembly. *Nat. Rev. Chem.* **2022**, *6*, 320–338. [CrossRef] [PubMed]
26. Mu, R.Q.; Zhu, D.Z.; Abdulmalik, S.; Wijekoon, S.; Wei, G.; Kumbar, S.G. Stimuli-responsive peptide assemblies: Design, self-assembly, modulation, and biomedical applications. *Bioact. Mater.* **2024**, *35*, 181–207. [CrossRef] [PubMed]
27. Zhou, Y.; Li, Q.Q.; Wu, Y.; Li, X.Y.; Zhou, Y.; Wang, Z.; Liang, H.; Ding, F.Q.; Hong, S.; Steinmetz, N.F.; et al. Molecularly stimuli-responsive self-assembled peptide nanoparticles for targeted imaging and therapy. *ACS Nano* **2023**, *17*, 8004–8025. [CrossRef]
28. Fernald, K.; Kurokawa, M. Evading apoptosis in cancer. *Trends Cell Biol.* **2013**, *23*, 620–633. [CrossRef]
29. Zitvogel, L.; Tesniere, A.; Kroemer, G. Cancer despite immunosurveillance: Immunoselection and immunosubversion. *Nat. Rev. Immunol.* **2006**, *6*, 715–727. [CrossRef]
30. Min, H.Y.; Lee, H.Y. Molecular targeted therapy for anticancer treatment. *Exp. Mol. Med.* **2022**, *54*, 1670–1694. [CrossRef]
31. Allen, G.M.; Lim, W.A. Rethinking cancer targeting strategies in the era of smart cell therapeutics. *Nat. Rev. Cancer* **2022**, *22*, 693–702. [CrossRef]
32. Wang, D.R.; Wu, X.L.; Sun, Y.L. Therapeutic targets and biomarkers of tumor immunotherapy: Response versus non-response. *Signal Transduct. Targted Ther.* **2022**, *7*, 331. [CrossRef]
33. Desgrosellier, J.S.; Cheresh, D.A. Integrins in cancer: Biological implications and therapeutic opportunities. *Nat. Rev. Cancer* **2010**, *10*, 9–22. [CrossRef] [PubMed]
34. Pierschbacher, M.D.; Ruoslahti, E. Cell attachment activity of fibronectin can be duplicated by small synthetic fragments of the molecule. *Nature* **1984**, *309*, 30–33. [CrossRef]
35. Pytela, R.; Pierschbacher, M.D.; Ruoslahti, E. Identification and isolation of a 140-kd cell-surface glycoprotein with properties expected of a fibronectin receptor. *Cell* **1985**, *40*, 191–198. [CrossRef]
36. Battistini, L.; Bugatti, K.; Sartori, A.; Curti, C.; Zanardi, F. RGD peptide-drug conjugates as effective dual targeting platforms: Recent advances. *Eur. J. Org. Chem.* **2021**, *2021*, 2506–2528. [CrossRef]
37. Kapp, T.G.; Rechenmacher, F.; Neubauer, S.; Maltsev, O.V.; Cavalcanti-Adam, E.A.; Zarka, R.; Reuning, U.; Notni, J.; Wester, H.J.; Mas-Moruno, C.; et al. A comprehensive evaluation of the activity and selectivity profile of ligands for RGD-binding integrins. *Sci. Rep.* **2017**, *7*, 39805. [CrossRef] [PubMed]
38. Javid, H.; Oryani, M.A.; Rezagholinejad, N.; Esparham, A.; Tajaldini, M.; Karimi-Shahri, M. RGD peptide in cancer targeting: Benefits, challenges, solutions, and possible integrin-RGD interactions. *Cancer Med-Us* **2024**, *13*, e6800. [CrossRef]
39. Smith, G.P. Filamentous fusion phage-novel expression vectors that display cloned antigens on the virion surface. *Science* **1985**, *228*, 1315–1317. [CrossRef]
40. Roberts, R.W.; Szostak, J.W. RNA-peptide fusions for the in vitro selection of peptides and proteins. *Proc. Natl. Acad. Sci. USA* **1997**, *94*, 12297–12302. [CrossRef]
41. Swain, S.M.; Shastry, M.; Hamilton, E. Targeting HER2-positive breast cancer: Advances and future directions. *Nat. Rev. Drug Discov.* **2023**, *22*, 101–126. [CrossRef] [PubMed]
42. Karasseva, N.G.; Glinsky, V.V.; Chen, N.X.; Komatireddy, R.; Quinn, T.P. Identification and characterization of peptides that bind human ErbB-2 selected from a bacteriophage display library. *J. Protein Chem.* **2002**, *21*, 287–296. [CrossRef] [PubMed]
43. Shadidi, M.; Sioud, M. Identification of novel carrier peptides for the specific delivery of therapeutics into cancer cells. *FASEB J.* **2002**, *17*, 256–258. [CrossRef] [PubMed]
44. Vinogradov, A.A.; Yin, Y.Z.; Suga, H. Macrocyclic peptides as drug candidates: Recent progress and remaining challenges. *J. Am. Chem. Soc.* **2019**, *141*, 4167–4181. [CrossRef] [PubMed]
45. Ji, X.J.; Nielsen, A.L.; Heinis, C. Cyclic peptides for drug development. *Angew. Chem. Int. Ed.* **2024**, *63*, e202308251. [CrossRef] [PubMed]
46. Dechantsreiter, M.A.; Planker, E.; Mathä, B.; Lohof, E.; Hölzemann, G.; Jonczyk, A.; Goodman, S.L.; Kessler, H. N-methylated cyclic RGD peptides as highly active and selective $\alpha\beta$ integrin antagonists. *J. Med. Chem.* **1999**, *42*, 3033–3040. [CrossRef] [PubMed]
47. Yu, Y.P.; Wang, Q.; Liu, Y.C.; Xie, Y. Molecular basis for the targeted binding of RGD-containing peptide to integrin $\alpha_V\beta_3$. *Biomaterials* **2014**, *35*, 1667–1675. [CrossRef] [PubMed]
48. Wu, C.L.; Leroux, J.C.; Gauthier, M.A. Twin disulfides for orthogonal disulfide pairing and the directed folding of multicyclic peptides. *Nat. Chem.* **2012**, *4*, 1045–1050. [CrossRef] [PubMed]
49. Lu, S.M.; Fan, S.H.; Xiao, S.L.; Li, J.J.; Zhang, S.L.; Wu, Y.P.; Kong, C.L.; Zhuang, J.; Liu, H.T.; Zhao, Y.B.; et al. Disulfide-directed multicyclic peptide libraries for the discovery of peptide ligands and drugs. *J. Am. Chem. Soc.* **2023**, *145*, 1964–1972. [CrossRef]

50. White, C.J.; Yudin, A.K. Contemporary strategies for peptide macrocyclization. *Nat. Chem.* **2011**, *3*, 509–524. [CrossRef]
51. Deyle, K.; Kong, X.D.; Heinis, C. Phage selection of cyclic peptides for application in research and drug development. *Acc. Chem. Res.* **2017**, *50*, 1866–1874. [CrossRef]
52. Heinis, C.; Rutherford, T.; Freund, S.; Winter, G. Phage-encoded combinatorial chemical libraries based on bicyclic peptides. *Nat. Chem. Biol.* **2009**, *5*, 502–507. [CrossRef] [PubMed]
53. Yamagishi, Y.; Shoji, I.; Miyagawa, S.; Kawakami, T.; Katoh, T.; Goto, Y.; Suga, H. Natural product-like macrocyclic-methyl-peptide inhibitors against a ubiquitin ligase uncovered from a ribosome-expressed de novo library. *Chem. Biol.* **2011**, *18*, 1562–1570. [CrossRef] [PubMed]
54. Huang, Y.C.; Wiedmann, M.M.; Suga, H. RNA display methods for the discovery of bioactive macrocycles. *Chem. Rev.* **2019**, *119*, 10360–10391. [CrossRef]
55. Goto, Y.; Katoh, T.; Suga, H. Flexizymes for genetic code reprogramming. *Nat. Protoc.* **2011**, *6*, 779–790. [CrossRef] [PubMed]
56. Worm, D.J.; Els-Heindl, S.; Beck-Sickinger, A.G. Targeting of peptide-binding receptors on cancer cells with peptide-drug conjugates. *Pept. Sci.* **2020**, *112*, e24171. [CrossRef]
57. Vadevoo, S.M.P.; Gurung, S.; Lee, H.S.; Gunassekaran, G.R.; Lee, S.M.; Yoon, J.W.; Lee, Y.K.; Lee, B.Y.H. Peptides as multifunctional players in cancer therapy. *Exp. Mol. Med.* **2023**, *55*, 1099–1109. [CrossRef]
58. Li, X.T.; Craven, T.W.; Levine, P.M. Cyclic peptide screening methods for preclinical drug discovery. *J. Med. Chem.* **2022**, *65*, 11913–11926. [CrossRef]
59. Nakajima, H.; Kim, Y.B.; Terano, H.; Yoshida, M.; Horinouchi, S. FR901228, a potent antitumor antibiotic, is a novel histone deacetylase inhibitor. *Exp. Cell Res.* **1998**, *241*, 126–133. [CrossRef]
60. Hennrich, U.; Kopka, K. Lutathera®: The First FDA- and EMA-approved radiopharmaceutical for peptide receptor radionuclide therapy. *Pharmaceuticals* **2019**, *12*, 114. [CrossRef]
61. Shi, J.; Fan, D.; Dong, C.; Liu, H.; Jia, B.; Zhao, H.; Jin, X.; Liu, Z.; Li, F.; Wang, F. Anti-tumor effect of integrin targeted ^{177}Lu-3PRGD$_2$ and combined therapy with endostar. *Theranostics* **2014**, *4*, 256–266. [CrossRef] [PubMed]
62. Loriot, Y.; Siefker-Radtke, A.O.; Friedlander, T.W.; Necchi, A.; Wei, A.Z.; Sridhar, S.S.; Garmezy, B.; Arroyo, S.; Gartside, E.; Liu, J.; et al. A phase 2/3 study of bicycle toxin conjugate BT8009 targeting nectin-4 in patients with locally advanced or metastatic urothelial cancer (la/mUC): Duravelo-2. *J. Clin. Oncol.* **2024**, *42*, TPS4619. [CrossRef]
63. Zhang, H.; Chen, S. Cyclic peptide drugs approved in the last two decades (2001–2021). *RSC Chem. Biol.* **2022**, *3*, 18–31. [CrossRef] [PubMed]
64. Fu, C.; Yu, L.; Miao, Y.; Liu, X.; Yu, Z.; Wei, M. Peptide-drug conjugates (PDCs): A novel trend of research and development on targeted therapy, hype or hope? *Acta Pharm. Sin. B* **2023**, *13*, 498–516. [CrossRef]
65. Wang, M.D.; Lv, G.T.; An, H.W.; Zhang, N.Y.; Wang, H. In situ self-assembly of bispecific peptide for cancer immunotherapy. *Angew. Chem. Int. Ed.* **2022**, *61*, e202113649. [CrossRef] [PubMed]
66. Shin, S.; O'Sullivan Coyne, G.H.; Rubinstein, L.V.; Takebe, N.; Wright, J.J.; Wilsker, D.; Ferry-Galow, K.V.; Karlovich, C.A.; Anderson, L.; Kuhlmann, L.; et al. Pilot study of CBX-12 pharmacodynamics in patients with advanced solid tumors. *J. Clin. Oncol.* **2024**, *42*, TPS3187. [CrossRef]
67. Wang, F.; Qu, L.; Ren, F.; Baghdasaryan, A.; Jiang, Y.; Hsu, R.; Liang, P.; Li, J.; Zhu, G.; Ma, Z.; et al. High-precision tumor resection down to few-cell level guided by NIR-IIb molecular fluorescence imaging. *Proc. Natl. Acad. Sci. USA* **2022**, *119*, e2123111119. [CrossRef] [PubMed]
68. Van der Heide, C.D.; Dalm, S.U. Radionuclide imaging and therapy directed towards the tumor microenvironment: A multi-cancer approach for personalized medicine. *Eur. J. Nucl. Med. Mol. Imaging* **2022**, *49*, 4616–4641. [CrossRef] [PubMed]
69. Xue, L.L.; Thatte, A.S.; Mai, D.; Haley, R.M.; Gong, N.Q.; Han, X.X.; Wang, K.; Sheppard, N.C.; June, C.H.; Mitchell, M.J. Responsive biomaterials: Optimizing control of cancer immunotherapy. *Nat. Rev. Mater.* **2024**, *9*, 100–118. [CrossRef]
70. Martínez-Reyes, I.; Chandel, N.S. Cancer metabolism: Looking forward. *Nat. Rev. Cancer* **2021**, *21*, 669–680. [CrossRef]
71. Chao, Y.; Liu, Z. Biomaterials tools to modulate the tumour microenvironment in immunotherapy. *Nat. Rev. Bioeng.* **2023**, *1*, 125–138. [CrossRef]
72. Quail, D.F.; Joyce, J.A. Microenvironmental regulation of tumor progression and metastasis. *Nat. Med.* **2013**, *19*, 1423–1437. [CrossRef] [PubMed]
73. Shahriari, M.; Zahiri, M.; Abnous, K.; Taghdisi, S.M.; Ramezani, M.; Alibolandi, M. Enzyme responsive drug delivery systems in cancer treatment. *J. Control Release* **2019**, *308*, 172–189. [CrossRef]
74. Lou, X.F.; Du, Y.Z.; Xu, X.L. Endogenous enzyme-responsive nanoplatforms for anti-tumor therapy. *Curr. Drug Targets* **2021**, *22*, 845–855. [CrossRef] [PubMed]
75. Maggi, M.; Scotti, C. Enzymes in metabolic anticancer therapy. *Adv. Exp. Med. Biol.* **2019**, *1148*, 173–199.
76. Li, M.Q.; Zhao, G.K.; Su, W.K.; Shuai, Q. Enzyme-responsive nanoparticles for anti-tumor drug delivery. *Front. Chem.* **2020**, *8*, 647. [CrossRef]
77. Zhao, Y.; Guo, Y.G.; Tang, L. Engineering cancer vaccines using stimuli-responsive biomaterials. *Nano Res.* **2018**, *11*, 5355–5371. [CrossRef]
78. Yang, Z.; Liang, G.; Xu, B. Enzymatic hydrogelation of small molecules. *Acc. Chem. Res.* **2008**, *41*, 315–326. [CrossRef] [PubMed]
79. He, H.J.; Tan, W.Y.; Guo, J.Q.; Yi, M.H.; Shy, A.N.; Xu, B. Enzymatic noncovalent synthesis. *Chem. Rev.* **2020**, *120*, 9994–10078. [CrossRef]

80. Yang, Z.M.; Gu, H.W.; Fu, D.G.; Gao, P.; Lam, J.K.; Xu, B. Enzymatic formation of supramolecular hydrogels. *Adv. Mater.* **2004**, *16*, 1440–1444. [CrossRef]
81. Kuang, Y.; Shi, J.; Li, J.; Yuan, D.; Alberti, K.A.; Xu, Q.; Xu, B. Pericellular hydrogel/nanonets inhibit cancer cells. *Angew. Chem. Int. Ed.* **2014**, *53*, 8104–8107. [CrossRef]
82. Chen, Y.M.; Li, W.; Wang, Z.Q.; Yu, Y.Y.; Li, J.; Ding, Y.H.; Hu, Z.W.; Liu, Q.; Yang, Z.M.; Gao, J. A transformable supramolecular bispecific cell engager for augmenting natural killer and T cell-based cancer immunotherapy. *Adv. Mater.* **2024**, *36*, e2306736. [CrossRef]
83. Zhang, W.; Hu, J.J.; Liu, R.; Dai, J.; Yuan, L.Z.; Liu, Y.H.; Chen, B.C.; Gong, M.X.; Xia, F.; Lou, X.D. A peptide-conjugated probe with cleavage-induced morphological change for treatment on tumor cell membrane. *Adv. Sci.* **2023**, *10*, 2207228. [CrossRef]
84. Liu, X.Y.; Zhan, W.J.; Gao, G.; Jiang, Q.C.; Zhang, X.P.; Zhang, H.B.; Sun, X.B.; Han, W.; Wu, F.G.; Liang, G.L. Apoptosis-amplified assembly of porphyrin nanofiber enhances photodynamic therapy of oral tumor. *J. Am. Chem. Soc.* **2023**, *145*, 7918–7930. [CrossRef] [PubMed]
85. Cheng, D.B.; Wang, D.; Gao, Y.J.; Wang, L.; Qiao, Z.Y.; Wang, H. Autocatalytic morphology transformation platform for targeted drug accumulation. *J. Am. Chem. Soc.* **2019**, *141*, 4406–4411. [CrossRef] [PubMed]
86. Cheung, E.C.; Vousden, K.H. The role of ROS in tumour development and progression. *Nat. Rev. Cancer* **2022**, *22*, 280–297. [CrossRef] [PubMed]
87. Curvello, R.; Kast, V.; Ordoñez-Morán, P.; Mata, A.; Loessner, D. Biomaterial-based platforms for tumour tissue engineering. *Nat. Rev. Mater.* **2023**, *8*, 314–330. [CrossRef]
88. De Angelis, B.; Depalo, N.; Petronella, F.; Quintarelli, C.; Curri, M.L.; Pani, R.; Calogero, A.; Locatelli, F.; De Sio, L. Stimuli-responsive nanoparticle-assisted immunotherapy: A new weapon against solid tumours. *J. Mater. Chem. B* **2020**, *8*, 1823–1840. [CrossRef]
89. Qin, J.L.; Sun, M.; Hu, W.; Cheng, J.J.; Fan, Z.; Du, J.Z. Stimuli-responsive hydrogels for cancer immunotherapy. *Polym. Chem.* **2023**, *14*, 793–802. [CrossRef]
90. Cheng, D.B.; Zhang, X.H.; Gao, Y.J.; Ji, L.; Hou, D.Y.; Wang, Z.Q.; Xu, W.H.; Qiao, Z.Y.; Wang, H. Endogenous reactive oxygen species-triggered morphology transformation for enhanced cooperative interaction with mitochondria. *J. Am. Chem. Soc.* **2019**, *141*, 7235–7239. [CrossRef]
91. Estrela, J.M.; Ortega, A.; Obrador, E. Glutathione in cancer biology and therapy. *Crit. Rev. Clin. Lab. Sci.* **2006**, *43*, 143–181. [CrossRef] [PubMed]
92. Niu, B.Y.; Liao, K.X.; Zhou, Y.X.; Wen, T.; Quan, G.L.; Pan, X.; Wu, C.B. Application of glutathione depletion in cancer therapy: Enhanced ROS-based therapy, ferroptosis, and chemotherapy. *Biomaterials* **2021**, *277*, 121110. [CrossRef] [PubMed]
93. Kennedy, L.; Sandhu, J.K.; Harper, M.E.; Cuperlovic-Culf, M. Role of glutathione in cancer: From mechanisms to therapies. *Biomolecules* **2020**, *10*, 1429. [CrossRef] [PubMed]
94. Bansal, A.; Simon, M.C. Glutathione metabolism in cancer progression and treatment resistance. *J. Cell Biol.* **2018**, *217*, 2291–2298. [CrossRef] [PubMed]
95. Liu, X.; Li, M.M.; Liu, J.Z.; Song, Y.Q.; Hu, B.B.; Wu, C.X.; Liu, A.A.; Zhou, H.; Long, J.F.; Shi, L.Q.; et al. Self-sorting peptide assemblies in living cells for simultaneous organelle targeting. *J. Am. Chem. Soc.* **2022**, *144*, 9312–9323. [CrossRef] [PubMed]
96. Wen, X.D.; Zhang, R.; Hu, Y.X.; Wu, L.Y.; Bai, H.; Song, D.F.; Wang, Y.F.; An, R.B.; Weng, J.H.; Zhang, S.R.; et al. Controlled sequential in situ self-assembly and disassembly of a fluorogenic cisplatin prodrug for cancer theranostics. *Nat. Commun.* **2023**, *14*, 800. [CrossRef] [PubMed]
97. Koltai, T. The pH paradigm in cancer. *Eur. J. Clin. Nutr.* **2020**, *74*, 14–19. [CrossRef] [PubMed]
98. Bogdanov, A.; Bogdanov, A.; Chubenko, V.; Volkov, N.; Moiseenko, F.; Moiseyenko, V. Tumor acidity: From hallmark of cancer to target of treatment. *Front. Oncol.* **2022**, *12*, 979154. [CrossRef]
99. Li, Z.; Zhu, Y.; Matson, J.B. pH-responsive self-assembling peptide-based biomaterials: Designs and applications. *ACS Appl. Bio Mater.* **2022**, *5*, 4635–4651. [CrossRef]
100. Yan, Y.F.; Ding, H.W. pH-responsive nanoparticles for cancer immunotherapy: A brief review. *Nanomaterials* **2020**, *10*, 1613. [CrossRef]
101. Manchun, S.; Dass, C.R.; Sriamornsak, P. Targeted therapy for cancer using pH-responsive nanocarrier systems. *Life Sci.* **2012**, *90*, 381–387. [CrossRef] [PubMed]
102. Pieszka, M.; Han, S.; Volkmann, C.; Graf, R.; Lieberwirth, I.; Landfester, K.; Ng, D.Y.W.; Weil, T. Controlled supramolecular assembly inside living cells by sequential multistaged chemical reactions. *J. Am. Chem. Soc.* **2020**, *142*, 15780–15789. [CrossRef]
103. Zhang, K.; Yang, P.P.; He, P.P.; Wen, S.F.; Zou, X.R.; Fan, Y.; Chen, Z.M.; Cao, H.; Yang, Z.; Yue, K.; et al. Peptide-based nanoparticles mimic fibrillogenesis of laminin in tumor vessels for precise embolization. *ACS Nano* **2020**, *14*, 7170–7180. [CrossRef] [PubMed]
104. Jia, H.R.; Zhu, Y.X.; Liu, X.Y.; Pan, G.Y.; Gao, G.; Sun, W.; Zhang, X.D.; Jiang, Y.W.; Wu, F.G. Construction of dually responsive nanotransformers with nanosphere-nanofiber-nanosphere transition for overcoming the size paradox of anticancer nanodrugs. *ACS Nano* **2019**, *13*, 11781–11792. [CrossRef]
105. Hunt, J.F.; Rath, P.; Rothschild, K.J.; Engelman, D.M. Spontaneous, pH-dependent membrane insertion of a transbilayer alpha-helix. *Biochemistry* **1997**, *36*, 15177–15192. [CrossRef] [PubMed]
106. Weerakkody, D.; Moshnikova, A.; Thakur, M.S.; Moshnikova, V.; Daniels, J.; Engelman, D.M.; Andreev, O.A.; Reshetnyak, Y.K. Family of pH (low) insertion peptides for tumor targeting. *Proc. Natl. Acad. Sci. USA* **2013**, *110*, 5834–5839. [CrossRef] [PubMed]

107. Reshetnyak, Y.K.; Andreev, O.A.; Engelman, D.M. Aiming the magic bullet: Targeted delivery of imaging and therapeutic agents to solid tumors by pHLIP peptides. *Front. Pharmacol.* **2024**, *15*, 1355893. [CrossRef]
108. Zhang, Y.; Zhang, Y.; Song, G.; He, Y.; Zhang, X.; Liu, Y.; Ju, H. A DNA-azobenzene nanopump fueled by upconversion luminescence for controllable intracellular drug release. *Angew. Chem. Int. Ed.* **2019**, *58*, 18207–18211. [CrossRef] [PubMed]
109. Wei, D.; Huang, Y.; Wang, B.; Ma, L.; Karges, J.; Xiao, H. Photo-reduction with NIR light of nucleus-targeting Pt^{IV} Nanoparticles for combined tumor-targeted chemotherapy and photodynamic immunotherapy. *Angew. Chem. Int. Ed.* **2022**, *61*, e202201486. [CrossRef]
110. Han, K.; Lei, Q.; Wang, S.B.; Hu, J.J.; Qiu, W.X.; Zhu, J.Y.; Yin, W.N.; Luo, X.; Zhang, X.Z. Dual-stage-light-guided tumor inhibition by mitochondria-targeted photodynamic therapy. *Adv. Funct. Mater.* **2015**, *25*, 2961–2971. [CrossRef]
111. Fuchs, N.; Zhang, L.; Calvo-Barreiro, L.; Kuncewicz, K.; Gabr, M. Inhibitors of immune checkpoints: Small molecule and peptide-based approaches. *J. Pers. Med.* **2024**, *14*, 68. [CrossRef] [PubMed]
112. Zhang, Y.; Liu, C.; Wu, C.; Song, L. Natural peptides for immunological regulation in cancer therapy: Mechanism, facts and perspectives. *Biomed. Pharmacother.* **2023**, *159*, 114257. [CrossRef] [PubMed]
113. Gabernet, G.; Müller, A.T.; Hiss, J.A.; Schneider, G. Membranolytic anticancer peptides. *MedChemComm* **2016**, *7*, 2232–2245. [CrossRef]
114. Wang, P.; Liu, J.; Zhu, X.; Kenry; Yan, Z.; Yan, J.; Jiang, J.; Fu, M.; Ge, J.; Zhu, Q.; et al. Modular synthesis of clickable peptides via late-stage maleimidation on C (7)-H tryptophan. *Nat. Commun.* **2023**, *14*, 3973. [CrossRef] [PubMed]
115. Bhardwaj, G.; O'Connor, J.; Rettie, S.; Huang, Y.H.; Ramelot, T.A.; Mulligan, V.K.; Alpkilic, G.G.; Palmer, J.; Bera, A.K.; Bick, M.J.; et al. Accurate de novo design of membrane-traversing macrocycles. *Cell* **2022**, *185*, 3520–3532. [CrossRef] [PubMed]
116. Salveson, P.J.; Moyer, A.P.; Said, M.Y.; Gökçe, G.; Li, X.; Kang, A.; Nguyen, H.; Bera, A.K.; Levine, P.M.; Bhardwaj, G.; et al. Expansive discovery of chemically diverse structured macrocyclic oligoamides. *Science* **2024**, *384*, 420–428. [CrossRef]
117. Cheng, D.; Chen, X.; Zhang, W.; Guo, P.; Xue, W.; Xia, J.; Wu, S.; Shi, J.; Ma, D.; Zuo, X.B.; et al. Design of multicomponent peptide fibrils with ordered and programmable compositional patterns. *Angew. Chem. Int. Ed.* **2023**, *62*, e202303684. [CrossRef] [PubMed]
118. Cheng, D.; Jia, F.; Jiang, Y.B.; Conticello, V.P.; Jiang, T. Assembly of peptide nanostructures with controllable sizes. *Nano Res.* **2024**, *17*, 151–161. [CrossRef]
119. Notin, P.; Rollins, N.; Gal, Y.; Sander, C.; Marks, D. Machine learning for functional protein design. *Nat. Biotechnol.* **2024**, *42*, 216–228. [CrossRef]
120. Min, J.W.; Rong, X.; Zhang, J.X.; Su, R.X.; Wang, Y.F.; Qi, W. Computational design of peptide assemblies. *J. Chem. Theory Comput.* **2024**, *20*, 532–550. [CrossRef] [PubMed]

targets

Review

Living Bacteriophage Engineering for Functional Material Synthesis, Bioanalytical Sensing and Disease Theranostics

Jing Zhang and Xuewen He *

The Key Lab of Health Chemistry and Molecular Diagnosis of Suzhou, College of Chemistry, Chemical Engineering and Materials Science, Soochow University, Suzhou 215123, China

* Correspondence: xheao@suda.edu.cn

Abstract: Facing the increasingly global crisis of antibiotic resistance, it is urgent to develop new antibacterial agents and methods. Simultaneously, as research progresses, the occurrence, development, and treatment of diseases, especially some malignant cancers, are found to be closely associated with the bacterial microenvironment, prompting us to reconsider the efficiency of existing antibacterial strategies for disease treatments. Bacteriophages have been employed as antibacterial agents for an extended period owing to their high biocompatibility and particular targetability toward the host bacterial strains. Nonetheless, they are almost neglected due to their slow and limited efficacy in antibacterial practice, especially in acute and severe infectious cases. In recent years, fantastic advancements in various biochemical technologies, such as bacteriophage display technology, genetic engineering, and chemical molecular engineering, have enabled scientists to conduct a broader range of modifications and transformations on the existing bacteriophages with inherited unique characteristics of themselves. As a result, a series of novel bacteriophage platforms are designed and fabricated with significantly enhanced properties and multiplied functionalities. These offer new avenues for combating infections caused by drug-resistant bacteria and treatment of malignancies that are associated with bacterial infections, holding great significance and potential in the innovative theranostic applications.

Keywords: bacteriophage; chemical modification; genetic engineering; bacterial targeting; bioanalysis; theranostics

1. Introduction

Bacteriophages, also called phages, are viruses with a unique ability to target and infect host bacteria, widely distributed on Earth [1]. Since being identified initially from *Staphylococcus* and *Shigella* bacteria in 1915 [2], bacteriophages have played a crucial role as natural antibacterial agents, contributing significantly to the advancement of biological research [3]. There are two major types of phages, i.e., lytic and temperate [4]. Phages are obligate intracellular parasites of hosts and have diverse life cycles. The life cycles include lytic, lysogenic, and pseudolysogenic cycles [5]. Lytic phages induce the lysis of host bacterial cells to release viral progeny [6]. Conversely, lysogenic or temperate phages integrate their nucleic acid (genome) into the host bacterial cell and replicate alongside the host, imparting new characteristics to the host bacteria [7]. Prompted by the growing crisis in antibiotic resistance, phage therapy is experiencing a resurgence as a potent solution against bacterial infections [8]. Although bacteriophages serve as widespread antibacterial agents, their clinical application still encounters a series of severe challenges. For instance, the effectiveness of individual bacteriophages is constrained, and obstacles like biofilms impede their penetration and weaken the antibacterial effectiveness [9]. Moreover, bacteriophages can potentially trigger adverse inflammatory responses [10]. Fortunately, genetic engineering and chemical alterations empower the modification of bacteriophages with enhanced robustness and multifunctionalities [11], expanding their host spectrum and amplifying their therapeutic capacity, presenting hopeful resolutions to these obstacles.

Bacteriophages consist mainly of proteins, forming the outer capsid, and protein-coding nucleic acids, which are protected within the capsid [12]. The protein capsid of bacteriophages contains various amino acids, each contributing distinct active functional groups [13]. Chemical modification techniques of phage rely on the interaction of functional groups on the capsid with active conjugates under specific solution pH and salt conditions [14]. By interacting with active sites like N-terminal alanine, lysine, and tyrosine, bacteriophages can acquire a variety of functionalities [15]. Bacteriophages have been tailored with various functional materials, including inorganic nanoparticles, organic molecules, fluorophores, DNA, antigens, and more, to align with their intended applications. Alternatively, genetic modification approaches entail the introduction of genes encoding amino acids, peptides, or protein sequences into either the bacteriophage genome or the capsid genes on the host plasmid, facilitating the generation of recombinant bacteriophages [16]. Utilizing advanced genome editing methods like recombination, CRISPR-Cas-assisted selection, or synthetic in vitro genome assembly allows for the precise design of particular genes or gene clusters within the bacteriophage genome, enabling the deliberate manipulation of functions such as merging, deleting, modifying, or diversifying [17]. In addition, bacteriophage display technology enables the swift and efficient identification of functional sites involved in protein–protein interactions, receptor binding, epitopes, isotopes, and antigens [18]. Through genetic engineering or chemical modification, bacteriophages acquire multifunctionality, paving the way for novel advancements in synthesizing functional materials, assembling and packing nanostructures, biosensing, biomedical applications, disease diagnosis, and treatment.

These genetic and chemical techniques for altering bacteriophages can be applied to develop diverse bacteriophage-based technologies [19]. These include the following: designing drugs and nanomaterials, synthesizing new proteins, detecting target molecules and pathogenic bacteria, constructing protein libraries, screening peptides or antibodies, treating diseases by killing bacteria, diagnosing diseases, and delivering drugs and genes [20]. Considering the close relationship of the microbiota environment with the occurrence, development, and therapeutics of cancer [21], the new-generation phage-based multifunctional platform has shown great promise in targeted bacterial killing-assisted modulation of the cancer immune environment and reversing cancer chemodrug resistance [22]. Further, bacteriophage-based nanocarriers demonstrate exceptional targeting precision, facilitating the precise delivery of drugs to specific lesion sites, thus enhancing the effectiveness, specificity, and safety of disease treatments, particularly in malignant cancer and brain disease [23]. Engineered bacteriophages can also overcome the challenges related to circulation metabolism and in vivo stability, prolonging their circulation time by evading attacks from the reticuloendothelial system and other bioactive molecules [24]. The gene of phages can be used to develop new DNA vaccines and antigen presentation systems because they can provide a highly organized and repetitive antigen presentation to immune cells, thus holding potential to rejuvenate the adaptive immune systems for combating the cancer cells. Phages allow new possibilities to target the specific molecular biomarkers of cancer cells. Phages can also be used as nanocarriers for delivery of imaging agents and therapeutic drugs/genes for cancer theranostics [25]. This review focused on the exploration of various techniques to engineer bacteriophages and expand their diverse applications across various domains, including templated synthesis of functional nanomaterials, guided assembly of nanostructures, bioanalytes assay, targeted bacterial elimination, assisted disease therapeutics, drug/gene delivery, etc., offering perspectives on the advancement of engineered bacteriophage-based technologies (Figure 1). Engineered phages that are chemically or genetically modified can be specifically localized at lesion sites (e.g., tumor tissue) that are infected by symbiotic target bacteria, which can not only contribute to their application in in vivo imaging but also serve for therapeutics by delivering gene/drug and act as immunomodulators to regulate the tumor microenvironment, thus achieving inhibition effects in the growth of tumors.

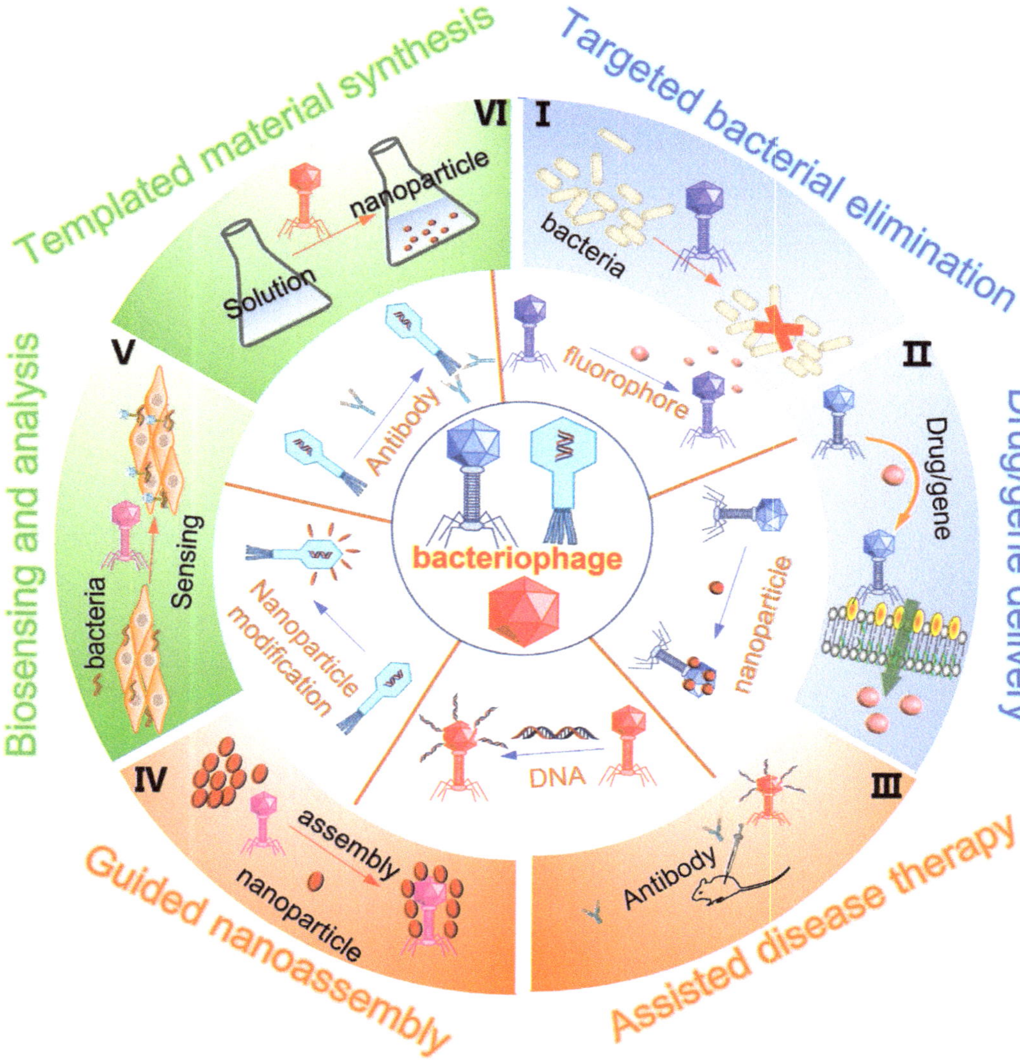

Figure 1. Overview of bacteriophage engineering for functional material synthesis, bioanalytical sensing, and disease theranostics. Diverse bacteriophages can undergo chemical modifications to incorporate functional materials like inorganic nanoparticles, organic polymer chains, fluorophore molecules, RNA/DNA, antibodies, etc. Meanwhile, they can undergo genetic engineering to integrate foreign peptides or proteins into their capsid proteins and undergo DNA recombination to aid the production of engineered bacteriophages. These engineered bacteriophages, with their multifunctionality, find applications in diverse fields such as templated nanoparticle synthesis, guided nanostructural assembly, bioanalyte assay, identification and eradication of pathogenic bacteria, assisted disease diagnosis and treatment, as well as drug and gene delivery.

2. The Modification of Bacteriophage

Bacteriophages serve as nano-sized probes for finely tuned disease diagnostics and as therapeutic agents for targeted treatment modalities. Given their bacteria-specific nature, phages refrain from infecting eukaryotic cells, thus ensuring safety. Moreover, they can be genetically engineered to home in on nanoparticles, cells, tissues, and organs, and can even be endowed with functional non-biological nanomaterials to enhance their diagnostic and therapeutic capabilities [26]. In recent years, extensive research endeavors have been undertaken to leverage phages as platforms in nanomedicine, capitalizing on their distinctive biological attributes. The versatility of phages as multifunctional platforms hinges significantly on the functionalization of their coat proteins with diverse functionalities. While genetic manipulation of these proteins stands out as a prominent approach for such functionalization, complementary chemical modification strategies also contribute to expanding the repertoire of materials, enabling precise, site-specific incorporation of multiple functionalities onto the phage [27].

2.1. Chemical Modification of Phage

Chemical modification stands as a pivotal approach in reshaping the structure and enhancing the functionality of bacteriophages. Within the proteinaceous capsids of these phages lie an array of amino acids, each offering diverse reactive functional groups encompassing carboxylic acids, amines, phenols, and thiols, presenting a rich landscape for molecular manipulation [28]. The abundance of nucleophilic functional groups further broadens the scope, enabling multiple amino acids to engage in chemical transformations. Through interactions with reactive sites including N-terminal alanine, lysine, aspartic/glutamic acid, cysteine, N-terminal serine/threonine, and tyrosine, phages can be endowed with an array of functionalities [29], ranging from DNA strands and antibodies to nanoparticles, fluorophores, and pharmaceutical agents, opening new avenues across biosensing, biomedicine, disease diagnosis, and treatment [30]. Especially, the phage can carry and deliver genes or drugs for the treatment of malignant tumors after chemical modification.

To precisely recognize, adhere, and kill bacteria at the infection sites, Mao et al. developed an innovative nano-enzyme-armed bacteriophage system, designated as phage@palladium (Pd), for combating bacterial infections (Figure 2a) [31]. The proposed phage@Pd retains the functionality of bacteriophages to achieve precise recognition and adhesion to host *Escherichia coli*. Meanwhile, the ultra-small Pd nano-enzymes with pronounced pH-dependent peroxidase-like activity can generate toxic hydroxyl radicals around bacteria in acidic and hydrogen peroxide-overexpressing infectious microenvironments. In contrast, under physiological conditions, the Pd nanoenzymes remained inert, leading to significant bacterial clearance at infection sites while excellent biocompatibility in healthy tissues. Additionally, phage@Pd can not only eradicate planktonic bacteria but also kill bacteria within biofilms in vitro. In in vivo models of acute bacterial pneumonia or subcutaneous abscesses, phage@Pd demonstrates significant activity in eliminating infections and promoting tissue recovery, suggesting a safe and effective antibacterial agent. In another example, Peng et al. coupled engineered chimeric M13 phage with photothermal-active gold nanorods and realized ablation of diverse bacterial strains [32]. Meanwhile, Wang et al. prepared a photocatalytic quantum-dot-equipped phage, abbreviated as QD@phage, showing excellent performance in combating infections caused by *Pseudomonas aeruginosa* [33].

Given the widespread use of filamentous bacteriophages as platforms for peptide and protein displaying, the ability to attach a large number of synthetic functional groups to their capsid proteins is of significant value to improve the performance of bacteriophages. Francis's group further decorated bacteriophages with polymer chains and imaging moieties (Figure 2b) [34]. After converting the N-terminal amines of the capsid proteins to ketones, further chemically specific modification via oxime formation with alkoxyamine groups can be achieved, as fluorescent moieties and up to 3000 polyethylene glycol molecules were attached on each bacteriophage capsid without significantly affecting the binding efficiency of bacteriophage-displayed antibody fragments to EGFR and HER2 receptors. The chemically modified fluorescent dyes, such as Alexa Fluor 488 and 647 C5-aminooxyacetamide (AF488/647-ONH2), can be easily substituted by imaging, MRI, PET, and other detection agents, allowing for various in vitro and in vivo biomedical applications. Additionally, filamentous bacteriophage M13 with N-terminal ketone can serve as an effective supramolecular scaffold for the assembly of multivalent CryA, a hyperpolarized ^{129}Xe contrast agent composed of many assembled latent A molecular cages (Figure 2c) [35]. The saturation of a large number of latent cages with xenon gas is chemically exchanged to a water-soluble xenon pool, leading to the effective generation of ultra-chemical exchange saturation transfer (CEST) contrast. The detection limit of this bacteriophage-based ultra-CEST agent can be as low as 230 fM, which represents the optimal single-dose sensitivity for magnetic resonance imaging agents.

As nucleic acid aptamers can serve as specific biotarget-recognizable ligands, an efficient oxidative coupling strategy was developed by Tong et al. to attach them to the surface of non-genomic bacteriophage capsid carriers (Figure 2d) [36]. Up to 60 DNA strands can be attached to each capsid without significant loss of base-pairing ability or protein stability. The specific binding capability was demonstrated by attaching aptamers that target tyrosine kinase receptors expressed on Jurkat T cells. With fluorescent dyes capsulated inside the capsid, the level of binding between phage capsids and Jurkat T cells was significantly enhanced. Confocal microscopy indicated that capsids were internalized and transported to lysosomes, suggesting the potential of that aptamer-modified bacteriophage in the application of targeted delivery and controlled release of cargos in living cells at lesion sites.

The bacteriophage T4 capsid is adorned with 155 copies of Hoc, a non-essential highly antigenic outer capsid protein. To induce effective immune responses and provide protection in the viral infection, Rao's group reported an in vitro assembly system that allows the surface display of human immunodeficiency virus (HIV) antigens, p24-gag, on the bacteriophage T4 capsid through Hoc–capsid interactions (Figure 2e) [37]. Fusion within the plasmid framework was constructed by splicing HIV genes to the 5′ or 3′ end of the Hoc gene. The Hoc fusion proteins were expressed, purified, assembled, and displayed on bacteriophage particles in the defined *E. coli* system. Single or multiple antigens were efficiently displayed, saturating all available capsid binding sites. In the absence of any external adjuvants, the displayed p24 demonstrated high immunogenicity in mice, eliciting strong p24-specific antibodies as well as Th1 and Th2 cell responses, providing new directions and insights for HIV vaccine development with the potential to broaden both cellular and humoral immune responses.

Phages have been modified by a range of functional moieties (e.g., fluorophores, nanoparticles, functional nucleic acids, antigens, and drugs) to serve their applications. The protein on the phage capsid can be coupled with functional molecules and materials by electrostatic adsorption or chemical reaction under appropriate pH values and ionic conditions to obtain chemically modified multifunctional phages. Modifying phage capsid protein is a promising method to obtain multifunctional phage. Flexible and diverse chemical modification strategies have expanded phage-based biomedical applications (Table 1).

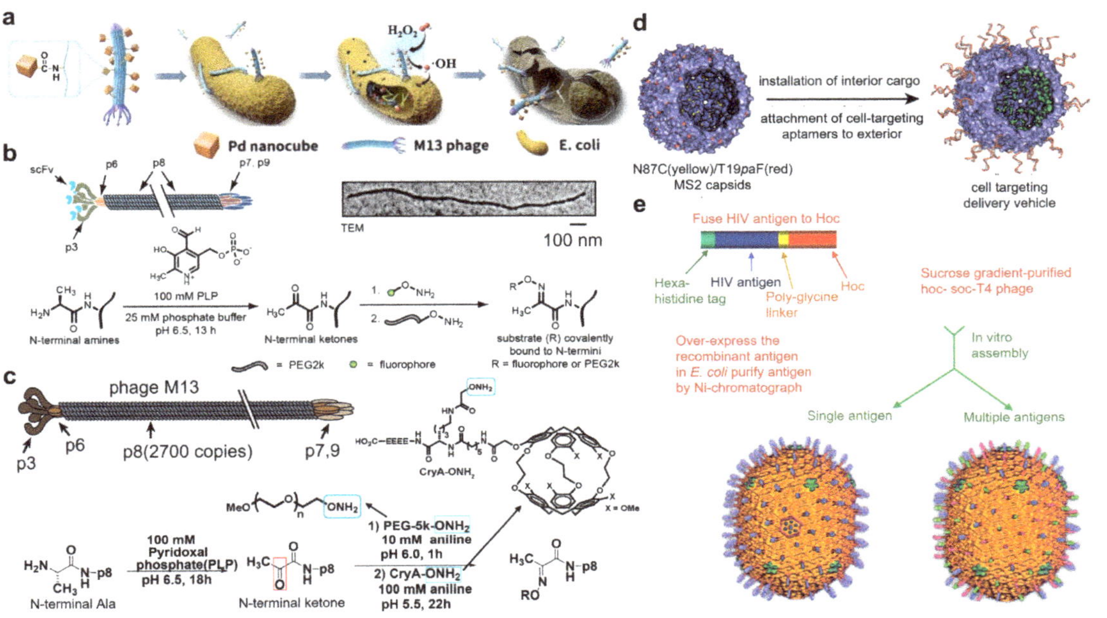

Figure 2. Chemical modification with diverse structure for reshaping the structures of bacteriophages. (**a**) The Pd nanocubes modified M13 phage (phage@Pd) for H_2O_2 catalytical hydrolysis and antibacterial study through the targeting and infection capabilities of phage toward the host bacteria [31]. Reproduced with permission from [31]; published by the Wiley Online Library, 2023. (**b**) Chemical modification of filamentous phage via transamination reaction. The N-termini were converted to ketone-bearing proteins, which were then reacted with aminooxy-functionalized fluorophores and aminooxy-functionalized PEG2k [34]. Reproduced with permission from [34]; published by the American Chemical Society, 2012. (**c**) Chemical modification of M13 phage. The N-termini of p8 proteins were transaminated to ketones along the capsids, which were then reacted with PEG-5k–ONH_2 chains and aminooxy-functionalized cryptophane-A cages (CryA–ONH_2) to construct M13-based biosensors [35]. Reproduced with permission from [35]; published by the Wiley Online Library, 2013. (**d**) Dual-surface modification of MS2 phage. For interior surfaces, an N87C mutation on the coat protein allowed for site-specific alkylation with up to 180 cargo molecules uploaded. For the exterior surface, a T19*pa*F mutation on the capsid allowed for the attachment of the phenylene diamine group-modified aptamer via a $NaIO_4$-mediated oxidative coupling reaction [36]. Reproduced with permission from [36]; published by the American Chemical Society, 2009. (**e**) Antigen spikes artificially fused to Hoc subunits. The left reconstruction showed blue spikes representing the single antigen display, and the right reconstruction showed blue, green, and pink spikes representing three antigen displays [37]. Reproduced with permission from [37]; published by the American Society for Microbiology, 2006.

Table 1. The modification of bacteriophage for application.

Modification Strategies	Modified Materials	Functionalization	Application	Reference
Amino-carboxyl group reaction	Pd nanocubes	Catalytic hydrolysis	Antibacterial study	Jin et al., 2023 [31]
N-succinimidyl-S-acetylthiopropionate (SATP)	Gold nanorods (AuNRs)	Photothermal	Antibacterial study	Peng et al., 2020 [32]
Avidin-biotin bioconjugation	Cd-based quantum dot	Photocatalytic localized reactive oxygen species (ROS)	Antibacterial study	Wang et al., 2022 [33]
N-termini reacted with aminooxy-functionalized fluorophores	Fluorophores and 2 kDa PEG	Fluorescence imaging	Characterization of breast cancer cells	Carrico et al., 2012 [34]
Site-specific protein bioconjugation	CryA cages	Hyperpolarized xenon MR contrast agent	NMR/MRI imaging	Stevens et al., 2012 [35]
Phenylene diamine-substituted oligonucleotides with aniline groups	Nucleic acid aptamers	Bind specific cellular targets	Targeted drug delivery of acid-labile prodrugs	Tong et al., 2009 [36]
Hoc–capsid interactions	HIV antigens	Highly immunogenic	HIV vaccine	Sathaliyawala et al., 2006 [37]

2.2. Genetic Modification of Phage

The distinguishing advantage of phages over other synthetic biomaterials lies in their capacity to present proteins or peptides on specific coat proteins through genetic modification [38]. Genetic engineering techniques involve the introduction of genes encoding alternative amino acids, peptides, or protein sequences into either the bacteriophage genomes or host plasmids containing capsid genes, facilitating the generation of recombinant phages via employing the well-established "phage display" method [26,39]. These recombinant capsid proteins then create additional sites for modification by enzymes or chemicals. For accommodating large, complex foreign protein fusions, recombinant capsid decoration proteins can be further assembled in vitro onto phages with accessible binding sites [40,41].

One method for engineering phages is homologous recombination (HR), which is a process involving the exchange of nucleotide sequences between two similar or identical molecules of double-stranded DNA (dsDNA) or single-stranded RNA (ssRNA) [42]. It serves crucial roles in DNA repair, DNA replication, and telomere maintenance. Regulated by the enzyme DNA helicase, HR has the capability to exchange up to 23 base pairs between two homologous DNA sequences, making it valuable for engineering phages within their bacterial hosts. This mechanism has been utilized for the introduction of foreign genes into phages (Figure 3a) [43]. Gene insertion, deletion, and substitution in phages through HR follow the same principles as bacterial recombination. In the development of recombinant phages, the desired gene sequence is initially incorporated into a replicative plasmid, followed by homologous integration of the desired phage sequences. Similarly, for engineering phage particles, bacterium containing donor plasmids is infected with phages, leading to the packaging of heterogeneous genes within phage particles [44].

The CRISPR-Cas system, found naturally in many prokaryotes, serves as an adaptive defense mechanism against invasive nucleic acids [45]. Comprising Cas proteins and the CRISPR array, this system has garnered attention for its potential applications. For instance, the type I-E CRISPR-Cas system has recently been employed as a counter-selection mechanism by Lv et al. for engineered T7 phages (Figure 3b) [46]. Through homologous recombination-mediated editing, the phage successfully excised the dispensable gene 1.7. This approach effectively removed nonrecombinant phage genomes containing gene 1.7 while preserving recombinant phage genomes lacking this gene. Additionally, the CRISPR/Cas II-A system can facilitate in vivo editing of the phage 2972 genome, enabling modifications such as point mutations, gene deletions, and DNA exchange [47]. With these promising genetic manipulations, the CRISPR/Cas strategy shows potential for broader applications across various phage genomes. The application of CRISPR-Cas-based techniques for phage engineering is limited to bacteria with either a native CRISPR-Cas system or the capacity for genetic transformation to express a functional heterologous CRISPR-Cas system. This limitation significantly hampers the manipulation of bacteriophages targeting bacteria lacking genetic manipulability [48].

Before introducing them into host bacterial cells, the genome of phages can undergo manipulation and modification in vitro. To eliminate overlaps between genetic segments, the refactoring process has been applied to T7 phages. In one study, 73 segments were divided into six sections using bracketing restriction sites to allow for changes within each section independently, without affecting others (Figure 3c) [43]. Subsequently, the refactoring of genomes after transformation into bacteria yielded a chimeric T7 phage genome composed of various engineered sections combined with viable phages. However, recombinant phages generated through the refactoring method displayed significantly smaller plaques compared to wild-type phages. Moreover, the refactoring approach to phage engineering necessitates large-scale DNA manipulation in vitro and modification of bacterial hosts with engineered genomes to retrieve viable phages. Consequently, this process yields only a limited number of engineered phages, particularly when applied to non-domesticated bacterial hosts.

The proliferation of phages within bacterial hosts can sometimes result in toxicity, leading to reduced efficiency of homologous recombination (HR), in vivo recombineering, and bacteriophage recombineering with electroporated DNA(BRED) genetic engineering techniques [49]. To address this issue, an intermediate host such as *Saccharomyces cerevisiae* can be employed, playing a vital role in genetic manipulation. In the approach of Ramirez-Chamorro's group, phages remain stable and non-toxic within yeast cells (Figure 3d) [43]. In the process of producing engineered phages, the genome of *S. cerevisiae* is utilized to capture the phage genome using a bacterial shuttle vector containing overhangs for recombination of the phage genome and vector. Subsequently, the newly assembled, modified, and propagated phage genomes in yeast are isolated and introduced into the host bacterium to recover recombinant phage particles. A similar strategy from Pires's group was employed in the creation of *Pseudomonas aeruginosa* phage vB_PaeP_PE3, where yeast-based assembly was utilized to delete approximately 48% of the hypothetical gene protein. During this process, the phage genome was PCR-amplified, excluding hypothetical protein sequences, and assembled in yeast before being isolated and incorporated into the host bacteria to recover engineered phages. Employing this approach, engineered phages can be assembled by deleting surplus parts of their genome through the incorporation of desired sequences [50,51].

Genetic modifications that introduce genes into the phage genome or capsid genes to replace amino acid, peptide, or protein sequences on the host plasmid can promote the production of recombinant phages. These recombinant capsid proteins provide additional sites for enzymes or chemicals to react with the capsid.

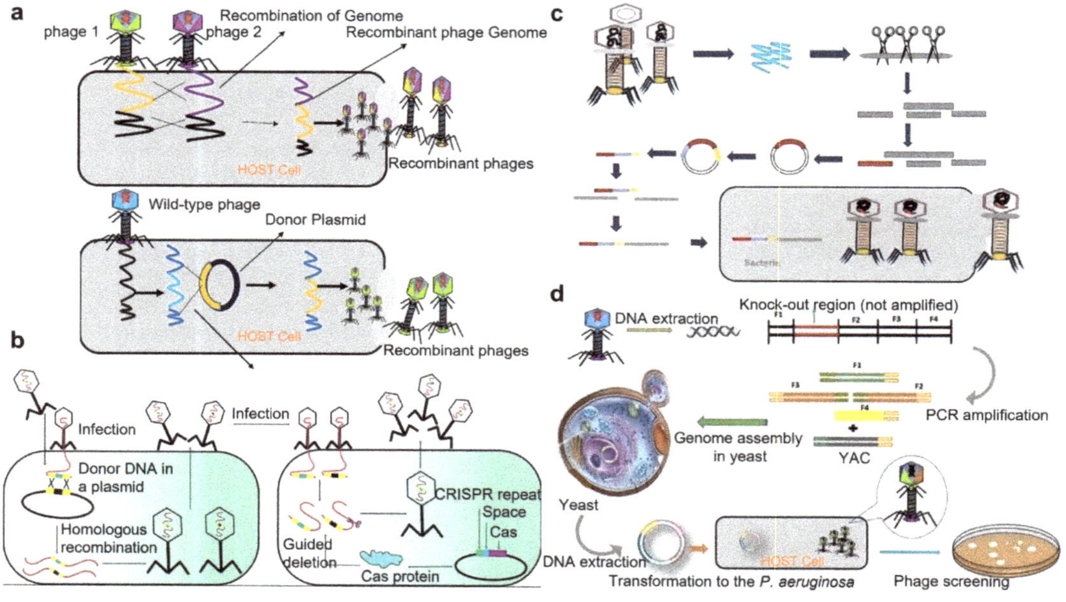

Figure 3. Genetic engineering techniques for modifying phages. (**a**) Developing engineered phages through homologous gene recombination. Phages 1 and 2 infect the same bacterial host, which results in the recombination of their genomes and forms new recombinant phages [43]. Reproduced with permission from [43]; published by Elsevier, ScienceDirect, 2023. (**b**) Phage genome engineering with CRISPR-Cas13a. The CRISPR-Cas13a selectively eliminated the harboring gene in the nonrecombinant phage genomes while sparing the recombinant phage genomes devoid of this gene [46]. Reproduced with permission from [46]; published by MDPI, 2023. (**c**) Rebuilding or refactoring the phage genome in vitro. The independent pieces of phage DNA via digesting the native restriction sites of purified phage DNA were further manipulated by sub-cloning. After release from the plasmid, the recombinant DNA section was ligated with the rest of the phage DNA and delivered to the host bacteria via electroporation, leading to the recovery of engineered phages [43]. Reproduced with permission from [43]; published by Elsevier, ScienceDirect, 2023. (**d**) Phage genome engineering by the yeast-based assembly. Phage genomic DNA is used as a template for the amplification of overlapped PCR products, except for the knockout region. For assembly of the genome, the genomic fragments are co-transformed to yeast cells along with a linearized YAC plasmid, and the phage genome is transformed into *P. aeruginosa* cells for the generation of engineered phages [43]. Reproduced with permission from [43]; published by Elsevier, ScienceDirect, 2023.

3. Phage Mediated Synthesis and Assembly

3.1. Phage Mediated Nanomaterial Synthesis

In the last decade, biotemplates, such as bacteriophages, have emerged as powerful tools for synthesizing inorganic materials [52]. Bacteriophages can undergo genetic modification to incorporate specific material-recognizing peptides, which can induce the nucleation of materials, primarily crystals, from an aqueous precursor solution onto the phage surface under mild conditions [53]. By employing phages of diverse shapes, controlled architectures of inorganic materials can be achieved, including one-dimensional nanowires, two-dimensional nanofilms, and three-dimensional scaffolds [54]. Phages serve as versatile building blocks, capable of organizing nanomaterials with various properties. Moreover, beyond utilizing the inherent structure of phages themselves, genetic engineering enables the display of peptides that confer affinity characteristics, serving as bio-interaction motifs or forming self-assembled nanomaterials [55]. Particularly, specific recognition peptides identified through phage library have been harnessed to template the growth of a wide array of inorganic nanocrystals and nanomaterials. By fusing these crystal-nucleating peptides onto multiple phage coat proteins, an assortment of intriguing nanomaterials and nanostructures can be constructed [56].

Mao et al. reported a phage-based scaffold for synthesizing semiconductor and magnetic materials including single-crystal ZnS, CdS, and chemically ordered CoPt and FePt nanowires (Figure 4a) [57]. Peptides exhibiting composition, size, and phase control during nanoparticle nucleation processes were expressed on the highly ordered filamentous coat of evolutionarily selected M13 bacteriophage. Annealing to remove the phage template promotes the growth of oriented crystalline aggregates, forming individual crystal nanowires. This was the first example to introduce the unique ability of substrate-specific peptides into the linear self-assembled filamentous structure of M13 phage for inorganic nanomaterial synthesis.

Ferroelectric materials, like tetragonal barium titanate ($BaTiO_3$), are widely utilized in various fields such as bioimaging, biosensing, and high-power switching devices. However, the conventional synthesis methods for tetragonal phase $BaTiO_3$ typically involve toxic organic reagents and high-temperature treatments, rendering them environmentally unfriendly and energy-inefficient. Therefore, Mao's group harnessed the phage display technique to devise a novel approach for producing $BaTiO_3$ nanowires [58]. As shown in Figure 4b, through biopanning of a phage-displayed random peptide library, a short $BaTiO_3$-binding/nucleating peptide, CRGATPMSC (named RS), was identified and genetically fused to the major coat protein (pVIII) of filamentous M13 phages to create pVIII-RS phages. These modified phages not only exhibited capability in binding with pre-synthesized $BaTiO_3$ crystals but also inducing the nucleation of uniform tetragonal $BaTiO_3$ nanocrystals at room temperature without the need for toxic reagents, resulting in one-dimensional polycrystalline $BaTiO_3$ nanowires. This method offers a sustainable approach for the eco-friendly synthesis of $BaTiO_3$ polycrystalline nanowires.

Besides nanowires being synthesized through bacteriophages, they were utilized to produce complex 3D structures of nanoparticles (NPs). M13 bacteriophage derivatives that display a ZnO-binding peptide (TMGANLGLKWPV) on either the pIII or pVIII coat protein were engineered and employed as biotemplates by Piotr Golec's team (Figure 4c) [59]. Depending on the type of recombinant phage used (M13-pIII-ZnO or M13-pVIII-ZnO), well-separated ZnO NPs or complex 3D structures of ZnO NPs with sizes of approximately 20–40 nm were synthesized at room temperature. The synthesized ZnO nanoparticles exhibited luminescence, emitting light near the short wavelength end of the visible region (around 400 nm). The presence of a very low-intensity emission band at 530 nm indicated that the ZnO material obtained had a low concentration of surface defects.

Besides these reports that focused on single modifications of the pIII minor coat protein to synthesize ZnS and CdS nanoparticles at the tip of the bacteriophage, Nam et al. modified both of the end minor coat proteins (pIII and pIX) with six histidine and HPQ (His–Pro–Gln) streptavidin binding peptide motifs [60]. Subsequently, these engineered phages were employed to fabricate nanoscale viral ring structures by leveraging the specific binding between chemically conjugated streptavidin and Ni-NTA linkers. Through expression of semiconductor and metal-binding peptides on the major coat proteins of phages, high-quality single crystal monodisperse ZnS nanowires with promising electro-optical and magnetic properties were produced through annealing processes. By simultaneously modifying pIII and pVIII coat proteins, a variety of nanostructures were successfully produced via the template of bacteriophage (Figure 4d) [61].

The unique recognition peptides of phages can direct mineralization processes for the synthesis of many useful nanoscale electronic and medical materials. The monodispersed and long rod-shaped phages, with specific recognition patterns, enable the synthesis of various nanomaterials such as ZnO, ZnS, and CdS quantum dots, which can be used in electronic, optical, and biotechnological applications.

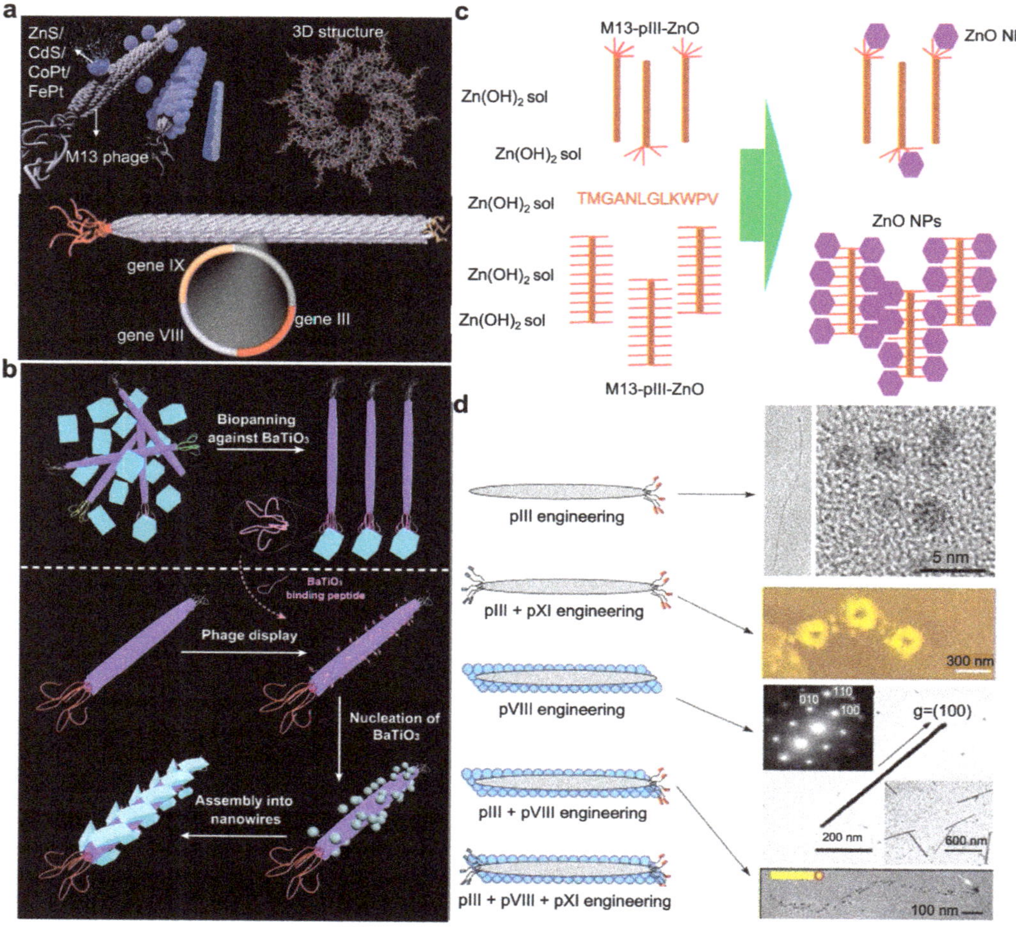

Figure 4. Phage-mediated nanomaterial synthesis. (**a**) The nucleation, ordering, and annealing of phage–nanoparticle assembly including II–VI semiconductors ZnS and CdS and ferromagnetic CoPt and FePt nanowire on M13 phage. The capsid and ends of M13 phages were genetically modified by encoding gP3, gP8, and gP9 peptides. The resulted symmetric phage allowed for ordering of the nucleated particles along the x, y, and z directions with the aid of the expressed peptides with rigidity and packing pattern [57]. Reproduced with permission from [57]; published by Science, 2004. (**b**) Using engineered phages for the synthesis of $BaTiO_3$ polycrystalline nanowires at room temperature. $BaTiO_3$-binding peptide was selected via biopanning and construction of the pVIII-RS phage via displaying of $BaTiO_3$-binding peptides on the side wall of phage, which can guide the nucleation of $BaTiO_3$ on phage to form polycrystalline nanowires [58]. Reproduced with permission from [58]; published by the American Chemical Society, 2016. (**c**) The modified phages acted as machinery for the synthesis of both the previously selected and identified ZnO-binding peptide and ZnO nanoparticles. The exposure of a ZnO-binding peptide to a phage mediated the growth of unique ZnO nanostructures [59]. Reproduced with permission from [59]; published by the American Chemical Society, 2016. (**d**) Genetic modification of the phage for programmed synthesis of ZnS nanoparticles, nanowires, viral rings, and arrayed hetero-nanostructures by engineered phages with display of pIII coat protein, pVIII coat protein, pIII/pIX dual-protein, and pIII/pVIII dual-protein, respectively [61]. Reproduced with permission from [61]; published by the American Chemical Society, 2005.

3.2. Phage-Guided Nanostructural Assembly

Previously, it was demonstrated that displaying semiconductor-binding peptides on phages can facilitate the production of single-crystalline semiconductor nanowires. This approach typically involves three key steps. Firstly, a target-binding peptide is identified through biopanning. Secondly, the peptide is genetically presented on the side wall of the phage. Finally, the resulting phage serves as a biotemplate for the assembly of nanoparticles [62]. Due to the capability of genetically fusing foreign peptides to bacteriophage coat proteins for generation of highly decorated phages and the significant influence of surface charges on bacteriophage assembly behavior, bacteriophages emerge as promising templates for controlling the assembly of nanostructures [63]. At each end of the phage nanocrystalline wires, specific receptors can be incorporated as fusions to terminal proteins. By programming the self-assembly of these constructs, complex nanowire arrays with diverse one-, two-, and three-dimensional geometries can be meticulously engineered.

Mao's group has introduced a system harnessing the M13 phage coats to nucleate and template II–VI semiconductor nanocrystals into well-oriented quantum dot nanowires (Figure 5a) [64]. The pVIII major coat protein was genetically engineered to showcase evolutionarily selected peptides, facilitating precise control over nanocrystal growth into phage–semiconductor hybrid nanowires with crystallographic alignment spanning micrometer scales. By incorporating semiconductor nucleating peptides onto the capsids of M13 phages, directed and phase-specific nanocrystal growth was achieved. This method seamlessly integrates the use of naturally ordered and assembled phage templates with the versatility of genetic manipulation for material synthesis, enabling the controllable assembly of semiconductor wires within supramolecular hybrid structures.

Later, the same group devised a versatile approach leveraging the surface proteins of biotemplates to catalyze the hydrolysis of organosilicate precursors and the ensuing polycondensation process to yield silica (Figure 5b) [65]. Meticulously organized mesoporous silica fibers featuring hexagonally arranged pores were fabricated by utilizing fd phage as biotemplate. By exhibiting peptides with varying charges on the side wall of filamentous bacteriophage, silica nanostructures were successfully controlled through manipulation of both bacteriophage surface charge density and the accompanying bacteriophage assembly behavior. Furthermore, the resultant mesoporous silica has been successfully utilized as a medium for fabricating three-dimensional arrays of PbS nanoparticles.

Building upon this foundation, Mao further expanded research efforts for genetic modification of phages. The filamentous M13 phage was anionized through the fusion of a negatively charged peptide to its major coat protein (pVIII) [66]. In the presence of cationic precursors like Ca^{2+} ions, the negatively charged phage underwent self-assembly into a nanofiber that promoted the formation of oriented nanocrystalline hydroxyapatite (HAP) (Figure 5c). In co-assembly, collagen fostered the formation of a hierarchical architecture akin to the extracellular matrix (ECM), while E8-displayed phage enhanced matrix mineralization. The arrangement and alignment of collagen and HAP dictated the superior mechanical properties of bone, holding promise for developing bone-mimetic biomaterials and bone tissue engineering.

Further, the same group introduced a novel VAM (Vascularized Bone Augmentation Matrix) strategy aimed at enhancing the formation of vascularized bone (Figure 5d) [67]. Through the incorporation of RGD-phage, the VAM exhibits the ability to regulate endothelial cell (EC) migration and adhesion, thereby promoting endothelialization. Simultaneously, it triggers osteoblastic differentiation of MSCs, consequently facilitating osteogenesis and angiogenesis in vivo. The critical role of engineered phage nanofibers, specifically RGD-phage, in stimulating both angiogenesis and osteogenesis marks a significant breakthrough in phage-based nanomedicine and regenerative medicine. The specific recognition capabilities of phages can guide the growth of inorganic semiconductors and bone-like biomaterials and facilitate osteogenic formation. Such functionalized phages can self-assemble to form highly organized hierarchies.

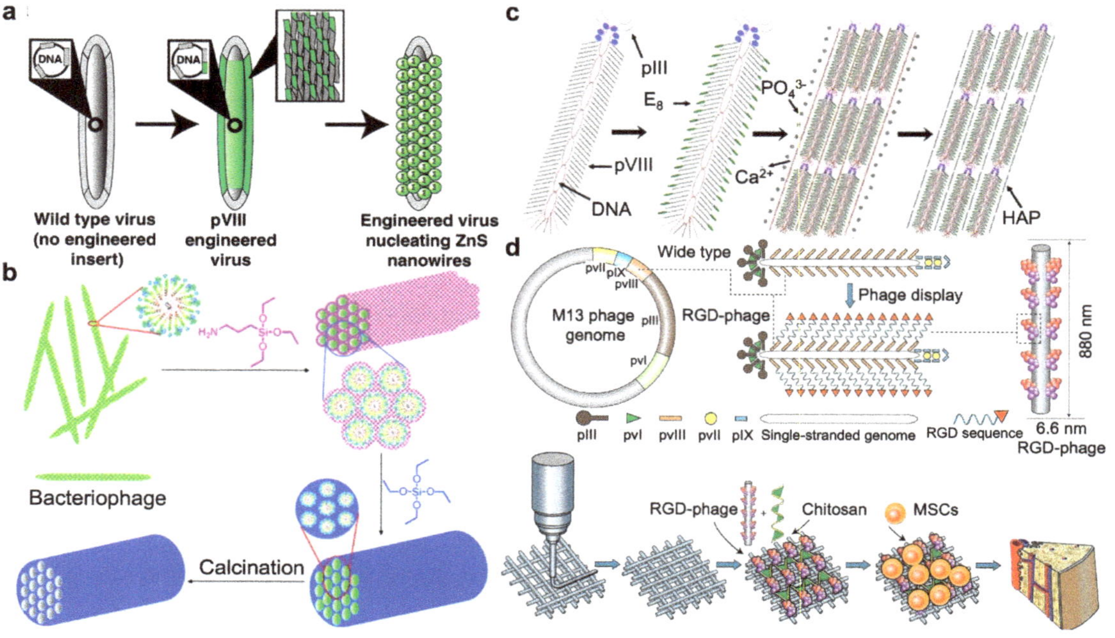

Figure 5. Phage-guided nanostructural assembly. (**a**) Directed ZnS nanocrystal synthesis on A7–pVIII-engineered phages at 0 °C [64]. Reproduced with permission from [64]; published by the National Academy of Sciences, 2003. (**b**) Templated silica nanostructure formation via monodispersed filamentous phages that self-assembled into highly ordered hexagonal lattices. Removal of the phage assembly through calcination led to the formation of mesoporous silica fibers with pore structures precisely defined by the phage assembly [65]. Reproduced with permission from [65]; published by the Wiley Online Library, 2012. (**c**) Controlled assembly of hydroxylapatite (HAP) on the wild-type M13 phage. Genetic fusion of E8 peptide to major coat protein (pVIII) on the side wall. Self-assembly of anionic phage into a nanofibrous structure (bundle) in the presence of Ca^{2+} ions. Mineralization of the phage bundle to form a fiber [66]. Reproduced with permission from [66]; published by the Wiley Online Library, 2010. (**d**) Bone generation via genetically modified M13 phage. RGD peptide was fused to the solvent-exposed terminal of the major coat protein (pVIII), constituting the side wall of filamentous phages. RGD-phage nanofibers (negatively charged) were integrated into a 3D printed bioceramic scaffold along with chitosan (positively charged). The resultant scaffold was seeded with MSCs and then implanted into bone defects, inducing the formation of new bone [67]. Reproduced with permission from [67]; published by the Wiley Online Library, 2014.

4. Bacteriophage in Precise Bioanalytes Assay

Molecular diagnostics endeavors to identify specific biomarkers for disease diagnosis, monitoring, risk assessment, and treatment determination. Although conventional enzyme-linked immunosorbent assays (ELISAs) and polymerase chain reaction (PCR) offer standard measurement techniques, the integration of nanotechnology into molecular diagnostics in recent years has also yielded numerous novel diagnostic approaches [68]. The nanoscale-sized phages, with their DNA enclosed within a protein capsid, present an opportunity for genetic engineering to display multiple peptides on the sidewall (pVIII) and tip (pIII) of the protein capsid [69]. Consequently, they have been extensively engineered to express nanobodies or mimotopes for selective recognition of target analytes or to interact with functional nanoparticles for serving as specific nanoprobes [70]. Thus, modified phage nanoprobes have great potential in enhancing the accuracy and reliability of diagnostic schemes. It can be expected that phages modified with specific peptides or functional materials can accurately target tumors and detect tumor biomarkers.

Bruce D. Hammock et al. introduced a new approach to immune-polymerase chain reaction (IPCR) known as phage anti-immunocomplex assay real-time PCR (PHAIA-PCR) for detecting small analytes, including 3-phenoxybenzoic acid and molinate (Figure 6a) [71]. The technique employed phage anti-immunocomplex assay technology, where a short peptide was presented on the M13 bacteriophage's surface specifically bound to the antibody–analyte complex. As the phage DNA can function as a versatile material for PCR, both universal amplification that targets the phagemid's common element and specific amplification when the real-time PCR probe is designed to bind the phagemid DNA can be achieved, leading to the significantly improved sensitivity. A tenfold increase in sensitivity was achieved via this PHAIA-PCRs method compared to traditional PHAIA. The successful assay conducted using agricultural drain water and human urine samples further demonstrated its reliability for monitoring human exposure or environmental contamination.

Counting disease biomarkers like miRNAs with the naked eye is very difficult in molecular diagnostics. Gu and Mao reported a highly sensitive method for quantifying miRNAs using T7 phage, which was modified to fluoresce and bound to a DNA-modified gold nanoparticle (GNP) in a one-to-one fashion after genetic engineering (Figure 6b) [72]. Target miRNAs then crosslinked the resulting phage–GNP complex with DNA-modified magnetic microparticles, forming a sandwich complex containing equimolar amounts of phage and miRNA. After release from the complex, the phages were grown into a macroscopic fluorescent plaque for visually counting. Thus, miRNA targets can be sensitively and conveniently quantified using only the naked eye, with detection limits down to aM level for miRNA biomarkers.

Traditional immunoassay-based detection methods face limitations due to the size effect of nanoprobes, which hinders the balance between molecular recognition and signal amplification. Tang's group presented a versatile approach by integrating immunoassay with a light-scattering signal of gold nanoparticles that were in situ grown on the M13 phage surface for quantitative detection of low-abundance analytes (Figure 6c) [73]. In the practical detection of ochratoxin A and alpha-fetoprotein, a detection limit of sub-femtomolar levels in real samples was achieved, showcasing four orders of magnitude enhancement in sensitivity compared to traditional phage-based ELISA methods. With the implementation of a biotin-streptavidin amplification scheme, further enhancement in sensitivity was realized, allowing the detection of severe acute respiratory syndrome coronavirus 2 spike protein down to the attomolar range.

Rapid and sensitive detection of pathogenic viruses is crucial for pandemic control. Hou et al. reported a swift and highly sensitive optical biosensing method for detecting avian influenza virus H9N2 (Figure 6d), utilizing a genetically modified filamentous M13 phage probe, M13@H9N2BP@AuBP, that borne an H9N2-binding peptide (H9N2BP) at its tip and a gold nanoparticle (AuNP)-binding peptide (AuBP) on its sidewall [74]. Simulation modeling demonstrated that the electric field in surface plasmon resonance (SPR) enhanced by 40-fold compared to conventional AuNPs, guaranteeing the experimentally ultrasensitive detection of H9N2 particles with a detection limit down to 6.3 copies/mL. It was even more sensitive than the quantitative polymerase chain reaction (qPCR). Further, after capturing H9N2 viruses on the sensor chip, the H9N2-binding phage nanofibers can be quantitatively converted into visible plaques, allowing for quantification through naked-eye observation. This phage-based biosensing strategy can be easily extended to the detection of other pathogens just by replacing the H9N2-binding peptides with other pathogen-binding peptides using phage display technology.

Phages have been developed as emerging nanoprobes for the detection of various biomarkers and pathogens. After displaying arbitrary sequences of exogenous peptides or antibodies, phages can specifically bind to targets, such as animal cells, bacteria, viruses, and protein molecules. Phage can also be co-modified with metal nanoparticles to detect targets by showing colorimetrical, fluorescence, and plasmonic signals.

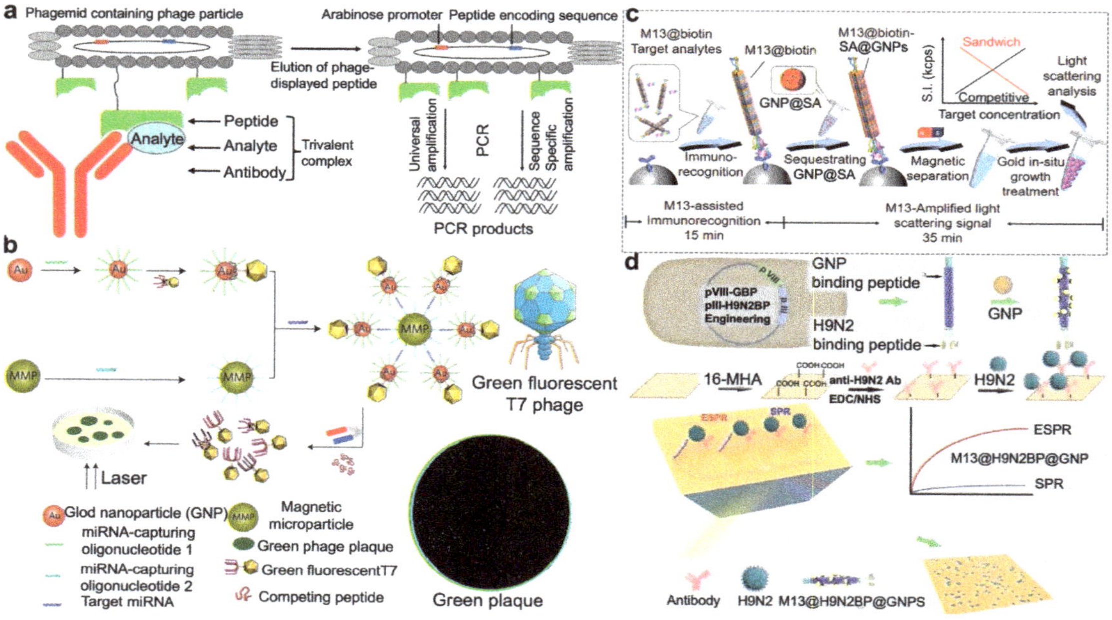

Figure 6. Modified phage for the selective recognition and analysis of biotargets. (**a**) Gene detection via PHAIA-PCR technology. The M13 phage with peptide binding to an immunocomplex of an antibody–analyte was eluted and immediately neutralized for PCR detection [71]. Reproduced with permission from [71]; published by the American Chemical Society, 2011. (**b**) A naked-eye-counting method for miRNA quantification with sensitivity comparable to PCR. The magnetic microparticle and the T7 phage–gold nanoparticle (GNP) binding probe were linked by cohybridization with target miRNA, forming a sandwich complex with miRNA number equal to the phage number. After magnetic separation, T7 phages were released from the GNPs by adding more competitive gold-binding peptides. Counting the number of fluorescent plaques formed by T7 phages by eye can quantify the miRNA [72]. Reproduced with permission from [72]; published by Springer Nature, 2015. (**c**) Immunoassay strategy via integrating designed M13 phage-assisted recognition with the in situ growth of gold nanoparticle-based light scattering signal for ultrasensitive quantitative detection of low-abundance ochratoxin A and alpha-fetoprotein analytes [73]. Reproduced with permission from [73]; published by the American Chemical Society, 2023. (**d**) Using phage@GNP nanofibers to detect H9N2 virus. M13 with H9N2 and GNP binding peptides can attach with GNP onto the sidewall to form the M13@H9N2BP@GNP nanofiber probe, showing an enhanced surface plasmon resonance (SPR) signal upon binding with H9N2 targets on the phage tail. SPR signals can be cross-validated with the blue-color phage plaques that were naked-eye discernable [74]. Reproduced with permission from [74]; published by Elsevier, ScienceDirect, 2023.

5. Bacteriophages in Bacterial Recognition and Detection

Conventional bacterial detection primarily relies on bacterial culture, which is cumbersome and time-consuming. PCR technology and ELISA-based immunoassay offer enhanced sensitivity and efficiency in bacterial detection yet necessitate specialized instruments and skilled personnel [75]. Recently, biosensors employing specific molecular recognition agents to target bacteria have emerged as promising alternatives. Phages utilize receptor-binding proteins at the distal end of the tail to recognize and interact with bacterial surface receptors such as lipopolysaccharides, teichoic acids, and porins. This unique ability allows phages to capture specific host bacteria from complex matrices [76]. More importantly, phage-based detection schemes offer the advantage of distinguishing between live and inactivated bacteria as phages replicate exclusively within viable hosts. This feature represents an improvement over DNA- or antibody-based detection assays, which may yield false positives if the pathogen has been successfully inactivated but its genetic material or epitopes remain detectable [77,78]. Thus, increasing studies have explored the utilization of phage-based schemes for the accurate and reliable detection of pathogenic bacteria.

To develop an inexpensive, rapid, and reliable method for detection of *E. coli* in drinking water or food, Chen et al. reported a phage-based magnetic separation technique for the swift detection of *E. coli* in drinking water as the broad host range specificity of T7 phage toward *E. coli* (Figure 7a) [79]. After capturing and isolating *E. coli* BL21 from the drinking water sample, phage-mediated lysis can liberate intrinsic β-galactosidase from the adhered bacteria, which can convert the yellow-color chlorophenol red-β-D-galactopyranoside to a red color. *E. coli* can be successfully detected within 2.5 h with a detection limit down to 1×10^4 CFU·mL^{-1}. This colorimetric change can be visually discerned, facilitating a straightforward and rapid quantification of target bacteria, particularly advantageous in resource-limited settings. By incorporating a pre-enrichment step in Luria–Bertani meida supplemented with isopropyl β-D-thiogalactopyranoside, even 10 CFU·mL^{-1} *E. coli* in drinking water can be detected after 6 h of pre-enrichment.

The prompt and precise identification of virulent foodborne pathogens is crucial for ensuring food safety. Gan's group introduces an ultra-sensitive and specific phage@DNAzyme probe for the detection of foodborne pathogens (Figure 7b) [80]. The carefully selected phage and functionalized DNAzyme enabled specific recognition of target foodborne pathogens at the strain level and efficient catalysis of copper(II)-based azide-alkyne cycloaddition (CuAAC) click reaction with fluorescent signal, respectively. Using virulent *E. coli* O157:H7 as the representative analyte, the phage@DNAzyme probes specifically bound to the captured *E. coli* O157:H7 and facilitated the click reaction to generate a visible fluorescent signal that can be quantified using a smartphone. A wide linear range from 10^2 to 10^8 CFU/mL was demonstrated with a detection limit of 50 CFU/mL, offering a novel approach for enhancing food safety.

Since single phages often encounter limitations in bacterial detection due to the high strain specificity of individual phages, Fu et al. prepared a cocktail comprising three *Klebsiella pneumoniae* (*K. pneumoniae*) phages to broaden the detection spectrum (Figure 7c) [81]. A total of 155 clinically isolated *K. pneumoniae* strains were used to evaluate the recognition spectrum. The recognition rate of 42.3–62.2% using a single phage can be significantly increased to 91.6% via phage cocktail, highlighting the complementarity of the recognition capabilities. Leveraging the wide-spectrum recognition capability of the phage cocktail, a fluorescence resonance energy transfer method was further established through labeling the phage cocktail with fluorescein isothiocyanate and Au nanoparticles with p-mercaptophenylboronic acid as energy donors and acceptors, respectively. The detection process for *K. pneumoniae* strains could be completed within 35 min, with a broad dynamic range of 5.0×10^2–1.0×10^7 CFU/mL, offering promising avenues for enhancing bacterial detection methodologies.

By introducing highly sensitive fluorescence imaging functionality, Gu et al. proposed a novel approach for detecting pathogenic bacteria via covalently conjugating the aggregation-induced emission photosensitizer (AIE-PS) TBTCP-PMB with a phage cocktail (Figure 7d) [82]. The host specificity of phages, the fluorescence imaging capability, and reactive oxygen species (ROS) of AIE-PS were synergized for bacteria detection and sterilization alongside favorable biocompatibility. Additionally, this approach enables rapid diagnosis of blood infections in clinical samples, offering a straightforward and precise method for pathogen detection and point-of-care diagnosis.

Rapid and reliable bacterial tests are essential for reducing the socio-economic burden associated with bacterial infections. Bacteriophages have high specificity to their host bacteria and can identify target bacteria quickly and accurately. Whether it is a single phage to recognize a specific genus of bacteria or a phage cocktail to identify a mixture of bacteria, it reflects the great potential of phage in bacterial detection. Thus, the particular specificity of phages and their widespread availability make them ideal tools for developing biosensors and bioassays for bacterial detection.

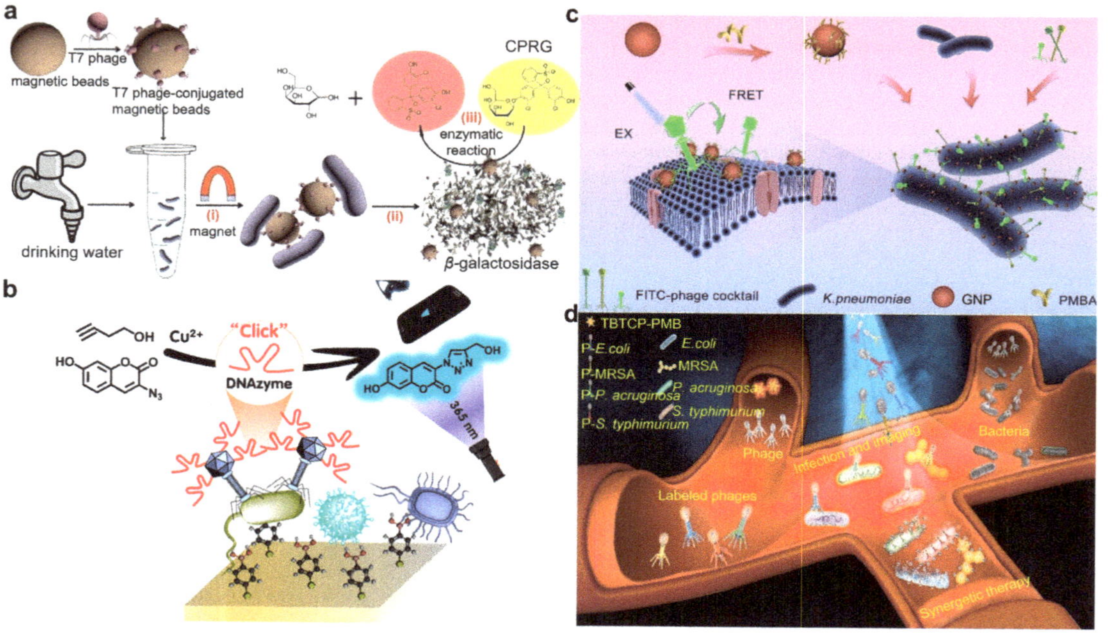

Figure 7. Phage for bacterial recognition and detection. (**a**) Detection of *E. coli* in drinking water using a T7 bacteriophage-conjugated magnetic probe. The separated *E. coli* released β-gal that can catalyze chlorophenol red-β-D-galactopyranoside (CPRG) hydrolysis to produce a colorimetric signal for readout [79]. Reproduced with permission from [79]; published by the American Chemical Society, 2015. (**b**) M13 phage@DNAzyme fluorescent probe for portable and ultrasensitive detection of bacteria. After selective conjugation with target *E. coli* O157:H7, the probe can efficiently catalyze the click reaction between 3-azido-7-hydroxycoumarin (AHC) and 3-butyn-1-ol (BOL) to generate a fluorescent product in the presence of Cu(II) [80]. Reproduced with permission from [80]; published by the American Chemical Society, 2023. (**c**) A phage cocktail composed of three phages for wide-spectrum detection of *K. pneumoniae* strains. Both the FITC-labeled phage cocktail and PMBA-modified GNPs were employed to conjugate *K. pneumoniae* to generate a bacterial complex as they recognized the different sites of the bacterial surface. In the present of *K. pneumoniae*, a decreased FL signal can be observed as a fluorescence resonance energy transfer (FRET) effect occurred between GNP and FITC, enabling the quantification of target bacteria [81]. Reproduced with permission from [81]; published by the American Chemical Society, 2023. (**d**) AIE-photosensitizer modified phage cocktails for the early diagnosis of sepsis. Phages were modified by AIE-active TBTCP-PMB through a simple nucleophilic substitution reaction between benzyl bromide functional groups and the sulfhydryl functional groups of proteins on the phage. The perfect and simultaneously maintained advantages of phages and PSs enabled the selective identification of bacterial species in 30 min through fluorescence imaging [82]. Reproduced with permission from [82]; published by the Wiley Online Library, 2023.

6. Bacteriophages in Targeted Bacterial Killing

Efficient Bacterial infections have long posed significant threats to public health, contributing to severe illness and persistently high mortality rates worldwide [83]. Since the discovery of penicillin, antibiotics have been the cornerstone of bacterial infection treatment [84]. However, the global misuse and overuse of antibiotics have fueled bacterial evolution, leading to the emergence of antibiotic-resistant strains. Phages that can selectively infect bacteria and cause lysis offer a promising alternative to antibiotics. They exhibit remarkable specificity towards their hosts and can adapt and evolve synchronously to infect even resistant bacteria [85]. Consequently, phages are increasingly recognized as a potent tool for antimicrobial therapy. Despite their potential, the efficacy of phages in combating infectious diseases, especially acute infections, is hindered by their relatively low bactericidal stability efficiency. To address this limitation, phages have been modified to enhance their functionality, enabling them to target and eradicate specific bacterial strains more effectively [86].

In light of the escalating global antibiotic resistance crisis, there is an urgent demand for novel agents exhibiting specific targeting toward bacterial strains and exceptional antibacterial efficacy. To address this challenge, Tang et al. introduce a groundbreaking approach that combines bacteriophage (PAP) with photodynamic inactivation (PDI)-active AIEgens (luminogens with aggregation-induced emission property) (Figure 8a) [87]. This strategy yields a unique AIE–PAP bioconjugate characterized by superior capabilities for targeted imaging and synergistic eradication of select bacterial species with maintained infection activity of the phage. Crucially, the PDI-active AIEgens function as potent in situ photosensitizers, generating high-efficiency reactive oxygen species (ROS) upon white light irradiation. Consequently, selective targeting and synergistic elimination of both antibiotic-sensitive and multi-drug-resistant (MDR) bacteria were achieved with excellent biocompatibility. This pioneering AIE–phage integrated approach diversifies the current arsenal of antibacterial agents and promises to catalyze the development of potent drug candidates for the future.

The misuse of antibiotics has spurred the rise of bacterial resistance, prompting the exploration of alternative treatments. Photodynamic antibacterial chemotherapy (PACT) holds promise but faces challenges such as inefficient ROS generation and lack of bacterial targeting. In response, Peng et al. reported a novel photodynamic antimicrobial agent (APNB) based on a cationic photosensitizer (NB) and bacteriophage (ABP) for precise bacterial eradication and effective biofilm ablation (Figure 8b) [88]. Through structural modification with a sulfur atom, NB exhibits enhanced ROS-production capability and near-infrared fluorescence. Furthermore, bacteriophages confer specific binding to pathogenic microorganisms and combinational therapeutic functionalities (PACT and phage therapy). Both in vitro and in vivo studies confirm APNB's efficacy against *A. baumannii* infection, with faster recovery compared to conventional treatments like ampicillin and polymyxin B in vivo. Additionally, the combination of bacteriophages and photosensitizers showed promising results in eradicating bacterial biofilms for the first time, offering a practical solution to bacterial infections.

Intracellular bacterial infections pose significant threats to human health, evading immune defenses and displaying notable drug resistance, presenting formidable challenges for treatment. He and Tang developed a versatile living phage nanoconjugate, integrating aggregation-induced emission luminogen (AIEgen) photosensitizers and nucleic acids onto a bacteriophage framework (forming MS2-DNA-AIEgen bioconjugates) (Figure 8c) [89]. These nanoconjugates efficiently penetrate mammalian cells, specifically identifying intracellular bacteria while emitting a detectable fluorescent signal. Leveraging the photodynamic properties of AIEgen photosensitizers and the bacteriophage's innate targeting and lysis abilities, effectively eliminate intracellular bacteria and restore infected cell activity. Furthermore, the engineered phage nanoconjugates accelerate wound healing in bacterially infected wounds observed in diabetic mouse models, enhancing immune activity within infected cells and in vivo, without evident toxicity. These multifunctional phage nanoconjugates employ AIEgen photosensitizers and spherical nucleic acids as a pioneering strategy for combating intracellular bacteria and offering potent avenues for theranostics applications in diseases associated with intracellular bacterial infections.

Increasing research results show that the occurrence and development of malignant tumors are closely related to the symbiotic bacteria in the tumor microenvironment. The treatment of tumor-associated bacteria by phages has become a promising avenue to augment the cancer therapy [90]. Emerging evidence underscores the role of gut microbiota in colorectal cancer (CRC) tumorigenesis, with symbiotic *Fusobacterium nucleatum* (*Fn*) selectively boosting immunosuppressive myeloid-derived suppressor cells (MDSCs), thereby impairing the host's anticancer immune response. A *Fn*-targeting M13 phage was identified by Zhang's group via phage display technology (Figure 8d) [91]. Subsequently, silver nanoparticles (AgNP) were electrostatically assembled on its surface capsid protein (forming M13@Ag) to achieve precise clearance of *Fn* and reshape the tumor-immune microenvironment. Further, antigen-presenting cells (APCs) can be activated to bolster the host immune system's response against CRC. When combined with immune checkpoint inhibitors (α-PD1) or chemotherapeutics (FOLFIRI), M13@Ag significantly extended overall survival in an orthotopic CRC model.

The engineered bacteriophage overcomes the limited killing ability of nude bacteriophage and achieves an augmented bactericidal effect assisted by photodynamic, photoacoustic, and photothermal effects. Whether it is sensitive or drug-resistant bacteria or even intracellular bacterial infections, excellent antibacterial performance can be achieved through engineered phages.

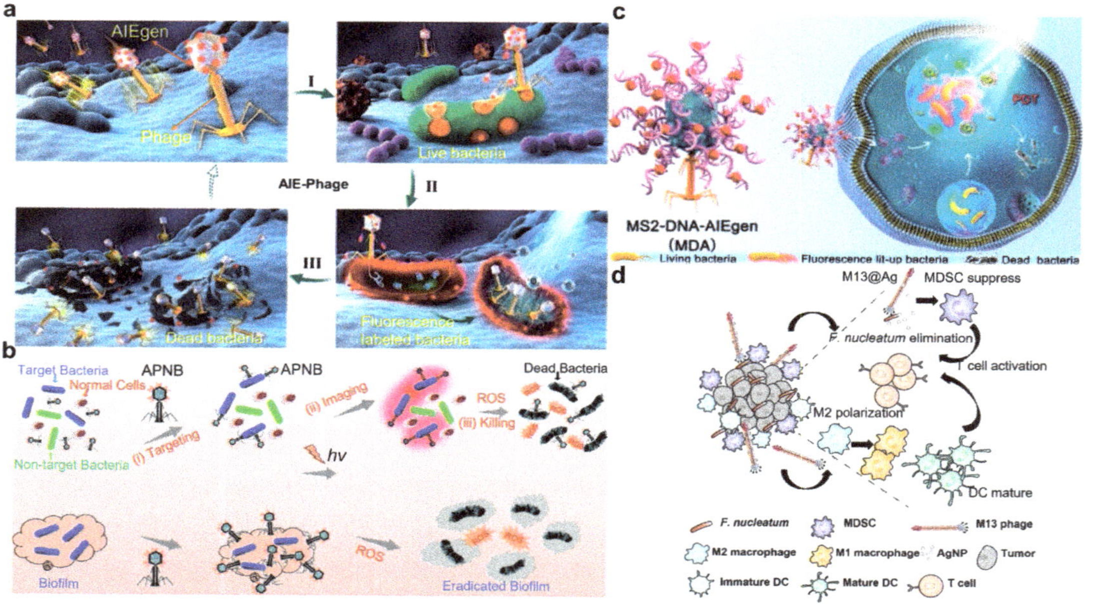

Figure 8. Modified phages to target and eradicate specific bacteria. (**a**) AIE photosensitizer modified phage for guided targeting, discriminative imaging, and synergistic killing of bacteria. Phage possessed particularly targetability toward the host bacteria that can be lit up by the AIE fluorescence and subsequently killed by the synergized effect of phage infection and photodynamic activity of AIE photosensitizers [87]. Reproduced with permission from [87]; published by the American Chemical Society, 2020. (**b**) The multi-functional antibacterial system (APNB) based on the ABP phage and Nile blue photosensitizer (NB) for the treatment of multi-drug-resistant *A. baumannii* and its biofilms [88]. Reproduced with permission from [88]; published by the Royal Society of Chemistry, 2021. (**c**) MS2-DNA-AIEgen bioconjugate for specific targeting and synergistic elimination of intracellular bacteria. After DNA modification on the surface, the MS2 phage was transferred to a type of spherical nucleic acid that can efficiently enter the infected immune cells and find the target bacteria. The fluorescence

property and photodynamic activity of AIE photosensitizers can subsequently label and eliminate the intracellular bacteria [89]. Reproduced with permission from [89]; published by the American Chemical Society, 2024. (**d**) Phage-based bio/abiotic hybrid system (M13@Ag) to regulate gut microbes for cancer-specific immune therapy. Silver (Ag) nanoparticles were electrostatically assembled on the surface of M13 phages (M13@Ag). After accumulation in the tumor microenvironment, AgNP can selectively kill the protumoral *F. nucleatum* under the guidance of M13 phages, which possessed exquisite recognition capability toward *F. nucleatum* and blocked the recruitment of immunosuppressive cells [91]. Reproduced with permission from [91]; published by Science, 2020.

7. Bacteriophage-Assisted Chemo/Radio/Immunotherapy

Emerging research highlights the significant role of the microbiota in both the development and progression of various types of cancer (e.g., colorectal cancers (CRCs) that exist in the gut with abundant bacteria). It has been demonstrated that bacteria can induce severe drug resistance in cancer chemotherapy and foster the immunosuppressive tumor microenvironment (TME) that impedes the immune cell responses [92,93]. While antibiotics have been proposed to regulate the microbiota, their broad spectrum of targets and negative impact on immunotherapy limit their efficacy. Alternatively, phages offer a promising approach to rebalance the tumor-immune microenvironment through precise manipulation of the gut microbiota and activation of innate immune cells [94]. In clinical practice, conventional cancer treatment modalities such as surgery, radiotherapy (RT), chemotherapy, endocrinotherapy, and targeted therapy are commonly employed. However, these approaches may induce resistance and severe side effects [95]. The integration of phage technology in cancer therapy presents a promising avenue for identifying tumor-targeting agents, carriers, and potential vaccines. Natural microorganisms containing pathogen-associated molecular patterns (PAMPs) effectively stimulate the host's innate immune response, surpassing synthetic material-based vaccines in robustness and efficacy [96].

The gut microbiota in humans is closely associated with the progression of colorectal cancer (CRC) and its response to treatment. By harnessing the higher concentration of pro-tumorigenic *Fusobacterium nucleatum* and the deficiency of anti-tumorigenic butyrate-producing bacteria in the fecal microbiota of CRC patients, Zhang's team discovered that oral or intravenous administration of irinotecan-loaded glucan nanoparticles with modification of bacteriophages targeting *Fusobacterium nucleatum* can effectively inhibit the growth of *Fusobacterium nucleatum* and significantly enhance the efficacy of chemotherapy for CRC in mice (Figure 9a) [97]. Additionally, the impact of bacteriophage-guided irinotecan nanoparticles on blood cell counts, immunoglobulin, and tissue protein levels, as well as liver and kidney function in piglets, was negligible. Utilizing this bacteriophage-guided nanotechnology to modulate the gut microbiota may inspire new approaches for treating colorectal cancer.

Mao et al. conjugated a photosensitizer to a phage for specifically targeting SKBR-3 breast cancer cells and consequently PDT via fusing a unique targeting peptide (VSSTQDFP) to each of the ~3900 copies of pVIII on the side wall of the phage (Figure 9b) [98]. A well-established photosensitizer, 9-ethenyl-14-ethyl-4,6,13,18-tetramethyl-20-oxo-3-phorbinepropanoic acid, commonly known as pyropheophorbid-a (PPa), was then partially modified on the N-terminus of the phage surface. Thus, the photosensitizers can be transported to the intended site, as the targeting specificity of the phage nanowire was well maintained through the free N terminus of the pVIII that is not conjugated to PPa. These phage-conjugated photosensitizers expand the clinical repertoire of target-specific PDT.

Zhang's group demonstrated the effectiveness of phages in immunotherapy via a bioactive vaccine platform termed HMP@Ag, which was composed by hybrid M13 phages and individualized tumor antigens (Figure 9c) [99]. The transportation of antigens to lymph nodes was facilitated, and the antigen-presenting cells (APCs) were activated via the Toll-like receptor 9 (TLR9) signaling pathway, leading to enhanced innate and adaptive immune responses. Serving as an adjuvant platform, hybrid M13 phages can easily deliver a variety of tumor-specific antigens by simple adsorption, thereby supporting the development of personalized cancer vaccines. Significantly, the HMP@Ag vaccine not only prevents tumor formation but also retards tumor growth in established (subcutaneous and orthotopic) and metastatic tumor-bearing models, particularly when used in combination with immune checkpoint blockade (ICB) therapy, with dramatically suppressed post-operative recurrence and concurrently promotes long-term immune memory. Furthermore, HMP@Ag induces a robust immune response targeting neoantigens in models of tumor-specific mutations, suggesting a potent tool for developing a bio-activated hybrid platform for personalized therapy.

The complexity and diversity within the tumor microenvironment have posed significant challenges to the efficacy of cancer treatments. Zhang et al. utilized an engineered M13 bacteriophage with immunogenic activity to remodel the tumor microenvironment (Figure 9d) [100]. Through chemical cross-linking and biomineralization, a living phage system capable of transforming the tumor microenvironment was fabricated. Through biomineralization, photothermal-active palladium nanoparticles (PdNPs) were in situ synthesized on the pVIII capsid protein of M13 phage. After chemical cross-linking and loading of NLG919 that aimed to inhibit the expression of the tryptophan metabolic enzyme indoleamine 2,3-dioxygenase 1 (IDO1), M13@Pd/NLG gel was formed. M13 bacteriophage not only acted as a carrier for cargo but also functioned as a self-immune adjuvant, effectively inducing immunogenic cell death in tumor cells and down-regulating IDO1 expression with good potential to reverse immunosuppression and significantly enhance the anti-breast cancer response.

The above studies confirm the efficacy and safety of using phage-based vectors as delivery tools for therapeutic genes and drugs in cancer therapy. Compared to the non-viral gene transfer methods, engineered phages show much higher efficiencies in gene delivery and expression in cancer cells with enhanced selectivity. However, as the interaction mechanisms between phage and eukaryotic cells are not illustrated clearly, it is necessary to carry out more in-depth investigations for further cancer diagnosis and treatment.

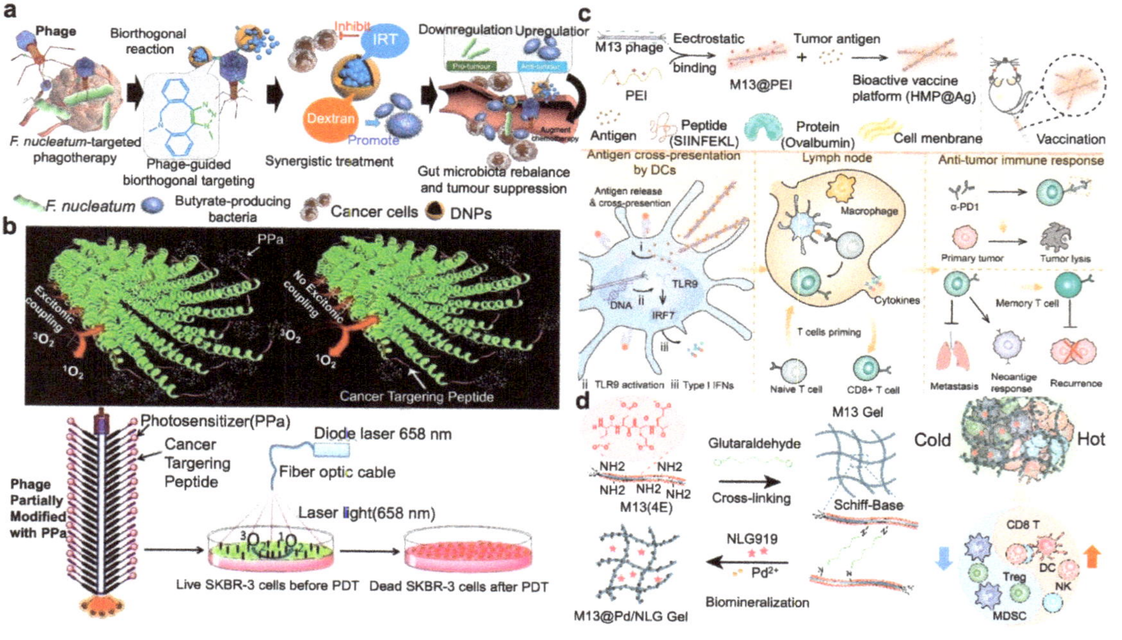

Figure 9. Phage-assisted chemo/radio/immunotherapy. (**a**) The phage-guided biotic–abiotic hybrid nanosystem for CRC therapeutics. Through bioorthogonal reaction, chemodrug IRT-loaded dextran

nanoparticles (IDNPs) were bound to the phages with the capability to target CRC and eliminate the pro-tumoral *Fusobacterium* and enhance the production of butyrate from anti-tumoral bacteria. The inhibition of pro-tumoral bacteria and the promotion of anti-tumoral bacteria together augmented the chemotherapeutic efficacy [97]. Reproduced with permission from [97]; published by Springer Nature, 2019. (**b**) Phage conjugated with photosensitizer for targeted photodynamic therapy (PDT). With modification of the major coat protein (pVIII) with PPa photosensitizer, the filamentous phage fd-tet nanowire was used as a vehicle for targeted delivery of photosensitizer to the SKBR-3 cells to realize cancer eradication [98]. Reproduced with permission from [98]; published by the Wiley Online Library, 2013. (**c**) Phage-based vaccine combined with immune checkpoint blockade (ICB) treatment for inhibition of primary and metastatic cancers. Tumor antigen (Ag) modified hybrid M13 phage (HMP@Ag)-based vaccine was internalized by DCs for antigen release and cross-presentation to promote DCs maturation, which further migrated to lymph nodes for activation and expansion of antigen-specific CD^{8+} T cells, representing a remarkable tumor recurrence suppression effect after surgery [99]. Reproduced with permission from [99]; published by Elsevier, ScienceDirect, 2022. (**d**) Construction of M13 phage-based self-adjuvant photothermal gel for antitumor therapy. The engineered M13 phage was chemically cross-linked through the Schiff base reaction to biomineralize Pd nanoparticles and load the inhibitor NLG919 of immunosuppressive factor indoleamine 2,3-dioxygenase 1 (IDO1), realizing photothermal and immunotherapy synergistically for robust antitumor response [100]. Reproduced with permission from [100]; published by the American Chemical Society, 2023.

8. Bacteriophages for Drug/Gene Delivery

As nanomedicine advances, an array of carriers has emerged for targeted drug delivery, including liposomes, dendrimers, polymers, micelles, virus-like particles, and even stem cells [101]. Phages stand out among these carriers due to their non-infectivity to mammalian cells, small genome size, simple structure, and ease of engineering. In addition, phage display technology offers an efficient strategy to modify phages by displaying target-specific peptides [102]. Moreover, in chemical drug delivery, filamentous phages offer superior pharmacokinetics and delivery efficiency compared to spherical nanoparticles as they can accommodate a larger payload of chemical drugs. For gene delivery, foreign genes can be inserted into the phage genome or loaded onto phage-mimetic nanoparticles. Thus, phages emerge as excellent candidates for drug and gene delivery applications [103].

With advancements in covalent modification techniques for phage capsids, Francis et al. can transform them into versatile carrier systems for drugs and imaging agents (Figure 10a) [104]. Two distinct modification strategies were employed to adorn the outer surface of genome-free MS2 capsids with PEG chains while incorporating 50–70 fluorescent dye inside to mimic drug cargo. Despite extensive modification, the capsids maintained their assembled state. The efficacy of the PEG coating in obstructing access to the capsid surface by polyclonal antibodies was as high as 90%. Biotin groups positioned at the distal ends of the polymer chains retained the ability to bind to streptavidin. Additionally, a modular approach was developed for attaching small-molecule targeting moieties to the PEG chains via an efficient oxime formation reaction, facilitating the development of robust and versatile platforms for therapeutic cargo delivery.

Wu et al. attached the water-soluble derivative of the chemotherapy drug paclitaxel to the capsid of MS2 phages via covalently bioconjugation (Figure 10b) [105]. Moreover, 180 copies of the MS2 protein monomer aggregate to create a hollow spherical carrier with a diameter of 27 nm and 32 pores of 2 nm wide that can facilitate the incorporation of small drug molecules onto the interior surface. Degradation and nonspecific interactions of the cargo with healthy tissue can be well avoided. Up to 65% modification of the MS2 monomer proteins was achieved, corresponding to approximately 110 molecules of taxol per capsid, or roughly 2% loading by weight. The modified capsid retains its capsid form and can efficiently release paclitaxel with excellent cytotoxicity toward the tested MCF-7 cancer cells.

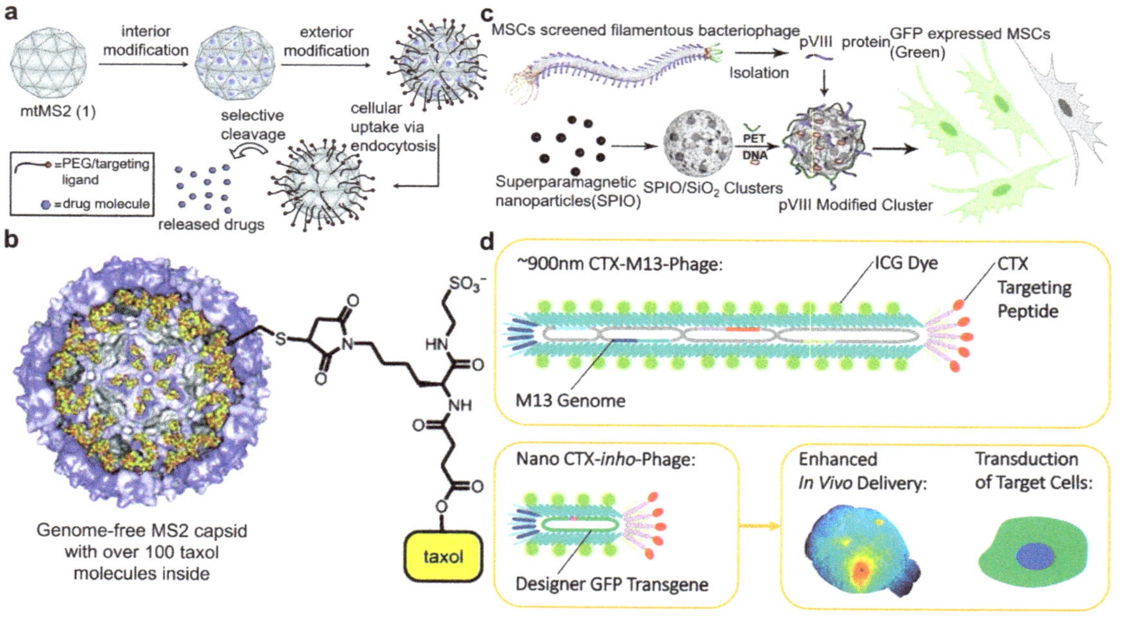

Figure 10. Phage for drug/gene delivery. (**a**) Construction of multifunctional drug delivery vessels derived from genome-free (mtMS2) phage capsids. Utilizing orthogonal coupling strategies to append over 180 PEG chains to the exterior surface of phage MS2, while decorating its interior surface with 50–70 fluorescein molecules as drug cargo mimic. The loaded drug can be efficiently released after cell uptake [104]. Reproduced with permission from [104]; published by the American Chemical Society, 2007. (**b**) Drug delivery system with a water-soluble derivative of the chemotherapeutic agent taxol conjugated to the capsids of MS2 phages. An amino acid on the interior of MS2 capsid was mutated to a cysteine residue (N87C) to provide a sulfhydryl group for the attachment of cargo. The modified capsids remained in their native form and released taxol when incubated with MCF-7 cells [105]. Reproduced with permission from [105]; published by the Wiley Online Library, 2009. (**c**) Phage capsids for gene delivery to mesenchymal stem cells (MSCs). The gene cargo was incorporated into the porous structure of SPIO-embedded magnetic/silica nanoclusters (MSNCs) that was modified with PEI and pVIII proteins extracted from the MSC-targeting phage. Bearing pVIII allowed the nanoclusters to mimic phage particles and target MSCs to achieve gene delivery [106]. Reproduced with permission from [106]; published by the Wiley Online Library, 2013. (**d**) CTX-M13 phage for gene delivery. After p3 capsid chlorotoxin fusion with the "inho" circular single-stranded DNA (cssDNA) gene packaging system, a miniatured chlorotoxin inho (CTX-inho) phage particle was produced that can target the GBM22 glioblastoma tumors in the brains of mice, achieving enhanced efficiency in gene delivery and target cell transduction [107]. Reproduced with permission from [107]; published by the American Chemical Society, 2022.

Non-viral gene therapy toward mesenchymal stem cells (MSCs) holds promise for treating various diseases, but MSCs are difficult to transfect using non-viral vectors. Mao's group has successfully engineered phage-borne rMSC-targeting pVIII proteins on the surface of gene-carrying fd-tet phage, resulting in the formation of virus-mimetic magnetic/silica nanoclusters (VMSNCs) (Figure 10c) [106]. These VMSNCs achieved a significantly improved transfection efficiency of approximately 40% in rMSCs compared to commercial transfection reagents. Furthermore, incorporating rMSC-targeting peptides significantly enhanced transfection efficiency (up to approximately 50%) with sustained gene expression for over 170 h, indicating enhanced stem cell-based gene therapy.

The M13 bacteriophage is a highly adaptable and genetically customizable nanocarrier for disease diagnosis and therapeutics. By fusing the p3 capsid with chlorotoxin and incorporating the "inho" circular single-stranded DNA (cssDNA) gene, Belcher's group successfully engineered miniature chlorotoxin inho (CTX-inho) phage particles with a length of at least 50 nm (Figure 10d) [107]. These particles can selectively target intracranial orthotopic patient-derived GBM22 glioblastoma tumors in mouse brains. Upon systemic administration, indocyanine green-conjugated CTX-inho phage accumulates in brain tumors by emitting out infrared fluorescence signals. Additionally, the transported cssDNA remained transcriptionally active when delivered to GBM22 glioma cells in vitro. Taking together the ability to adjust capsid display, surface loading, phage length, and cssDNA gene content, the recombinant M13 phage particle can be engineered to be an optimal delivery platform.

Chemical drugs can be modified on the phage surface, and gene drugs can also be inserted into the phage genome by recombinant DNA technology. Meanwhile, specific peptides/proteins displayed on the phage surface are able to bind to functional materials, giving them specific targeting and huge drug loading capabilities. What is more, phage peptides or proteins can be directly self-assembled into nanovehicles for self-guided drug delivery.

9. Conclusions and Prospectives

Although numerous approaches have demonstrated potential in modifying and armoring bacteriophages that bring considerable strides in the applications of bioanalysis and disease theranostics, certain constraints persist and imped their continuous advancement and utilization. First, the screening process for phages is relatively complicated, time-consuming, and labor-intensive, leading to the lack of a standardized phage library. Second, long-term examination of the genetic stability of the engineered bacteriophages that acquired through genetic modification and transformation is necessary. Despite the encouraging outcomes of applying the existing generation of engineered bacteriophages, the durability of their genetic alterations and inheritability of properties and functions are remaining unresolved problems. Thirdly, safety is the pivotal issue in the real clinical translation. Apart from a few experiments that have demonstrated the safety of engineered bacteriophages in vitro and in cell-level, persistent challenges still impede the progress of applying them in animal-level practices. These obstacles necessitate us to gain a deeper understanding of the immune response of modified bacteriophages and carry out thorough evaluations of the potential risks linked to their in vivo and clinic applications. Consequently, further research is imperative to probe the interactions between bacteriophages and the bodies, aiming to clearly illustrate the stability and degradability of engineered bacteriophages in diverse environments.

In sum, this review outlines the capabilities of chemical modification and genetic engineering to create engineered bacteriophages with enhanced properties and multiplied functionalities. Accompanying the advancements in phage display technology and molecular engineering, armored bacteriophages are extensively utilized across diverse fields. This review summarizes the outstanding achievements of engineered bacteriophages in recent years, including the synthesis and assembly of nanomaterials, detection of biotargets and pathogenic bacteria, screening of peptides or antibodies, identification and elimination of bacteria, diagnosis and treatment of diseases, as well as drug and gene delivery, offering promising avenues for future clinical research endeavors.

Author Contributions: Conceptualization, X.H.; resources, J.Z.; writing—original draft preparation, J.Z.; writing—review and editing, X.H.; supervision, X.H.; funding acquisition, X.H. All authors have read and agreed to the published version of the manuscript.

Funding: This research was funded by the National Natural Science Foundation of China, grant numbers 22274106 and 22104104; the Natural Science Foundation of Jiangsu Province, grant number

BK20210701; and the Program of Suzhou Innovation and Entrepreneurship Leading Talents grant number ZXL2022513. The APC was funded by Soochow University.

Acknowledgments: The authors acknowledge funding from the National Natural Science Foundation of China (22274106 and 22104104), the Natural Science Foundation of Jiangsu Province (BK20210701), the Program of Suzhou Innovation and Entrepreneurship Leading Talents (ZXL2022513), and startup funds from Soochow University.

Conflicts of Interest: The authors declare no conflicts of interest.

References

1. Hendrix, R.W. Bacteriophages: Evolution of the majority. *Theor. Popul. Biol.* **2002**, *61*, 471–480. [CrossRef] [PubMed]
2. Twort, F.W. Further investigations on the nature of ultra-microscopic viruses and their cultivation. *J. Hyg.* **1936**, *36*, 204–235. [CrossRef] [PubMed]
3. Sharma, S.; Chatterjee, S.; Datta, S.; Prasad, R.; Dubey, D.; Prasad, R.K.; Vairale, M.G. Bacteriophages and its applications: An overview. *Folia Microbiol.* **2017**, *62*, 17–55. [CrossRef] [PubMed]
4. Yue, H.; Li, Y.; Yang, M.; Mao, C. T7 phage as an emerging nanobiomaterial with genetically tunable target specificity. *Adv. Sci.* **2022**, *9*, 2103645. [CrossRef] [PubMed]
5. Zhang, M.; Zhang, T.; Yu, M.; Chen, Y.-L.; Jin, M. The life cycle transitions of temperate phages: Regulating factors and potential ecological implications. *Viruses* **2022**, *14*, 1904. [CrossRef] [PubMed]
6. Young, R.; Wang, I.-N.; Roof, W.D. Phages will out: Strategies of host cell lysis. *Trends Microbiol.* **2000**, *8*, 120–128. [CrossRef] [PubMed]
7. Sunderland, K.S.; Yang, M.; Mao, C. Phage-enabled nanomedicine: From probes to therapeutics in precision medicine. *Angew. Chem. Int. Ed.* **2017**, *56*, 1964–1992. [CrossRef]
8. Hesse, S.; Adhya, S. Phage therapy in the twenty-first century: Facing the decline of the antibiotic era; is it finally time for the age of the phage? *Annu. Rev. Microbiol.* **2019**, *73*, 155–174. [CrossRef] [PubMed]
9. Pires, D.P.; Meneses, L.; Brandão, A.C.; Azeredo, J. An overview of the current state of phage therapy for the treatment of biofilm-related infections. *Curr. Opin. Virol.* **2022**, *53*, 101209. [CrossRef] [PubMed]
10. Champagne-Jorgensen, K.; Luong, T.; Darby, T.; Roach, D.R. Immunogenicity of bacteriophages. *Trends Microbiol.* **2023**, *31*, 1058–1071. [CrossRef] [PubMed]
11. Kortright, K.E.; Chan, B.K.; Koff, J.L.; Turner, P.E. Phage therapy: A renewed approach to combat antibiotic-resistant bacteria. *Cell Host Microbe* **2019**, *25*, 219–232. [CrossRef]
12. Mochizuki, K.; Matsukura, L.; Ito, Y.; Miyashita, N.; Taki, M. A medium-firm drug-candidate library of cryptand-like structures on t7 phage: Design and selection of a strong binder for hsp90. *Org. Biomol. Chem.* **2021**, *19*, 146–150. [CrossRef]
13. Schmidt, A.; Kellermann, J.; Lottspeich, F. A novel strategy for quantitative proteomics using isotope-coded protein labels. *Proteomics* **2005**, *5*, 4–15. [CrossRef] [PubMed]
14. Bernard, J.M.L.; Francis, M.B. Chemical strategies for the covalent modification of filamentous phage. *Front. Microbiol.* **2014**, *5*, 734. [CrossRef]
15. Abello, N.; Kerstjens, H.A.M.; Postma, D.S.; Bischoff, R. Selective acylation of primary amines in peptides and proteins. *J. Proteome Res.* **2007**, *6*, 4770–4776. [CrossRef]
16. Carmody, C.M.; Goddard, J.M.; Nugen, S.R. Bacteriophage capsid modification by genetic and chemical methods. *Bioconj. Chem.* **2021**, *32*, 466–481. [CrossRef] [PubMed]
17. Marinelli, L.J.; Piuri, M.; Swigoňová, Z.; Balachandran, A.; Oldfield, L.M.; van Kessel, J.C.; Hatfull, G.F. Bred: A simple and powerful tool for constructing mutant and recombinant bacteriophage genomes. *PLoS ONE* **2008**, *3*, e3957. [CrossRef]
18. Tan, Y.; Tian, T.; Liu, W.; Zhu, Z.; Yang, C.J. Advance in phage display technology for bioanalysis. *Biotechnol. J.* **2016**, *11*, 732–745. [CrossRef] [PubMed]
19. Peng, H.; Chen, I.A. Phage engineering and the evolutionary arms race. *Curr. Opin. Biotechnol.* **2021**, *68*, 23–29. [CrossRef] [PubMed]
20. Ragothaman, M.; Yoo, S.Y. Engineered phage-based cancer vaccines: Current advances and future directions. *Vaccines* **2023**, *11*, 919. [CrossRef] [PubMed]
21. Veeranarayanan, S.; Azam, A.H.; Kiga, K.; Watanabe, S.; Cui, L. Bacteriophages as solid tumor theragnostic agents. *Int. J. Mol. Sci.* **2022**, *23*, 402. [CrossRef] [PubMed]
22. Li, X.-T.; Peng, S.-Y.; Feng, S.-M.; Bao, T.-Y.; Li, S.-Z.; Li, S.-Y. Recent progress in phage-based nanoplatforms for tumor therapy. *Small* **2024**, *20*, 2307111. [CrossRef] [PubMed]
23. Farr, R.; Choi, D.S.; Lee, S.-W. Phage-based nanomaterials for biomedical applications. *Acta Biomater.* **2014**, *10*, 1741–1750. [CrossRef]
24. Chang, C.; Guo, W.; Yu, X.; Guo, C.; Zhou, N.; Guo, X.; Huang, R.-L.; Li, Q.; Zhu, Y. Engineered m13 phage as a novel therapeutic bionanomaterial for clinical applications: From tissue regeneration to cancer therapy. *Mater. Today Bio* **2023**, *20*, 100612. [CrossRef] [PubMed]

25. Rehman, S.; Ali, Z.; Khan, M.; Bostan, N.; Naseem, S. The dawn of phage therapy. *Rev. Med. Virol.* **2019**, *29*, e2041. [CrossRef] [PubMed]
26. Meile, S.; Du, J.; Dunne, M.; Kilcher, S.; Loessner, M.J. Engineering therapeutic phages for enhanced antibacterial efficacy. *Curr. Opin. Virol.* **2022**, *52*, 182–191. [CrossRef] [PubMed]
27. Stacey, H.J.; De Soir, S.; Jones, J.D. The safety and efficacy of phage therapy: A systematic review of clinical and safety trials. *Antibiotics* **2022**, *11*, 1340. [CrossRef] [PubMed]
28. Fujiwara, D.; Mihara, K.; Takayama, R.; Nakamura, Y.; Ueda, M.; Tsumuraya, T.; Fujii, I. Chemical modification of phage-displayed helix-loop-helix peptides to construct kinase-focused libraries. *ChemBioChem* **2021**, *22*, 3406–3409. [CrossRef] [PubMed]
29. Branston, S.; Stanley, E.; Ward, J.; Keshavarz-Moore, E. Study of robustness of filamentous bacteriophages for industrial applications. *Biotechnol. Bioeng.* **2011**, *108*, 1468–1472. [CrossRef] [PubMed]
30. Zheng, M.; Gao, J. Phage display of two distinct warheads to inhibit challenging proteins. *ACS Chem. Biol.* **2023**, *18*, 2259–2266. [CrossRef] [PubMed]
31. Jin, L.; Cao, F.; Gao, Y.; Zhang, C.; Qian, Z.; Zhang, J.; Mao, Z. Microenvironment-activated nanozyme-armed bacteriophages efficiently combat bacterial infection. *Adv. Mater.* **2023**, *35*, 2301349. [CrossRef] [PubMed]
32. Peng, H.; Borg, R.E.; Dow, L.P.; Pruitt, B.L.; Chen, I.A. Controlled phage therapy by photothermal ablation of specific bacterial species using gold nanorods targeted by chimeric phages. *Proc. Natl. Acad. Sci. USA* **2020**, *117*, 1951–1961. [CrossRef] [PubMed]
33. Wang, L.; Fan, X.; Gonzalez Moreno, M.; Tkhilaishvili, T.; Du, W.; Zhang, X.; Nie, C.; Trampuz, A.; Haag, R. Photocatalytic quantum dot-armed bacteriophage for combating drug-resistant bacterial infection. *Adv. Sci.* **2022**, *9*, 2105668. [CrossRef]
34. Carrico, Z.M.; Farkas, M.E.; Zhou, Y.; Hsiao, S.C.; Marks, J.D.; Chokhawala, H.; Clark, D.S.; Francis, M.B. N-terminal labeling of filamentous phage to create cancer marker imaging agents. *ACS Nano* **2012**, *6*, 6675–6680. [CrossRef] [PubMed]
35. Stevens, T.K.; Palaniappan, K.K.; Ramirez, R.M.; Francis, M.B.; Wemmer, D.E.; Pines, A. Hypercest detection of a 129xe-based contrast agent composed of cryptophane-a molecular cages on a bacteriophage scaffold. *Magn. Reson. Med.* **2013**, *69*, 1245–1252. [CrossRef] [PubMed]
36. Tong, G.J.; Hsiao, S.C.; Carrico, Z.M.; Francis, M.B. Viral capsid DNA aptamer conjugates as multivalent cell-targeting vehicles. *J. Am. Chem. Soc.* **2009**, *131*, 11174–11178. [CrossRef]
37. Sathaliyawala, T.; Rao, M.; Maclean, D.M.; Birx, D.L.; Alving, C.R.; Rao, V.B. Assembly of human immunodeficiency virus (hiv) antigens on bacteriophage t4: A novel in vitro approach to construct multicomponent hiv vaccines. *J. Virol.* **2006**, *80*, 7688–7698. [CrossRef]
38. Guan, J.; Oromí-Bosch, A.; Mendoza, S.D.; Karambelkar, S.; Berry, J.D.; Bondy-Denomy, J. Bacteriophage genome engineering with CRISPR–Cas13a. *Nat. Microbiol.* **2022**, *7*, 1956–1966. [CrossRef] [PubMed]
39. Paul, B.; Montoya, G. CRISPR-Cas12a: Functional overview and applications. *Biomed. J.* **2020**, *43*, 8–17. [CrossRef] [PubMed]
40. Park, J.Y.; Moon, B.Y.; Park, J.W.; Thornton, J.A.; Park, Y.H.; Seo, K.S. Genetic engineering of a temperate phage-based delivery system for CRISPR/Cas9 antimicrobials against staphylococcus aureus. *Sci. Rep.* **2017**, *7*, 44929. [CrossRef] [PubMed]
41. Braun, P.; Raab, R.; Bugert, J.J.; Braun, S. Recombinant reporter phage rTUN1::*nLuc* enables rapid detection and real-time antibiotic susceptibility testing of klebsiella pneumoniae K64 strains. *ACS Sens.* **2023**, *8*, 630–639. [CrossRef] [PubMed]
42. Marinelli, L.J.; Hatfull, G.F.; Piuri, M. Recombineering. *Bacteriophage* **2012**, *2*, 5–14. [CrossRef] [PubMed]
43. Hussain, W.; Yang, X.; Ullah, M.; Wang, H.; Aziz, A.; Xu, F.; Asif, M.; Ullah, M.W.; Wang, S. Genetic engineering of bacteriophages: Key concepts, strategies, and applications. *Biotechnol. Adv.* **2023**, *64*, 108116. [CrossRef]
44. Ramirez-Chamorro, L.; Boulanger, P.; Rossier, O. Strategies for bacteriophage T5 mutagenesis: Expanding the toolbox for phage genome engineering. *Front. Microbiol.* **2021**, *12*, 667332. [CrossRef] [PubMed]
45. Martel, B.; Moineau, S. CRISPR-Cas: An efficient tool for genome engineering of virulent bacteriophages. *Nucleic Acids Res.* **2014**, *42*, 9504–9513. [CrossRef] [PubMed]
46. Lv, S.; Wang, Y.; Jiang, K.; Guo, X.; Zhang, J.; Zhou, F.; Li, Q.; Jiang, Y.; Yang, C.; Teng, T. Genetic engineering and biosynthesis technology: Keys to unlocking the chains of phage therapy. *Viruses* **2023**, *15*, 1736. [CrossRef] [PubMed]
47. Shivram, H.; Cress, B.F.; Knott, G.J.; Doudna, J.A. Controlling and enhancing CRISPR systems. *Nat. Chem. Biol.* **2021**, *17*, 10–19. [CrossRef] [PubMed]
48. Malone, L.M.; Birkholz, N.; Fineran, P.C. Conquering CRISPR: How phages overcome bacterial adaptive immunity. *Curr. Opin. Biotechnol.* **2021**, *68*, 30–36. [CrossRef] [PubMed]
49. da Silva, J.L.; Piuri, M.; Broussard, G.; Marinelli, L.J.; Bastos, G.M.; Hirata, R.D.C.; Hatfull, G.F.; Hirata, M.H. Application of BRED technology to construct recombinant D29 reporter phage expressing EGFP. *FEMS Microbiol. Lett.* **2013**, *344*, 166–172. [CrossRef] [PubMed]
50. Jaschke, P.R.; Lieberman, E.K.; Rodriguez, J.; Sierra, A.; Endy, D. A fully decompressed synthetic bacteriophage øX174 genome assembled and archived in yeast. *Virol. J.* **2012**, *434*, 278–284. [CrossRef]
51. Assad-Garcia, N.; D'Souza, R.; Buzzeo, R.; Tripathi, A.; Oldfield, L.M.; Vashee, S.; Fouts, D.E. Cross-genus "boot-up" of synthetic bacteriophage in staphylococcus aureus by using a new and efficient DNA transformation method. *Appl. Environ. Microbiol.* **2022**, *88*, e01486-21. [CrossRef] [PubMed]
52. Dhanker, R.; Hussain, T.; Tyagi, P.; Singh, K.J.; Kamble, S.S. The emerging trend of bio-engineering approaches for microbial nanomaterial synthesis and its applications. *Front. Microbiol.* **2021**, *12*, 638003. [CrossRef]

53. Asar, M.; Newton-Northup, J.; Soendergaard, M. Improving pharmacokinetics of peptides using phage display. *Viruses* **2024**, *16*, 570. [CrossRef]
54. Zhang, H.; Xu, Y.; Huang, Q.; Yi, C.; Xiao, T.; Li, Q. Natural phage nanoparticle-mediated real-time immuno-PCR for ultrasensitive detection of protein marker. *Chem. Commun.* **2013**, *49*, 3778–3780. [CrossRef] [PubMed]
55. Lee, S.-H.; Moody, I.; Zeng, Z.; Fleischer, E.B.; Weiss, G.A.; Shea, K.J. Synthesis of a high affinity complementary peptide–polymer nanoparticle (NP) pair using phage display. *ACS Appl. Bio Mater.* **2021**, *4*, 2704–2712. [CrossRef] [PubMed]
56. Bush, J.; Singh, S.; Vargas, M.; Oktay, E.; Hu, C.-H.; Veneziano, R. Synthesis of DNA origami scaffolds: Current and emerging strategies. *Molecules* **2020**, *25*, 3386. [CrossRef] [PubMed]
57. Mao, C.; Solis, D.J.; Reiss, B.D.; Kottmann, S.T.; Sweeney, R.Y.; Hayhurst, A.; Georgiou, G.; Iverson, B.; Belcher, A.M. Virus-based toolkit for the directed synthesis of magnetic and semiconducting nanowires. *Science* **2004**, *303*, 213–217. [CrossRef] [PubMed]
58. Li, Y.; Cao, B.; Yang, M.; Zhu, Y.; Suh, J.; Mao, C. Identification of novel short $BaTiO_3$-binding/nucleating peptides for phage-templated in situ synthesis of $BaTiO_3$ polycrystalline nanowires at room temperature. *ACS Appl. Mater. Interfaces* **2016**, *8*, 30714–30721. [CrossRef] [PubMed]
59. Żelechowska, K.; Karczewska-Golec, J.; Karczewski, J.; Łoś, M.; Kłonkowski, A.M.; Węgrzyn, G.; Golec, P. Phage-directed synthesis of photoluminescent zinc oxide nanoparticles under benign conditions. *Bioconjugate Chem.* **2016**, *27*, 1999–2006. [CrossRef] [PubMed]
60. Nam, K.T.; Peelle, B.R.; Lee, S.-W.; Belcher, A.M. Genetically driven assembly of nanorings based on the M13 virus. *Nano Lett.* **2004**, *4*, 23–27. [CrossRef]
61. Merzlyak, A.; Lee, S.-W. Phage as templates for hybrid materials and mediators for nanomaterial synthesis. *Curr. Opin. Chem. Biol.* **2006**, *10*, 246–252. [CrossRef]
62. Tang, J.; Fraden, S. Isotropic-cholesteric phase transition in colloidal suspensions of filamentous bacteriophage *fd*. *Liq. Cryst.* **1995**, *19*, 459–467. [CrossRef]
63. Moon, J.-S.; Choi, E.J.; Jeong, N.-N.; Sohn, J.-R.; Han, D.-W.; Oh, J.-W. Research progress of M13 bacteriophage-based biosensors. *J. Nanomater.* **2019**, *9*, 1448. [CrossRef] [PubMed]
64. Mao, C.; Flynn, C.E.; Hayhurst, A.; Sweeney, R.; Qi, J.; Georgiou, G.; Iverson, B.; Belcher, A.M. Viral assembly of oriented quantum dot nanowires. *Proc. Natl. Acad. Sci. USA* **2003**, *100*, 6946–6951. [CrossRef] [PubMed]
65. Mao, C.; Wang, F.; Cao, B. Controlling nanostructures of mesoporous silica fibers by supramolecular assembly of genetically modifiable bacteriophages. *Angew. Chem. Int. Ed.* **2012**, *51*, 6411–6415. [CrossRef] [PubMed]
66. He, T.; Abbineni, G.; Cao, B.; Mao, C. Nanofibrous bio-inorganic hybrid structures formed through self-assembly and oriented mineralization of genetically engineered phage nanofibers. *Small* **2010**, *6*, 2230–2235. [CrossRef] [PubMed]
67. Wang, J.; Yang, M.; Zhu, Y.; Wang, L.; Tomsia, A.P.; Mao, C. Phage nanofibers induce vascularized osteogenesis in 3D printed bone scaffolds. *Adv. Mater.* **2014**, *26*, 4961–4966. [CrossRef] [PubMed]
68. Taton, T.A.; Mirkin, C.A.; Letsinger, R.L. Scanometric DNA array detection with nanoparticle probes. *Science* **2000**, *289*, 1757–1760. [CrossRef]
69. Wang, L.; Li, H.; Wang, X.; Yang, X.; Tian, C.; Sun, D.; Liu, L.; Li, J. Modification of low-energy surfaces using bicyclic peptides discovered by phage display. *J. Am. Chem. Soc.* **2023**, *145*, 17613–17620. [CrossRef]
70. Guliy, O.I.; Evstigneeva, S.S.; Khanadeev, V.A.; Dykman, L.A. Antibody phage display technology for sensor-based virus detection: Current status and future prospects. *Biosensors* **2023**, *13*, 640. [CrossRef] [PubMed]
71. Kim, H.-J.; McCoy, M.; Gee, S.J.; González-Sapienza, G.G.; Hammock, B.D. Noncompetitive phage anti-immunocomplex real-time polymerase chain reaction for sensitive detection of small molecules. *Anal. Chem.* **2011**, *83*, 246–253. [CrossRef] [PubMed]
72. Zhou, X.; Cao, P.; Zhu, Y.; Lu, W.; Gu, N.; Mao, C. Phage-mediated counting by the naked eye of mirna molecules at attomolar concentrations in a petri dish. *Nat. Mater.* **2015**, *14*, 1058–1064. [CrossRef]
73. Fang, H.; Zhou, Y.; Ma, Y.; Chen, Q.; Tong, W.; Zhan, S.; Guo, Q.; Xiong, Y.; Tang, B.Z.; Huang, X. M13 bacteriophage-assisted recognition and signal spatiotemporal separation enabling ultrasensitive light scattering immunoassay. *ACS Nano* **2023**, *17*, 18596–18607. [CrossRef]
74. Hou, J.; Qian, X.; Xu, Y.; Guo, Z.; Thierry, B.; Yang, C.-T.; Zhou, X.; Mao, C. Rapid and reliable ultrasensitive detection of pathogenic H9N2 viruses through virus-binding phage nanofibers decorated with gold nanoparticles. *Biosens. Bioelectron.* **2023**, *237*, 115423. [CrossRef] [PubMed]
75. Ma, Y.-X.; Wang, C.-Y.; Li, Y.-Y.; Li, J.; Wan, Q.-Q.; Chen, J.-H.; Tay, F.R.; Niu, L.-N. Considerations and caveats in combating ESKAPE pathogens against nosocomial infections. *Adv. Sci.* **2020**, *7*, 1901872. [CrossRef] [PubMed]
76. Nobrega, F.L.; Vlot, M.; de Jonge, P.A.; Dreesens, L.L.; Beaumont, H.J.E.; Lavigne, R.; Dutilh, B.E.; Brouns, S.J.J. Targeting mechanisms of tailed bacteriophages. *Nat. Rev. Microbiol.* **2018**, *16*, 760–773. [CrossRef] [PubMed]
77. Paczesny, J.; Richter, Ł.; Hołyst, R. Recent progress in the detection of bacteria using bacteriophages: A review. *Viruses* **2020**, *12*, 845. [CrossRef] [PubMed]
78. Richter, Ł.; Janczuk-Richter, M.; Niedziółka-Jönsson, J.; Paczesny, J.; Hołyst, R. Recent advances in bacteriophage-based methods for bacteria detection. *Drug Discov. Today* **2018**, *23*, 448–455. [CrossRef] [PubMed]
79. Chen, J.; Alcaine, S.D.; Jiang, Z.; Rotello, V.M.; Nugen, S.R. Detection of Escherichia coli in drinking water using T7 bacteriophage-conjugated magnetic probe. *Anal. Chem.* **2015**, *87*, 8977–8984. [CrossRef]

80. You, H.; Wang, M.; Wang, S.; Xu, J.; Hu, S.; Li, T.; Yu, Z.; Tang, D.; Gan, N. Ultrasensitive and specific phage@DNAzyme probe-triggered fluorescent click chemistry for on-site detection of foodborne pathogens using a smartphone. *Anal. Chem.* **2023**, *95*, 11211–11218. [CrossRef] [PubMed]
81. Yang, H.; Gou, Z.; Yuan, H.; Yue, X.; Lu, S.; Fu, Z. Employment of the phage cocktail as a species-specific recognition agent for wide-spectrum detection of bacterial strains. *Anal. Chem.* **2023**, *95*, 10008–10016. [CrossRef]
82. Wu, M.-Y.; Chen, L.; Chen, Q.; Hu, R.; Xu, X.; Wang, Y.; Li, J.; Feng, S.; Dong, C.; Zhang, X.-L.; et al. Engineered phage with aggregation-induced emission photosensitizer in cocktail therapy against sepsis. *Adv. Mater.* **2023**, *35*, 2208578. [CrossRef] [PubMed]
83. Brown, E.D.; Wright, G.D. Antibacterial drug discovery in the resistance era. *Nature* **2016**, *529*, 336–343. [CrossRef] [PubMed]
84. Kupferschmidt, K. Resistance fighters. *Science* **2016**, *352*, 758–761. [CrossRef] [PubMed]
85. Thiel, K. Old dogma, new tricks—21st century phage therapy. *Nat. Biotechnol.* **2004**, *22*, 31–36. [CrossRef] [PubMed]
86. Wang, M.; Zhang, J.; Wei, J.; Jiang, L.; Jiang, L.; Sun, Y.; Zeng, Z.; Wang, Z. Phage-inspired strategies to combat antibacterial resistance. *Crit. Rev. Microbiol.* **2024**, *50*, 196–211. [CrossRef] [PubMed]
87. He, X.; Yang, Y.; Guo, Y.; Lu, S.; Du, Y.; Li, J.-J.; Zhang, X.; Leung, N.L.C.; Zhao, Z.; Niu, G.; et al. Phage-guided targeting, discriminative imaging, and synergistic killing of bacteria by AIE bioconjugates. *J. Am. Chem. Soc.* **2020**, *142*, 3959–3969. [CrossRef] [PubMed]
88. Ran, B.; Yuan, Y.; Xia, W.; Li, M.; Yao, Q.; Wang, Z.; Wang, L.; Li, X.; Xu, Y.; Peng, X. A photo-sensitizable phage for multidrug-resistant Acinetobacter baumannii therapy and biofilm ablation. *Chem. Sci.* **2021**, *12*, 1054–1061. [CrossRef]
89. Zhang, J.; He, X.; Tang, B.Z. Aggregation-induced emission-armored living bacteriophage-DNA nanobioconjugates for targeting, imaging, and efficient elimination of intracellular bacterial infection. *ACS Nano* **2024**, *18*, 3199–3213. [CrossRef] [PubMed]
90. Jiang, T.; Yang, T.; Chen, Y.; Miao, Y.; Xu, Y.; Jiang, H.; Yang, M.; Mao, C. Emulating interactions between microorganisms and tumor microenvironment to develop cancer theranostics. *Theranostics* **2022**, *12*, 2833–2859. [CrossRef]
91. Dong, X.; Pan, P.; Zheng, D.-W.; Bao, P.; Zeng, X.; Zhang, X.-Z. Bioinorganic hybrid bacteriophage for modulation of intestinal microbiota to remodel tumor-immune microenvironment against colorectal cancer. *Sci. Adv.* **2020**, *6*, eaba1590. [CrossRef] [PubMed]
92. Geller, L.T.; Barzily-Rokni, M.; Danino, T.; Jonas, O.H.; Shental, N.; Nejman, D.; Gavert, N.; Zwang, Y.; Cooper, Z.A.; Shee, K.; et al. Potential role of intratumor bacteria in mediating tumor resistance to the chemotherapeutic drug gemcitabine. *Science* **2017**, *357*, 1156–1160. [CrossRef] [PubMed]
93. Armstrong, H.; Bording-Jorgensen, M.; Dijk, S.; Wine, E. The complex interplay between chronic inflammation, the microbiome, and cancer: Understanding disease progression and what we can do to prevent it. *Cancers* **2018**, *10*, 83. [CrossRef] [PubMed]
94. Federici, S.; Kredo-Russo, S.; Valdés-Mas, R.; Kviatcovsky, D.; Weinstock, E.; Matiuhin, Y.; Silberberg, Y.; Atarashi, K.; Furuichi, M.; Oka, A.; et al. Targeted suppression of human IBD-associated gut microbiota commensals by phage consortia for treatment of intestinal inflammation. *Cell* **2022**, *185*, 2879–2898.e2824. [CrossRef] [PubMed]
95. Baudino, A.T. Targeted cancer therapy: The next generation of cancer treatment. *Curr. Drug Discov. Technol.* **2015**, *12*, 3–20. [CrossRef]
96. Chen, X.; Mendes, B.G.; Alves, B.S.; Duan, Y. Chapter four—Phage therapy in gut microbiome. In *Progress in Molecular Biology and Translational Science*; Academic Press: Cambridge, MA, USA, 2023; Volume 201, pp. 93–118.
97. Zheng, D.-W.; Dong, X.; Pan, P.; Chen, K.-W.; Fan, J.-X.; Cheng, S.-X.; Zhang, X.-Z. Phage-guided modulation of the gut microbiota of mouse models of colorectal cancer augments their responses to chemotherapy. *Nat. Biomed. Eng.* **2019**, *3*, 717–728. [CrossRef] [PubMed]
98. Gandra, N.; Abbineni, G.; Qu, X.; Huai, Y.; Wang, L.; Mao, C. Bacteriophage bionanowire as a carrier for both cancer-targeting peptides and photosensitizers and its use in selective cancer cell killing by photodynamic therapy. *Small* **2013**, *9*, 215–221. [CrossRef]
99. Dong, X.; Pan, P.; Ye, J.-J.; Zhang, Q.-L.; Zhang, X.-Z. Hybrid M13 bacteriophage-based vaccine platform for personalized cancer immunotherapy. *Biomaterials* **2022**, *289*, 121763. [CrossRef] [PubMed]
100. Dong, X.; Pan, P.; Zhang, Q.; Ye, J.-J.; Zhang, X.-Z. Engineered living bacteriophage-enabled self-adjuvanting hydrogel for remodeling tumor microenvironment and cancer therapy. *Nano Lett.* **2023**, *23*, 1219–1228. [CrossRef]
101. Qamar, Z.; Qizilbash, F.F.; Iqubal, K.M.; Ali, A.; Narang, K.J.; Ali, J.; Baboota, S. Nano-based drug delivery system: Recent strategies for the treatment of ocular disease and future perspective. *Recent. Pat. Drug Deliv. Formul.* **2019**, *13*, 246–254. [CrossRef] [PubMed]
102. Sioud, M. Phage display libraries: From binders to targeted drug delivery and human therapeutics. *Mol. Biotechnol.* **2019**, *61*, 286–303. [CrossRef] [PubMed]
103. Kalarical Janardhanan, S.; Narayan, S.; Abbineni, G.; Hayhurst, A.; Mao, C. Architectonics of phage-liposome nanowebs as optimized photosensitizer vehicles for photodynamic cancer therapy. *Mol. Cancer Ther.* **2010**, *9*, 2524–2535. [CrossRef] [PubMed]
104. Kovacs, E.W.; Hooker, J.M.; Romanini, D.W.; Holder, P.G.; Berry, K.E.; Francis, M.B. Dual-surface-modified bacteriophage MS2 as an ideal scaffold for a viral capsid-based drug delivery system. *Bioconjugate Chem.* **2007**, *18*, 1140–1147. [CrossRef] [PubMed]
105. Wu, W.; Hsiao, S.C.; Carrico, Z.M.; Francis, M.B. Genome-free viral capsids as multivalent carriers for taxol delivery. *Angew. Chem. Int. Ed.* **2009**, *48*, 9493–9497. [CrossRef]

106. Gandra, N.; Wang, D.-D.; Zhu, Y.; Mao, C. Virus-mimetic cytoplasm-cleavable magnetic/silica nanoclusters for enhanced gene delivery to mesenchymal stem cells. *Angew. Chem. Int. Ed.* **2013**, *52*, 11278–11281. [CrossRef]
107. Tsedev, U.; Lin, C.-W.; Hess, G.T.; Sarkaria, J.N.; Lam, F.C.; Belcher, A.M. Phage particles of controlled length and genome for in vivo targeted glioblastoma imaging and therapeutic delivery. *ACS Nano* **2022**, *16*, 11676–11691. [CrossRef]

MDPI AG
Grosspeteranlage 5
4052 Basel
Switzerland
Tel.: +41 61 683 77 34

Targets Editorial Office
E-mail: targets@mdpi.com
www.mdpi.com/journal/targets

www.ingramcontent.com/pod-product-compliance
Lightning Source LLC
Chambersburg PA
CBHW061619210926
51850CB00009B/2377
* 9 7 8 3 7 2 5 8 5 0 4 9 5 *